AF335568

Lunar Base Agriculture:
Soils for Plant Growth

Lunar Base Agriculture: Soils for Plant Growth

Editors
D. W. Ming
D. L. Henninger

Editorial Committee
D. W. Ming
D. L. Henninger
K. S. Brady

Editor-in-Chief ASA
G. H. Heichel

Editor-in-Chief CSSA
C. W. Stuber

Editor-in-Chief SSSA
D. E. Kissel

Managing Editor
S. H. Mickelson

American Society of Agronomy, Inc.
Crop Science Society of America, Inc.
Soil Science Society of America, Inc.
Madison, Wisconsin, USA

1989

American Society of Agronomy, Inc.
Crop Science Society of America, Inc.
Soil Science Society of America, Inc.
677 South Segoe Road, Madison, WI 53711 USA

Library of Congress Cataloging-in-Publication Data

Lunar base agriculture.

Papers from a NASA-sponsored workshop held June 1–2, 1987 in
Houston, Tex.
   1. Lunar bases—Congresses. 2. Lunar soil—Congresses. I. Ming,
D. W. (Douglas W.) II. Henninger, D. L. (Donald L.) III. United
States. National Aeronautics and Space Administration.
TL799.M6L82   1989   631.4′799′1   89-15141
ISBN 0-89118-100-8

Printed in the United States of America

Plate 1. The surface geology of the Moon is dominated by two primary units. The heavily cratered, light-colored regions are the lunar highlands and the smooth dark regions are lunar maria. The lunar highlands are the oldest surface features of the Moon. The maria were formed by basaltic lava flows, mostly after the primary flux of meteorites that bombarded the highlands (Lick Observatory photograph).

Plate 2. This photograph taken from the lunar surface shows the footprint left behind by Apollo 11 astronaut Neil Armstrong (NASA photo AS11–49–5877).

Plate 3. A core tube is being driven with a hammer by astronaut "Buzz" Aldrin on Apollo 11 (NASA photo AS11–40–5964).

Plate 4. In this Apollo 16 photograph, astronaut John Young is jumping up and saluting the American flag in front of the lunar module and rover. In the 1/6 gravity environment on the lunar surface, John Young is able to jump nearly a meter above the surface while wearing a suit that weighed approximately 100 kg on Earth (NASA photo AS16–113–18339).

Plate 5. This photograph was taken in the Lunar Sample Laboratory at the NASA Johnson Space Center and shows a typical Apollo 17 "soil" breccia. This sample is one of many samples collected on the lunar surface (approximately 382 kg of lunar samples were collected) and brought back to Earth during the six Apollo missions (NASA photo S73–15443).

CONTENTS

FOREWORD

The American Society of Agronomy, the Crop Science Society of America, and the Soil Science Society of America are pleased to publish *Lunar Base Agriculture: Soils for Plant Growth*.

Whether or not plants will be grown on the Moon has not been answered; however, it is a possible scenario that will continue to be discussed by the planetary community. Consequently, this book will be a major source of information for many years to those seeking ways to produce food in space, particularly if the Moon serves as an outpost to launch human exploration to our inner solar system.

The Societies thank the organizers of the symposium, the editors, and authors of *Lunar Base Agriculture: Soils for Plant Growth*.

E. C. A. Runge, *president*
American Society of Agronomy

C. O. Qualset, *president*
Crop Science Society of America

J. J. Mortvedt, *president*
Soil Science Society of America

PREFACE

The National Aeronautics and Space Administration (NASA) is considering a step into a new frontier: extending permanent human presence beyond low Earth orbit. As part of the planning process, selected missions are being evaluated as case studies, including (i) the human exploration of Mars and its satellites, Phobos and Deimos, (ii) the establishment of a lunar science outpost, and (iii) the evolutionary expansion of humans into our inner solar system. Evolutionary expansion will be a step-by-step program to open the inner planets for exploration, space science research, on-site resource development, and permanent human presence. The first step will probably be the establishment of a lunar base. Some reasons to return to the Moon have been suggested, including (i) scientific research, (ii) utilization of lunar resources, and (iii) attainment of self-sufficiency in the lunar environment as a first step in planetary exploration. A self-sufficient lunar base will require the use of on-site resources. Of primary concern will be the production of food at a lunar base. No doubt, lunar materials will play an important role in the development of lunar base agriculture. The study of lunar surface materials and how they might react as a soil, source of plant nutrients, or both opens a new frontier for researchers in the agricultural community.

This publication, *Lunar Base Agriculture: Soils for Plant Growth*, represents a collection of papers by soil and plant scientists, along with experts in planetary sciences, concerned with the future role of agriculture at a lunar base. More than 70 scientists representing more than 15 universities, 6 federal agencies, and 8 industries were invited to a NASA-sponsored workshop held 1 to 2 June 1987 in Houston, TX. The goal was to identify a course of research dealing with the interaction of lunar resources and agricultural systems. This publication addresses that goal and contains sections on (i) lunar base scenarios, (ii) the lunar environment, (iii) chemical and physical considerations for a lunar-derived soil, (iv) biological considerations for a lunar-derived soil, (v) current research in controlled ecological life support systems, and (vi) future research needs for plant growth at a lunar base.

We are indebted to the authors for their outstanding contributions. The editorial committee expresses its appreciation to all the anonymous reviewers who contributed suggestions for improvement of the chapters. Special thanks goes to Sherri Mickelson, David Kral, and the rest of the staff at ASA Headquarters for their conscientious and painstaking job of style editing and typesetting the final copy for publication. We hope that these chapters will stimulate interdisciplinary cooperation among soil and plant scientists and the planetary community in the area of lunar base agriculture that will undoubtedly become an important topic in the years to come.

D. W. Ming, *co-editor, chair*
NASA Johnson Space Center
Houston, TX

D. L. Henninger, *co-editor*
NASA Johnson Space Center
Houston, TX

K. S. Brady, *co-editor*
Tennessee Valley Authority
Muscle Shoals, AL

CONTRIBUTORS

D. B. Alexander
Postdoctoral Research Associate, Soil and Crop Sciences Department, Texas A&M University, College Station, TX 77843

E. R. Allen
Graduate Research Assistant, Soil and Crop Sciences Department, Texas A&M University, College Station, TX 77843

M. M. Averner
Discipline Scientist, NASA-CELSS Program, NASA Headquarters, Washington, DC 20546

D. L. Bubenheim
Research Scientist, NASA-Ames Research Center, Life Science Division, Moffett Field, CA 94035

Bruce G. Bugbee
Assistant Professor, Plant Science Department, Utah State University, Logan, UT 84322

R. B. Corey
Professor, Department of Soil Science, University of Wisconsin, Madison, WI 53706

L. R. Drees
Research Associate, Soil and Crop Sciences Department, Texas A&M University, College Station, TX 77843

Michael B. Duke
Chief, Solar System Exploration Division, NASA Johnson Space Center, Houston, TX 77058

Henry L. Ehrlich
Professor of Biology, Rensselaer Polytechnic Institute, Troy, NY 12180

Kyle O. Fairchild
Manager for Technology and Enterprise, New Initiatives Office, NASA Johnson Space Center, Houston, TX 77058

J. Gale
Professor, Department of Botany, Hebrew University of Jerusalem, Jerusalem 91904, Israel

P. A. Helmke
Professor, Department of Soil Science, University of Wisconsin, Madison, WI 53706

Donald L. Henninger
Space Scientist, Solar System Exploration Division, NASA Johnson Space Center, Houston, TX 77058

L. R. Hossner
Professor, Soil and Crop Sciences Department, Texas A&M University, College Station, TX 77843

D. H. Hubbell
Professor of Soil Microbiology, Department of Soil Science, University of Florida, Gainesville, FL 32611

W. M. Knott, III
Bioscience Officer, NASA, Life Sciences Research Office, Kennedy Space Center, FL 32899

R. D. MacElroy
Research Scientist, NASA-Ames Research Center, Life Science Division, Moffett Field, CA 94035

David S. McKay
Manager, Space Resources Utilization Office, NASA Johnson Space Center, Houston, TX 77058

Wendell W. Mendell
Chief Scientist, Lunar Base Studies, NASA Johnson Space Center, Houston, TX 77058

Douglas W. Ming
Space Scientist, Solar System Exploration Division, NASA Johnson Space Center, Houston, TX 77058

R. P. Prince
Environmental Engineer, NASA Life Sciences Research Office, Kennedy Space Center, FL 32899

Barney B. Roberts   Manager, Planet Surface and Human Systems Office, New Initiatives Office, NASA Johnson Space Center, Houston, TX 77058

Frank B. Salisbury   Professor of Plant Physiology, Plant Science Department, Utah State University, Logan, UT 84322

G. Stotzky   Professor, Biology Department, New York University, New York, NY 10003

G. Jeffrey Taylor   Assistant Director, Institute of Meteoritics, University of New Mexico, Albuquerque, NM 87131

T. W. Tibbitts   Director of Biotron and Professor of Horticulture, Biotron and Horticulture Department, University of Wisconsin, Madison, WI 53706

J. Tremor   Formerly Research Scientist, Ecosystem Science and Technology Branch, Ames Research Center. Currently Manager, Science Department, Lockheed Engineering and Sciences Company, Western Programs Office, Moffett Field, CA 94035

Gene Whitney   Research Geologist, U.S. Geological Survey, Denver Federal Center, Denver, CO 80225

L. P. Wilding   Professor of Pedology, Soil and Crop Sciences Department, Texas A&M University, College Station, TX 77843

D.A. Zuberer   Associate Professor, Soil Microbiology, Soil and Crop Sciences Department, Texas A&M University, College Station, TX 77843

Conversion Factors for SI and non-SI Units

Conversion Factors for SI and non-SI Units

To convert Column 1 into Column 2, multiply by	Column 1 SI Unit	Column 2 non-SI Unit	To convert Column 2 into Column 1, multiply by
		Length	
0.621	kilometer, km (10^3 m)	mile, mi	1.609
1.094	meter, m	yard, yd	0.914
3.28	meter, m	foot, ft	0.304
1.0	micrometer, μm (10^{-6} m)	micron, μ	1.0
3.94×10^{-2}	millimeter, mm (10^{-3} m)	inch, in	25.4
10	nanometer, nm (10^{-9} m)	Angstrom, Å	0.1
		Area	
2.47	hectare, ha	acre	0.405
247	square kilometer, km^2 $(10^3$ m$)^2$	acre	4.05×10^{-3}
0.386	square kilometer, km^2 $(10^3$ m$)^2$	square mile, mi^2	2.590
2.47×10^{-4}	square meter, m^2	acre	4.05×10^3
10.76	square meter, m^2	square foot, ft^2	9.29×10^{-2}
1.55×10^{-3}	square millimeter, mm^2 $(10^{-3}$ m$)^2$	square inch, in^2	645
		Volume	
9.73×10^{-3}	cubic meter, m^3	acre-inch	102.8
35.3	cubic meter, m^3	cubic foot, ft^3	2.83×10^{-2}
6.10×10^4	cubic meter, m^3	cubic inch, in^3	1.64×10^{-5}
2.84×10^{-2}	liter, L (10^{-3} m^3)	bushel, bu	35.24
1.057	liter, L (10^{-3} m^3)	quart (liquid), qt	0.946
3.53×10^{-2}	liter, L (10^{-3} m^3)	cubic foot, ft^3	28.3
0.265	liter, L (10^{-3} m^3)	gallon	3.78
33.78	liter, L (10^{-3} m^3)	ounce (fluid), oz	2.96×10^{-2}
2.11	liter, L (10^{-3} m^3)	pint (fluid), pt	0.473

Mass

2.20×10^{-3}	gram, g (10^{-3} kg)	pound, lb	454
3.52×10^{-2}	gram, g (10^{-3} kg)	ounce (avdp), oz	28.4
2.205	kilogram, kg	pound, lb	0.454
0.01	kilogram, kg	quintal (metric), q	100
1.10×10^{-3}	kilogram, kg	ton (2000 lb), ton	907
1.102	megagram, Mg (tonne)	ton (U.S.), ton	0.907
1.102	tonne, t	ton (U.S.), ton	0.907

Yield and Rate

0.893	kilogram per hectare, kg ha^{-1}	pound per acre, lb acre^{-1}	1.12
7.77×10^{-2}	kilogram per cubic meter, kg m^{-3}	pound per bushel, bu^{-1}	12.87
1.49×10^{-2}	kilogram per hectare, kg ha^{-1}	bushel per acre, 60 lb	67.19
1.59×10^{-2}	kilogram per hectare, kg ha^{-1}	bushel per acre, 56 lb	62.71
1.86×10^{-2}	kilogram per hectare, kg ha^{-1}	bushel per acre, 48 lb	53.75
0.107	liter per hectare, L ha^{-1}	gallon per acre	9.35
893	tonnes per hectare, t ha^{-1}	pound per acre, lb acre^{-1}	1.12×10^{-3}
893	megagram per hectare, Mg ha^{-1}	pound per acre, lb acre^{-1}	1.12×10^{-3}
0.446	megagram per hectare, Mg ha^{-1}	ton (2000 lb) per acre, ton acre^{-1}	2.24
2.24	meter per second, m s^{-1}	mile per hour	0.447

Specific Surface

10	square meter per kilogram, m^2 kg^{-1}	square centimeter per gram, cm^2 g^{-1}	0.1
1000	square meter per kilogram, m^2 kg^{-1}	square millimeter per gram, mm^2 g^{-1}	0.001

Pressure

9.90	megapascal, MPa (10^6 Pa)	atmosphere	0.101
10	megapascal, MPa (10^6 Pa)	bar	0.1
1.00	megagram per cubic meter, Mg m^{-3}	gram per cubic centimeter, g cm^{-3}	1.00
2.09×10^{-2}	pascal, Pa	pound per square foot, lb ft^{-2}	47.9
1.45×10^{-4}	pascal, Pa	pound per square inch, lb in^{-2}	6.90×10^3

continued on next page

Conversion Factors for SI and non-SI Units

To convert Column 1 into Column 2, multiply by	Column 1 SI Unit	Column 2 non-SI Unit	To convert Column 2 into Column 1, multiply by
	Temperature		
$1.00\ (K - 273)$	Kelvin, K	Celsius, °C	$1.00\ (°C + 273)$
$(9/5\ °C) + 32$	Celsius, °C	Fahrenheit, °F	$5/9\ (°F - 32)$
	Energy, Work, Quantity of Heat		
9.52×10^{-4}	joule, J	British thermal unit, Btu	1.05×10^{3}
0.239	joule, J	calorie, cal	4.19
10^{7}	joule, J	erg	10^{-7}
0.735	joule, J	foot-pound	1.36
2.387×10^{-5}	joule per square meter, $J\ m^{-2}$	calorie per square centimeter (langley)	4.19×10^{4}
10^{5}	newton, N	dyne	10^{-5}
1.43×10^{-3}	watt per square meter, $W\ m^{-2}$	calorie per square centimeter minute (irradiance), $cal\ cm^{-2}\ min^{-1}$	698
	Transpiration and Photosynthesis		
3.60×10^{-2}	milligram per square meter second, $mg\ m^{-2}\ s^{-1}$	gram per square decimeter hour, $g\ dm^{-2}\ h^{-1}$	27.8
5.56×10^{-3}	milligram (H_2O) per square meter second, $mg\ m^{-2}\ s^{-1}$	micromole (H_2O) per square centimeter second, $\mu mol\ cm^{-2}\ s^{-1}$	180
10^{-4}	milligram per square meter second, $mg\ m^{-2}\ s^{-1}$	milligram per square centimeter second, $mg\ cm^{-2}\ s^{-1}$	10^{4}
35.97	milligram per square meter second, $mg\ m^{-2}\ s^{-1}$	milligram per square decimeter hour, $mg\ dm^{-2}\ h^{-1}$	2.78×10^{-2}
	Plane Angle		
57.3	radian, rad	degrees (angle), °	1.75×10^{-2}

Electrical Conductivity, Electricity, and Magnetism

10	siemen per meter, S m^{-1}	millimho per centimeter, mmho cm^{-1}	0.1
10^4	tesla, T	gauss, G	10^{-4}

Water Measurement

9.73×10^{-3}	cubic meter, m^3	acre-inches, acre-in	102.8
9.81×10^{-3}	cubic meter per hour, m^3 h^{-1}	cubic feet per second, ft^3 s^{-1}	101.9
4.40	cubic meter per hour, m^3 h^{-1}	U.S. gallons per minute, gal min^{-1}	0.227
8.11	hectare-meters, ha-m	acre-feet, acre-ft	0.123
97.28	hectare-meters, ha-m	acre-inches, acre-in	1.03×10^{-2}
8.1×10^{-2}	hectare-centimeters, ha-cm	acre-feet, acre-ft	12.33

Concentrations

1	centimole per kilogram, cmol kg^{-1} (ion exchange capacity)	milliequivalents per 100 grams, meq 100 g^{-1}	1
0.1	gram per kilogram, g kg^{-1}	percent, %	10
1	milligram per kilogram, mg kg^{-1}	parts per million, ppm	1

Radioactivity

2.7×10^{-11}	bequerel, Bq	curie, Ci	3.7×10^{10}
2.7×10^{-2}	bequerel per kilogram, Bq kg^{-1}	picocurie per gram, pCi g^{-1}	37
100	gray, Gy (absorbed dose)	rad, rd	0.01
100	sievert, Sv (equivalent dose)	rem (roentgen equivalent man)	0.01

Plant Nutrient Conversion

	Elemental	*Oxide*	
2.29	P	P$_2$O$_5$	0.437
1.20	K	K$_2$O	0.830
1.39	Ca	CaO	0.715
1.66	Mg	MgO	0.602

1 Options for the Human Settlement of the Moon and Mars

Kyle O. Fairchild and Barney B. Roberts

NASA—Johnson Space Center
Houston, Texas

People have pondered visiting and living on the Moon and Mars since the dawn of civilization. We are only now entering into the era where such dreams are realizable in the foreseeable future. The nation, through the National Aeronautics and Space Administration (NASA), is seriously considering the steps that would be required to establish permanent habitation on the Moon or Mars. Instead of brief sorties into space, NASA is defining options for evolutionary growth that begins with continuous human presence at a low Earth orbit (LEO) space station and culminates in permanent facilities on the Moon and Mars. How this can best be achieved is a matter of some controversy, subject to technical, economic, political, and cultural pressures.

INTRODUCTION

The most recent resurgence of interest in lunar and Mars programs was heralded by the President's National Commission on Space (NCOS, 1986). NCOS has recommended a bold evolutionary approach that considers the solar system as humanity's extended home and envisions space development that will be built on an ever-growing and outwardly expanding infrastructure. This expansion would begin with routine operations in (LEO), then extend to include lunar operations and a lunar base, and finally establish at least semi-permanent habitation on Mars. NASA's response to this report was articulated in the Ride Task Group report (Ride, 1987), "Leadership and America's Future in Space." This report outlines four options for space development: (i) Mission to Planet Earth; (ii) Exploration of the Solar System; (iii) Outpost of the Moon, and (iv) Humans to Mars.

In 1987, the NASA Administrator created the Office of Exploration (OEXP) at NASA Headquarters in Washington, DC with the charter to define options for human development of space after the Space Station Freedom is operational. The unmanned options for space development are under consideration in the NASA Headquarters Office of Space Science and Applications (OSSA). Interest in human exploration of space was further bolstered on 5 Jan. 1988, when the Presidential Directive on National Space Policy commissioned NASA to set a long-range goal "to expand human presence and activity beyond Earth orbit into the solar system."

Further space activities will be driven by a complex balance of evolving exigencies and priorities, both internal and external to the space development communities. In 1988, the OEXP has taken the step to define a set of options that span the spectrum of development options; from Apollo-type missions to Mars and Phobos (one of the two moons of Mars); to human-tended lunar observatories and scientific lunar excursions, to an evolutionary approach that would establish a lunar outpost, supplying resources and planetary surface technology and experience in support of a manned Mars mission. In coming years these scenarios or "case studies" as they are known, will be further defined and expanded, with the goal of recommending a scenario to the President in 1992—500 yr after Columbus discovered the Americas.

The focus of this chapter will be the evolutionary approach to space development, including lunar and martian bases, since it is this scenario in which lunar and Mars soils have the greatest applicability. There also appears to be a growing consensus that this approach is in the best long-term interests of the nation. However, the other case studies will provide a sense of perspective. Although not directly related to lunar soils, some understanding of these case studies can provide insight into the case study generation and evaluation process. This may help to provide some guidance for potential development pathways in lunar and Mars soils research.

1988 CASE STUDIES

Four case studies were developed with three overall strategies in mind. These strategies have been grouped into (i) human expeditions, (ii) science outposts, and (iii) evolutionary expansion (Fig. 1–1).

The human expeditions strategy places emphasis on highly visible, near-term "firsts" by sending humans to Mars or to one of the two moons of Mars. These expeditions are similar in scope and objectives to the Apollo program, with infrastructure development only done to the degree necessary to support one or two short-duration trips. Two such expeditionary scenarios were developed for the 1988 case studies, one to the martian moon Phobos and the other to the surface of Mars.

Science outposts, the second strategy, emphasizes scientific exploration as well as investigation of technologies and operations needed for permanent habitation. The lunar observatories case study is driven primarily by this strategy.

The evolutionary expansion strategy would explore and settle the inner solar system in a series of steps, with continued development of technologies, experience, and infrastructure. This is in contrast to the way space development has traditionally been done in the USA. Although there was some technological and experiential inheritance from one large program to another,

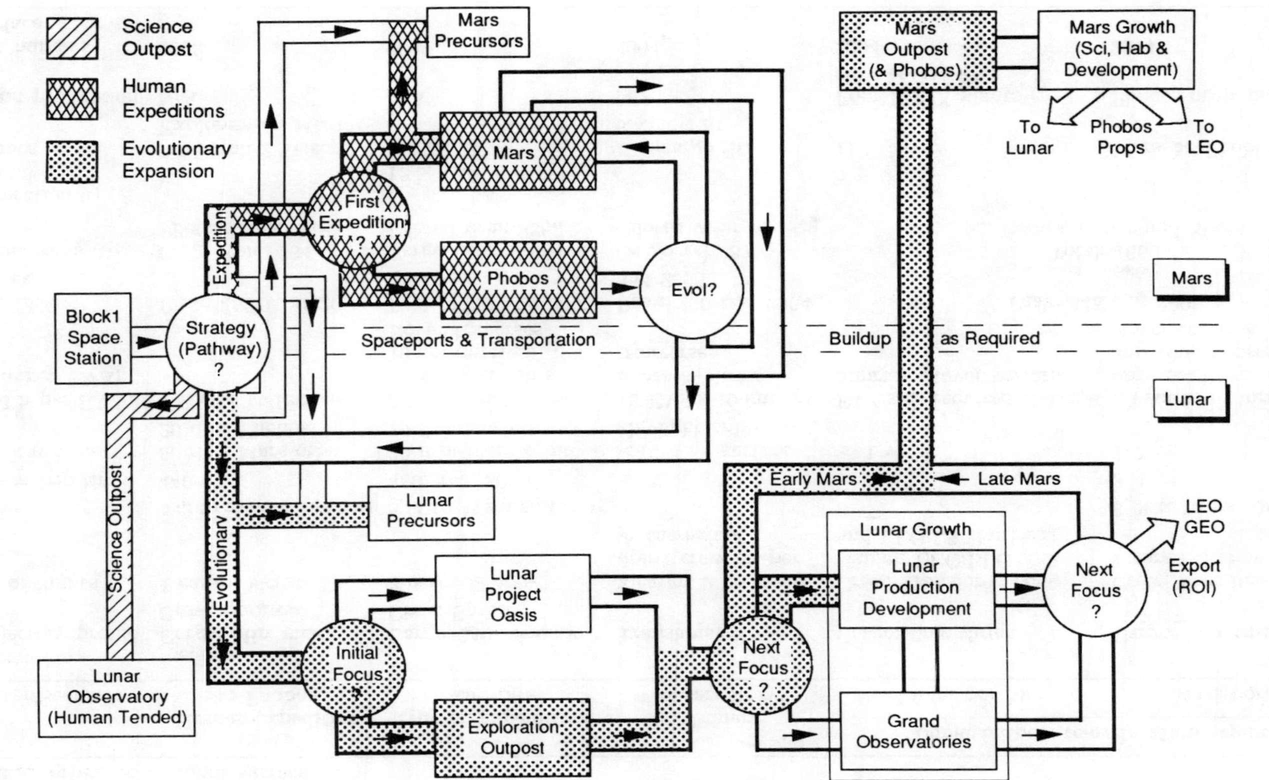

Human exploration roadmap.

Fig. 1-1. Roadmap of human exploration options (LEO = low Earth orbit, GEO = geosynchronous Earth orbit, ROI = return on investment).

Table 1-1. Summary of case study key parameters (EVA = extravehicular activity, LEO = low Earth orbit, LLOX = lunar liquid O_2, LLO = low lunar orbit, LS = lunar surface).

Descriptor/Scenario	Human expedition to Phobos	Human expeditions to Mars	Lunar observatory	Lunar outpost-to-early Mars evolution	
				Lunar portion	Mars portion
Transportation					
—Trajectory profile	Cargo: Min. energy Crew: Sprint	Cargo: Min. energy Crew: Sprint	Translunar	Cargo: Low thrust	Cargo: Low thrust
—No. of flights	1 cargo, 1 crew	3 cargo, 3 crew	2 cargo, 2 crew (set-up): 1 crew flt. per yr thereafter	Crew: Translunar continuing LEO/LLO and LLO/LS shuttles	Crew: Near fuel min. 1 cargo, 3 crew
Crew size	4 (2 to Phobos surface)	8 (4 to Mars surface)	4	8	8 (8 to Mars surface)
Total crew trip time	440 d	440, 440, 500 d	≤ 20 d	≤ 1 yr	3–4 yr.
Surface stay time	30 d in Mars orbit 20 d at Phobos	30 d in Mars orbit 20 d on surface	<14 d on surface (daylight only)	≤ 1 yr	1–2 yr
EVAs (6 h per EVA; two cre per EVA)	4 EVAs at Phobos	4 EVAs at Moons 10 EVA at Mars 10-km unpress. Rover traverses	12 EVAs 10-km unpress. Rover tranverses	EVAs as required 10-km unpress. Rover traverse	EVAs as required 10-km unpress. Rover traverse and 100-km press. Rover traverse
Mass to LEO peak year	Peak: 4531 t at 2002	Peak: 1770 t at 2006	Peak: 250 t at 2004 and 2005	Peak: 345 t at 2005	
Propellant mass (t)	Cargo veh.: 234 Pilotted veh.: 318	Cargo veh.: 1796 Piloted veh.: 3363	Cargo veh. 87 Piloted veh.: 96	Total: 1660 (all flights, Lunar and Mars)	
User allocation (t) —Orbital —Surface	N/A 7 t Total: 2 Teleop. Explorers on Mars	12.5, 12.5, 6 15, 15, 15	7 17.5/cargo flt. 6.5/crew flt.	3.3 112	12 Mars–58 Phobos–10
Propellant production	N/A	N/A	N/A	Four LLOX plants (40 t each)	Phobos prop. plant (86 t)
Year-1st humans to surface	2003	2007	2004	2004	2014

the Mercury, Gemini, Apollo, and Shuttle programs were developed largely independent of each other, as was the unmanned development that preceded them.

The 1988 cast studies have been documented by NASA (NASA Office of Exploration, 1988a, b, c). The bulk of the descriptions of these case studies is taken from Vol. I, the Executive Summary of the Exploration Studies Technical Report (NASA Office of Exploration, 1988a). Table 1–1 summarizes the most salient features of the four 1988 case studies. The relative level of resource commitment is measured generally in terms of the mass needed in LEO. Unfortunately, this can be somewhat misleading. For example, it does not adequately consider the buildup of infrastructure needed on Earth, nor technologial and cultural considerations.

Human Expedition to Phobos (Case Study 1)

The primary objective of this mission is to establish leadership in the human exploration of the solar system. To that end, baseline vehicles are designed for minimum dependence on advanced technology, and human presence is extended only to Mars orbit and the surface of Phobos. In this case study, the first human beings will arrive at the martian moon Phobos to explore, conduct resource surveys, and establish a science station. Secondary objectives are to conduct enhanced robotic exploration of Mars from orbit, using rovers, penetrators, balloons, and sample collectors; and to return samples of Mars and Phobos to Earth for detailed analyses. The expedition to Phobos combined human exploration objectives with those of previously studied robotic Mars Rover and Sample Return missions, but allows different approaches to exploration of Mars because of the capability for nearly real-time teleoperation of robotic systems from the vicinity of Mars.

Phobos Expedition Mission Description

The mission scenario, shown in Fig. 1–2 and 1–3, employs a "split-sprint" trajectory: a cargo vehicle carrying the Phobos and Deimos exploration equipment, Mars rovers, and the crew's return propellant will be launched on a minimum-energy trajectory in February 2001. Upon arrival, this vehicle will be placed in Mars orbit to await the piloted flight. In August 2002, approximately 72 wk after the first launch, a second vehicle carrying a crew of four will be launched to Mars on a high-energy sprint-class trajectory, requiring about 36 wk to reach Phobos.

Upon arrival in Mars orbit, the piloted vehicle will rendezvous with the cargo vehicle. Two crew members will transfer to a Phobos Excursion Vehicle and depart for a 20 d exploration of the martian moon. During that time, the crew on Phobos will make observations, conduct experiments, and gather samples, during a total of 24 h of extravehicular activity. The two crew members who remain in the orbiting vehicle will teleoperate, or remotely control, rovers that will gather samples from the Mars surface. After spending 30

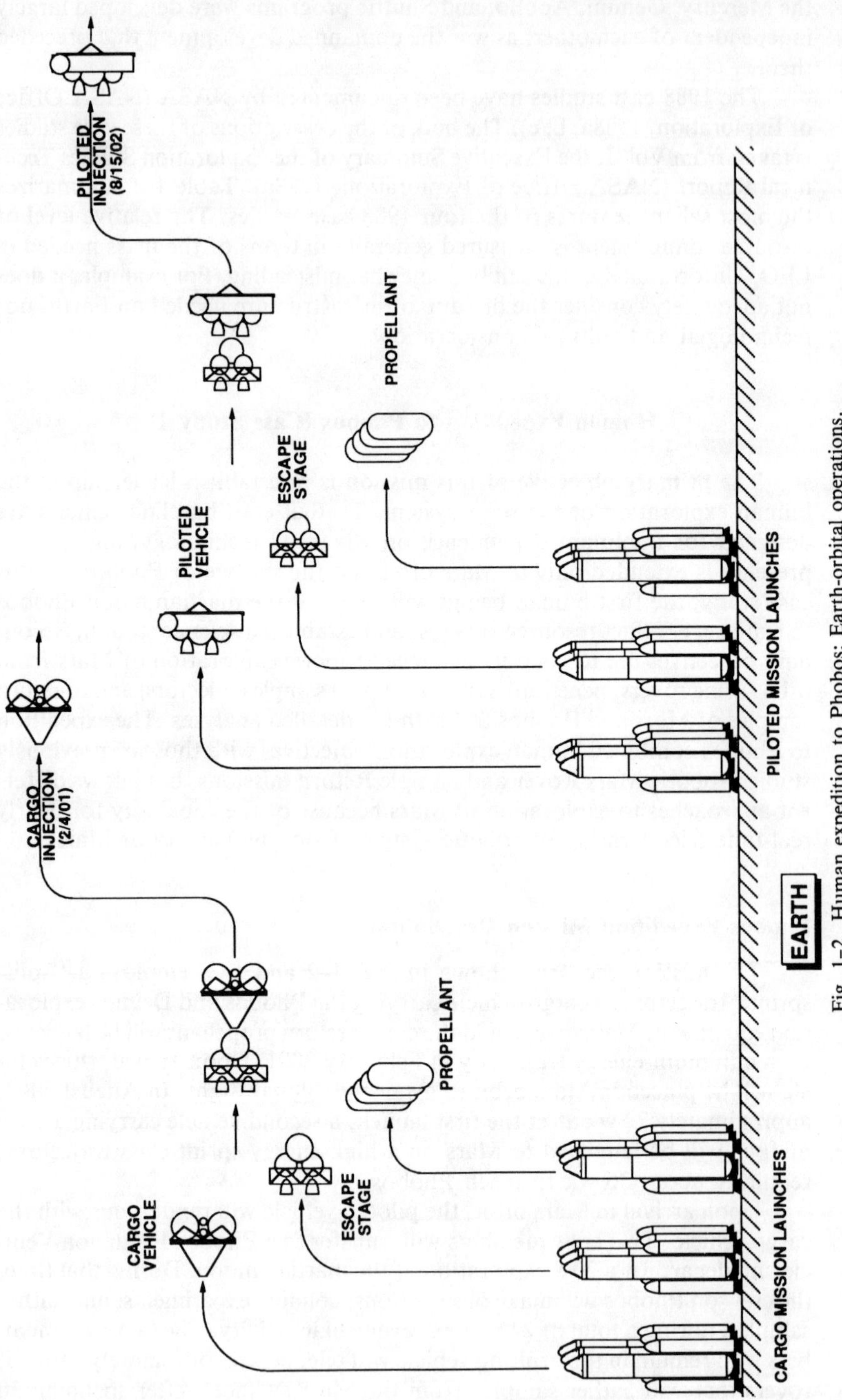

Fig. 1-2. Human expedition to Phobos: Earth-orbital operations.

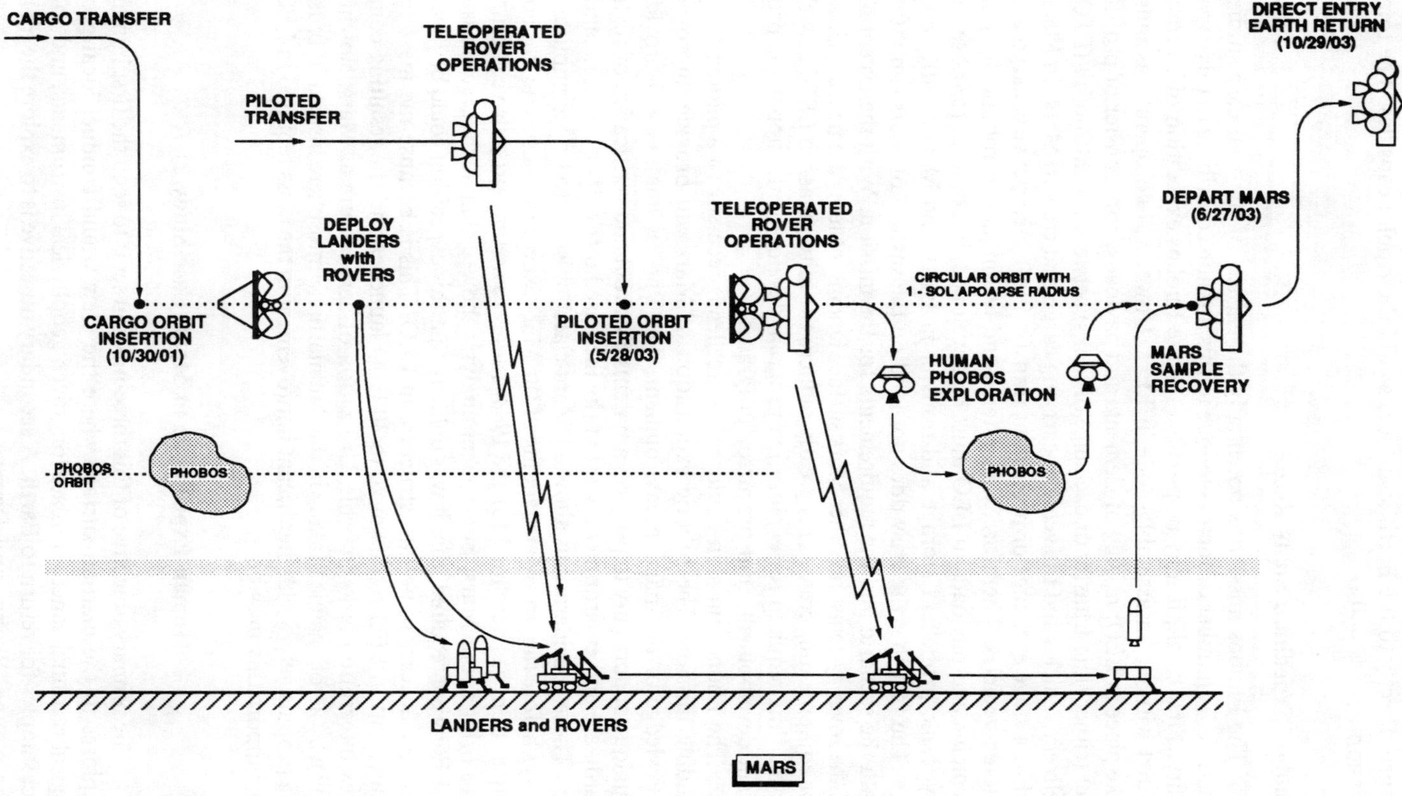

Fig. 1-3. Human expedition to Phobos: Mars/Phobos orbital operations.

d in the martian system, the crew will spend 16 wk returning to Earth and return directly to the Earth's surface, as did the Apollo capsules. The total mission time is 440 d.

Phobos Expedition Study Results

The Phobos mission is potentially the earliest of the four case studies possible. A number of factors unique to this mission contribute to this capability. First of all, it may be possible for the Phobos expedition to be completed without an assembly node in LEO. However, two operations must take place in LEO: mating of elements and payloads, and transfer of propellant (either fluid transfer or exchange of tanks) between Earth-to-orbit (ETO) delivery vehicles and the vehicles carrying cargo and crew to Mars. In Mars' orbit, a stage exchange or propellant transfer is effected between cargo and piloted vehicles. Therefore, the systems and techniques to robotically join elements and payloads in LEO must be developed, in addition to those for cryogenic propellant storage and transfer in Earth and Mars orbit.

The fact that the crew does not land on the surface of Mars simplifies both the scenario and the requirements for the mission. With the exception of the rover systems, no other Mars surface landing systems are needed, either for equipment or crew. This greatly reduces the initial mass to LEO, as well as the time required for exploration program development, supporting program development, and precursor programs.

The Phobos mission could be an excellent precursor to a piloted Mars landing mission. The robotic exploration of Mars will provide improved knowledge of the martian environment. The Phobos mission will provide a unique opportunity to perform a systems checkout and verification of flight hardware and environment without the increased difficulty of a Mars landing. These considerations allow a "Mars-class" mission to be accomplished 4.5 yr before the first Mars landing of the Mars expedition case. The Phobos Expedition was baselined for the 1988 studies to assume propulsive capture into orbit about Mars. Subsequent analysis showed that such large masses in LEO were required, that it was unlikely that this expedition could be flown without at least some infrastructure in LEO. This is because the mass requirement in LEO results in 20 to 30 ETO launches and the resulting complex integration tasks in orbit. Utilization of aerocapture at Mars, slowing down by aerodynamic braking in the thin martian atmosphere instead of braking propulsively, yields significant improvement in the mass required in LEO to support this mission.

Human Expeditions to Mars (Case Study 2)

The primary objective of this three-mission set is to send the first human explores to the martian surface where the crew would conduct local geological reconnaissance, emplace long-lived geophysical instruments, and collect samples for return to Earth. A secondary objective is to explore the martian moons, Phobos and Deimos.

Mars Expedition Mission Description

The transportation strategy employed for each of the three missions will be a split/sprint trajectory as shown in Fig. 1–4 and 1–5. For the first expedition, a cargo transport carrying the landing vehicle (including Mars surface habitat and exploration equipment and the ascent vehicle), and the Earth-return propellant will be launched on a minimum-energy trajectory in September 2005. Upon arrival at Mars, this vehicle will be placed in Mars orbit to await the piloted flight. In December 2006, 60 wk after the first launch, a vehicle carrying eight crew members will be launched to Mars on a high-energy, sprint-class trajectory.

Upon arrival at Mars, the piloted vehicle will rendezvous with the cargo vehicle in Mars orbit. Four crew members will transfer to the Mars orbit. Four crew members will transfer to the Mars Lander Vehicle and depart for a 20 d exploration of the Martian surface. The four remaining crew members will perform the propellant transfer from the cargo to the piloted vehicle, conduct Mars-orbital science, and monitor and assist the activities on the surface of Mars. After 30 d in the martian system, the surface crew will rendezvous with the orbiting piloted vehicle to depart for Earth, arriving about 20 wk later. The total length of the mission will be 440 d.

Cargo/piloted vehicle pairs will again be launched to Mars during the next two launch opportunities (2009 and 2011). The third piloted flight, in 2011, will have a total round-trip flight time of 500 d. This longer flight time will be necessary to avoid prohibitive mass penalties associated with the sprint trajectory in 2011. Piloted excursions to Phobos and Deimos are envisioned as part of the first two Mars expeditions. Each of the three Mars landing missions will also visit a different site on the Martian surface.

Mars Expedition Study Results

The Mars expeditions will deliver a crew of eight to Mars, with four landing on the surface; but arrivals will begin almost 5 yr later than the Phobos expedition. This difference results from the fact that the Mars case is of a much larger scale, with increased dependence on infrastructure and new technologies. The Mars expeditions will require significant LEO infrastructure and a substantial degree of on-orbit assembly operations at a LEO transportation node.

The Mars expeditions are much more complicated than the Phobos Expedition from several standpoints. First, landing on Mars itself increases mission difficulty for at least two reasons: (i) the vehicles that transport the crew to the martian surface and return them to Mars orbit are major factors in the significant increase in the mass in LEO for these missions, as compared to Phobos; and (ii) the crew must endure changing gravity environments. The necessity of crew operations in a one-third gravity environment on Mars after an extended period in zero gravity is a life science issue that needs resolution. The fact that the Mars expeditions launched to Mars over three successive opportunities introduces significant orbital mechanics complexities into the analysis. Substantial variations in propellant requirements

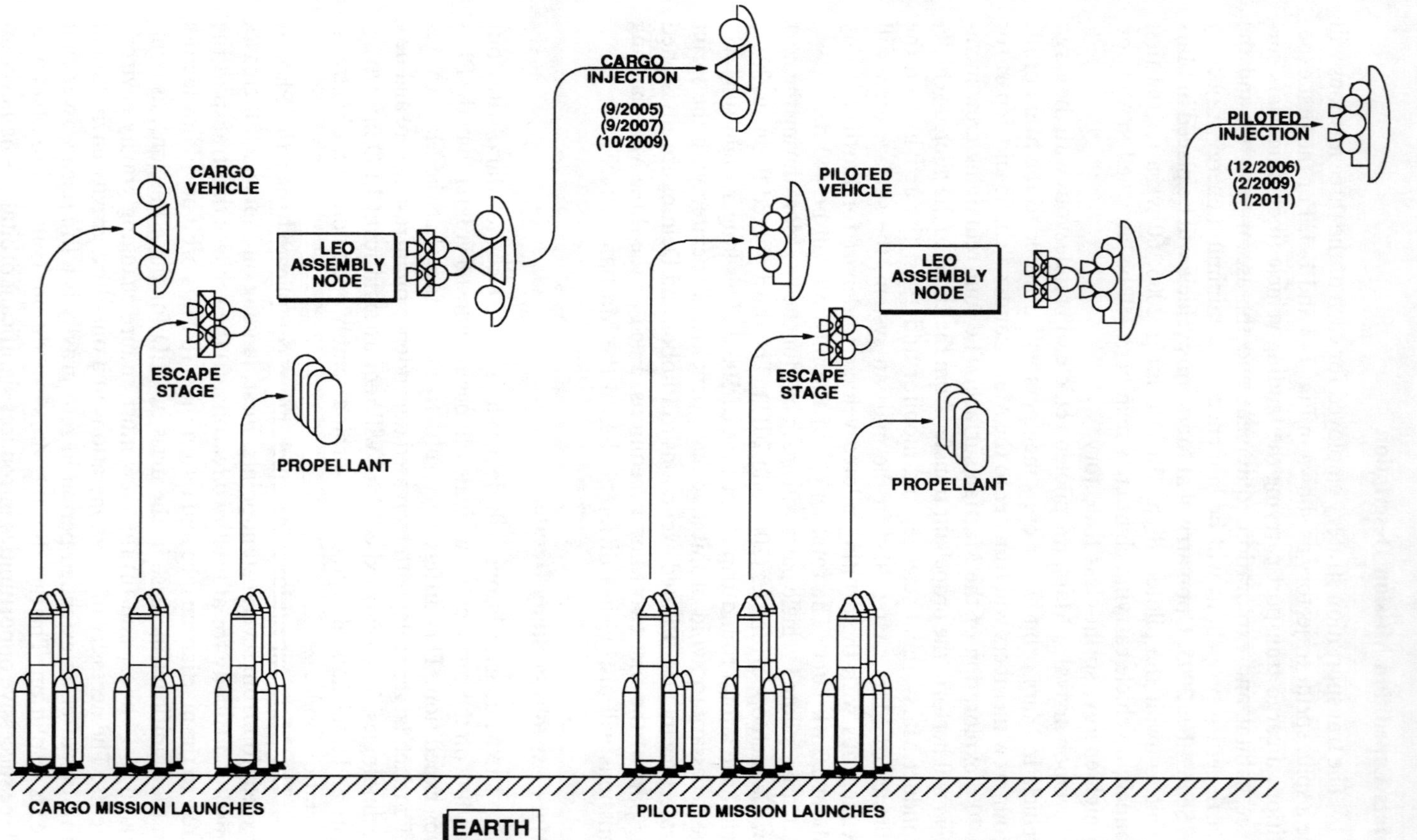

Fig. 1-4. Human expeditions to Mars: Earth-orbital operations (LEO = low Earth orbit).

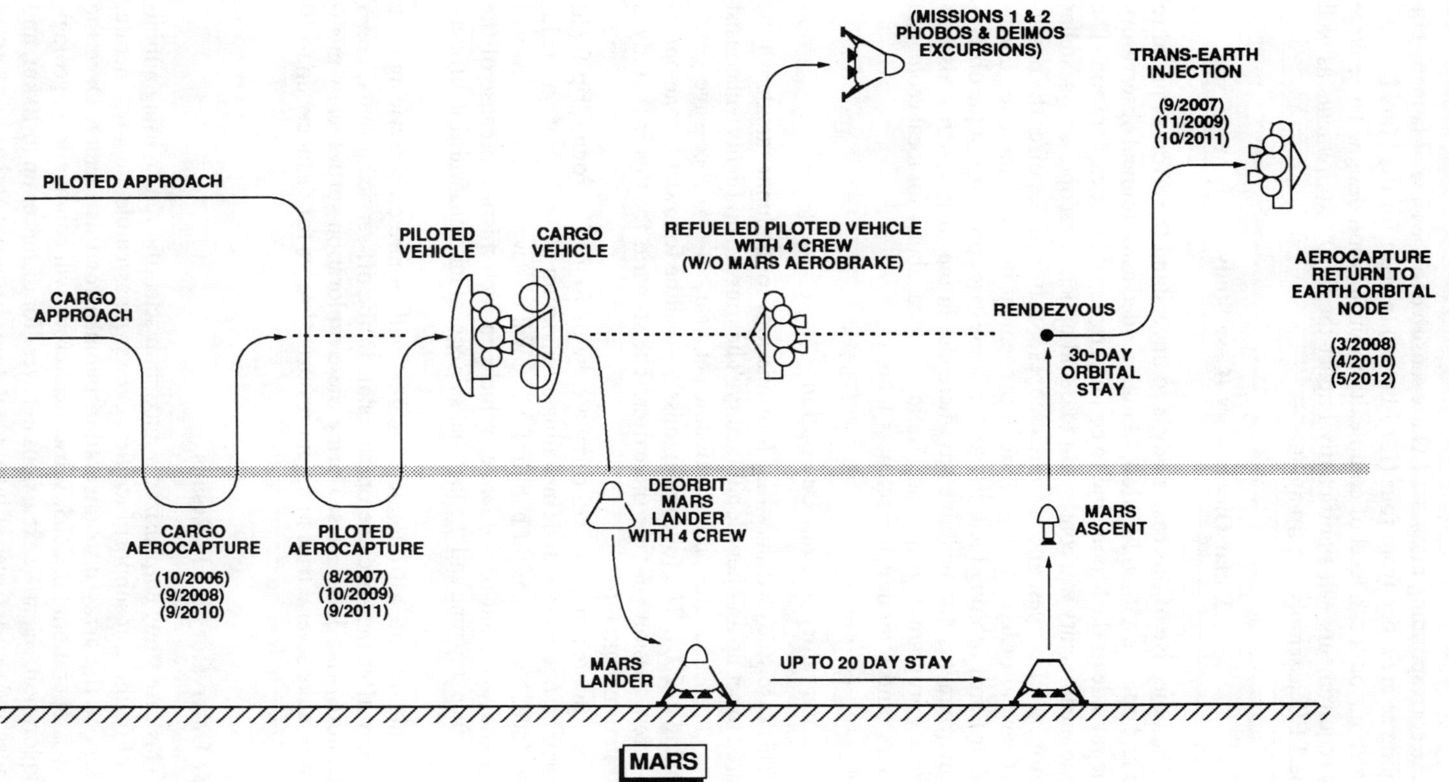

Fig. 1-5. Human expeditions to Mars: Mars-orbital/surface operations.

and the corresponding mass in LEO are sensitive to the time of launch. The doubling of crew size from four (for Phobos) to eight for the Mars Expedition also introduces a level of complexity to the mission design. These large mass requirements will significantly impact the ETO requirements as well as the LEO assembly requirements.

Lunar Observatory (Case Study 3)

The objective of this case study is to understand the effort required to build and operate a long-duration, human-tended astronomical observatory on the far side of the Moon, and to conduct regional lunar exploration. The astronomical facility will consist of radio and optical telescopic arrays, stellar monitoring telescopes, and radio telescopes. Such facilities offer the potential of several orders of magnitude improvement in resolution over Earth-based or Earth-orbiting facilities, and in some cases provide unique observing environments not available anywhere else in the solar system. Also included is a program of geophysical stations, the capability for local geological traverses, and a modest life sciences laboratory.

Lunar Observatory Mission Description

This case study assumes that four missions to the lunar farside will be required to set up an operational facility. The four set-up flights will consist of one cargo and one piloted mission per year, in two successive years, beginning in 2004. The four set-up missions will be followed by one operational crew mission per year thereafter. The scenario for this case study is illustrated in Fig. 1–6.

Each piloted mission will carry a crew of four. The round-trip flight time will be fewer than 20 d, including a maximum of 14 d spent on the lunar surface. No permanent habitat facility will be developed. The crew will live in and work out of the lander vehicle on each mission because of the short surface stay time and the fact that subsequent missions will visit different sites.

Nominally, the astronomical facilities will require crew servicing once every 3 yr after they become operational. In the off-servicing years, crews will explore other lunar sites. During these exploration sorties, crew members will make several trips in an unpressurized rover for distances up to 10 km.

Lunar Observatory Study Results

This case study emphasized a maximum scientific return using a minimum of permanent support facilities. Crews will assemble, deploy, operate, and service the array instrumentation planned for this facility. Once the facility is operational in 2005, subsequent crews will conduct local geological exploration, scientific excursions in rovers for distances up to 10 km, and upgrades and sensor/receiver instrumentation changeout at the observatory.

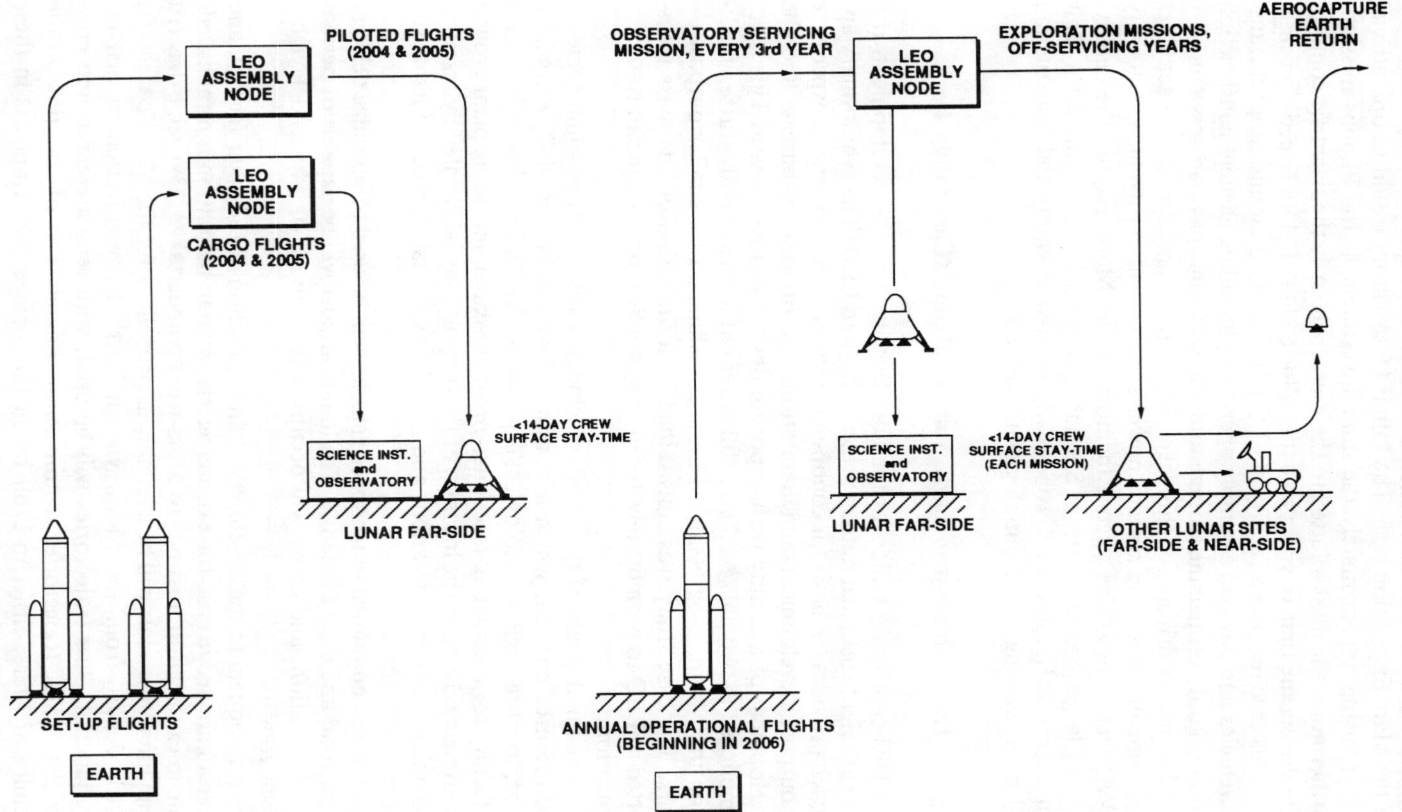

Fig. 1–6. Lunar observatory (LEO = low Earth orbit).

This scenario can be accomplished for less than one-fourth the total mass for the Mars Expedition case. The Lunar Observatory would be operational in 2 yr, using approximately the same investment as the Phobos mission. Furthermore, the mass allotted to the users (primarily the scientific community) during the first 10 yr is twice that allotted in the Mars Expedition case.

This case requires a facility in LEO to house the crew and support transfer vehicles and payload assembly operation, including element construction and checkout, propellant storage and transfer, and payload servicing.

The major drivers for this scenario are the planetary surface activities requirements, including surface power systems and extravehicular activity (EVA) technology. The science facilities on the Moon require the deployment of large, complex arrays. Special equipment is also required for their emplacement. In this regard, robotically assisted assembly and construction will be promising new technologies to investigate.

Lunar Outpost to Early Mars Evolution (Case Study 4)

This case study builds a capability that leads to the development of a self-sufficient sustained human presence beyond LEO. The evolutionary approach provides the basis for continuing technology advancement, experience in outpost development and habitation, use of local resources, and the development of a facility with opportunities for further growth. This is accomplished in two phases: the establishment of a permanently staffed facility on the Moon, progressing to the establishment of a similar outpost on Mars. The case study was constrained by a limitation on the mass transported to LEO to promote new technology applications and in situ resources utilization.

The lunar phase of the mission includes development of a lunar science and resource outpost, which is dominated by a lunar liquid O_2 extraction plant, local-to-regional geological exploration, and a life sciences laboratory facility for conducting fractional-gravity research, including plant growth experiments. Because the location of the outpost may be dictated by resource and operational considerations and not observational science, a farside site is not mandatory.

Cargo and piloted lunar space vehicles will be used to optimize delivery of payload and crew exchange. The lunar outpost will be capable of permanent habitation, and crews will occupy it for periods of 24 wk to 1 yr between rotations.

Subsequent to the development and operation of the lunar facility, and after a knowledge base for extraterrestrial human habitation is established, human exploration missions to Mars will be undertaken. Oxygen extracted from lunar materials will be available as propellant for the Mars spacecraft, which depart from the Moon via an Earth flyby injection maneuver. Conjunction-class trajectories will be used, with separate cargo and crew vehicles. These trajectories require approximately 1 yr of stay time in the vicinity of Mars, either in orbit or on the surface. The emphasis in these scenarios is the local-to-regional geological exploration of the surface of Mars, using piloted and robotic mobility systems, and the exploration of Phobos and Deimos, with the objective of establishing the capability to extract propellant from one of the moons to support subsequent missions. The study envisions three missions to the Mars system, each to a different site, in preparation for the establishment of a permanent outpost on Mars.

Lunar/Mars Evolution Mission Description

Beginning early in the next century (2004), a series of piloted and cargo flights would embark for the Moon. As illustrated in Fig. 1–7, the crew will transfer to the Moon aboard chemically propelled transfer vehicles. Several

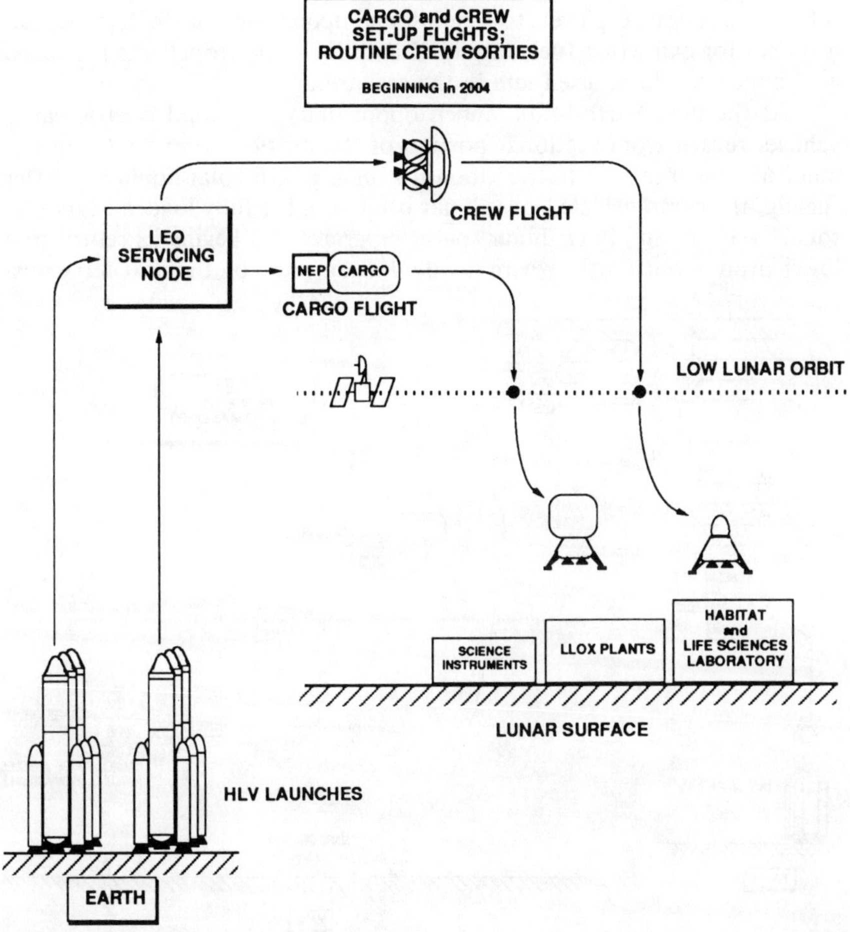

Fig. 1–7. Lunar outpost to early Mars evolution: Lunar operations (LEO = low Earth orbit, LLOX = lunar liquid oxygen, NEP = nuclear electric propulsion, HLV = heavy launch vehicle).

years will be spent in constructing a permanently staffed surface facility. Experience will be accumulated in all aspects of long-duration human planetary exploration missions: life sciences, psychological effects and human dynamics, utilization of local resources, and scientific exploration. One goal of the base is to extract O_2 from lunar soil which will be needed for subsequent Mars flights.

In approximately 2010, the branch to Mars would take place. The nominal scenario for this phase is depicted in Fig. 1–8, 1–9, and 1–10. The specific timing is left open, but in general would occur when the lunar capability is sufficient to provide enough propellant to enable the Mars mission. First, the electric cargo vehicle would carry the Mars surface equipment, excursion modules for transportation between Mars and Phobos, and various types of scientific equipment to the Mars system. As the spacecraft approaches the Mars system, it would drop off communications satellites in synchronous orbit, send robotic explorers to Deimos, and upon arrival at Phobos, deposit a system for extracting fuel. Liquid H and liquid O_2 propellants produced on Phobos would be used late in the scenario.

At the next Earth-Mars launch opportunity, a second electric cargo vehicle, reused from the lunar portion of the mission, would take an unmanned crew transport to the Moon for fueling with lunar liquid O_2. After fueling, the cargo vehicle leaves lunar orbit with the fully loaded Mars personnel vehicle, and in *cis*-lunar space, separates and begins its return to a lower orbit about Earth, where it would await reuse on the next ferry mis-

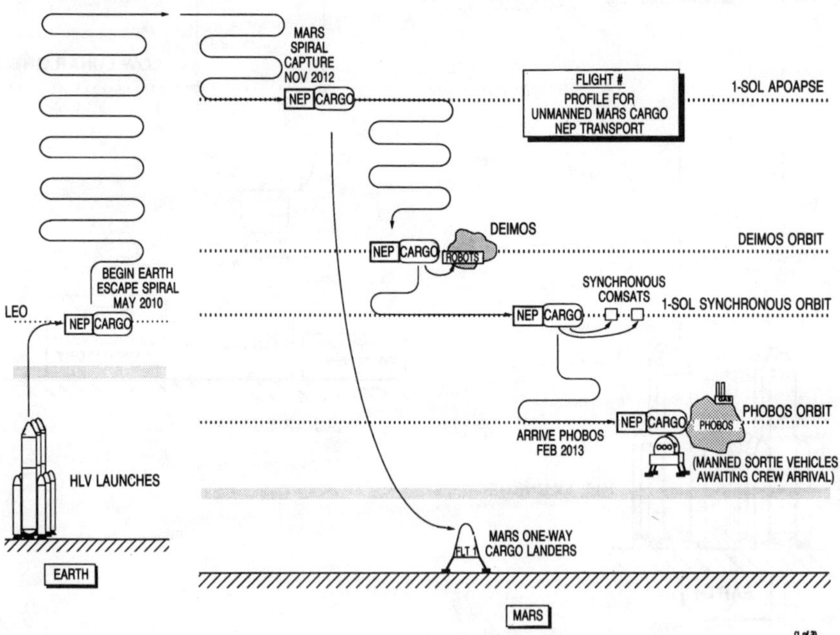

Fig. 1–8. Lunar outpost to early Mars evolution: Flight no. 1 Mars operations (LEO = low Earth orbit, NEP = nuclear electric propulsion, HLV = heavy launch vehicle).

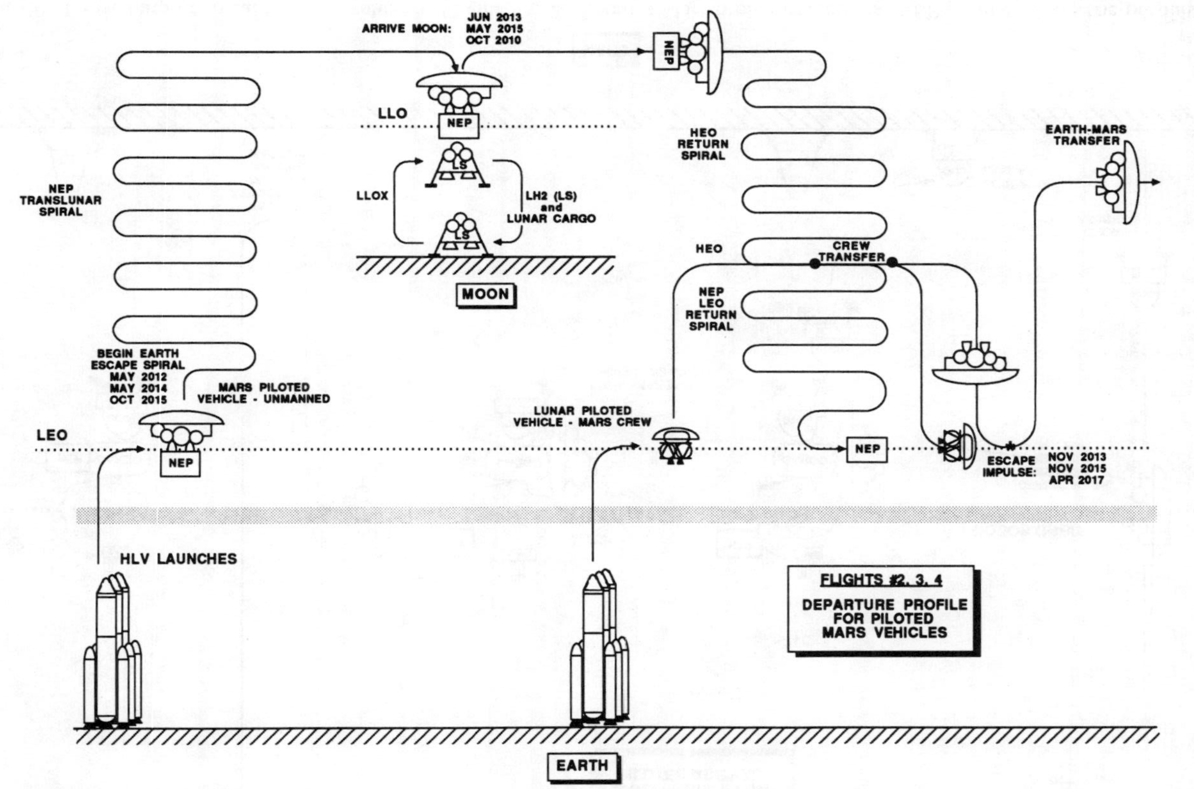

Fig. 1-9. Lunar outpost to early Mars evolution: Flights no. 2, 3, and 4 Mars operations (LEO = low-Earth orbit, LLOX = lunar liquid O_2, LH2 = lunar hydrogen, NEP = nuclear electric propulsion, LLO = low lunar orbit, HLV = heavy launch vehicle, HEO = high Earth orbit, LS = lunar surface).

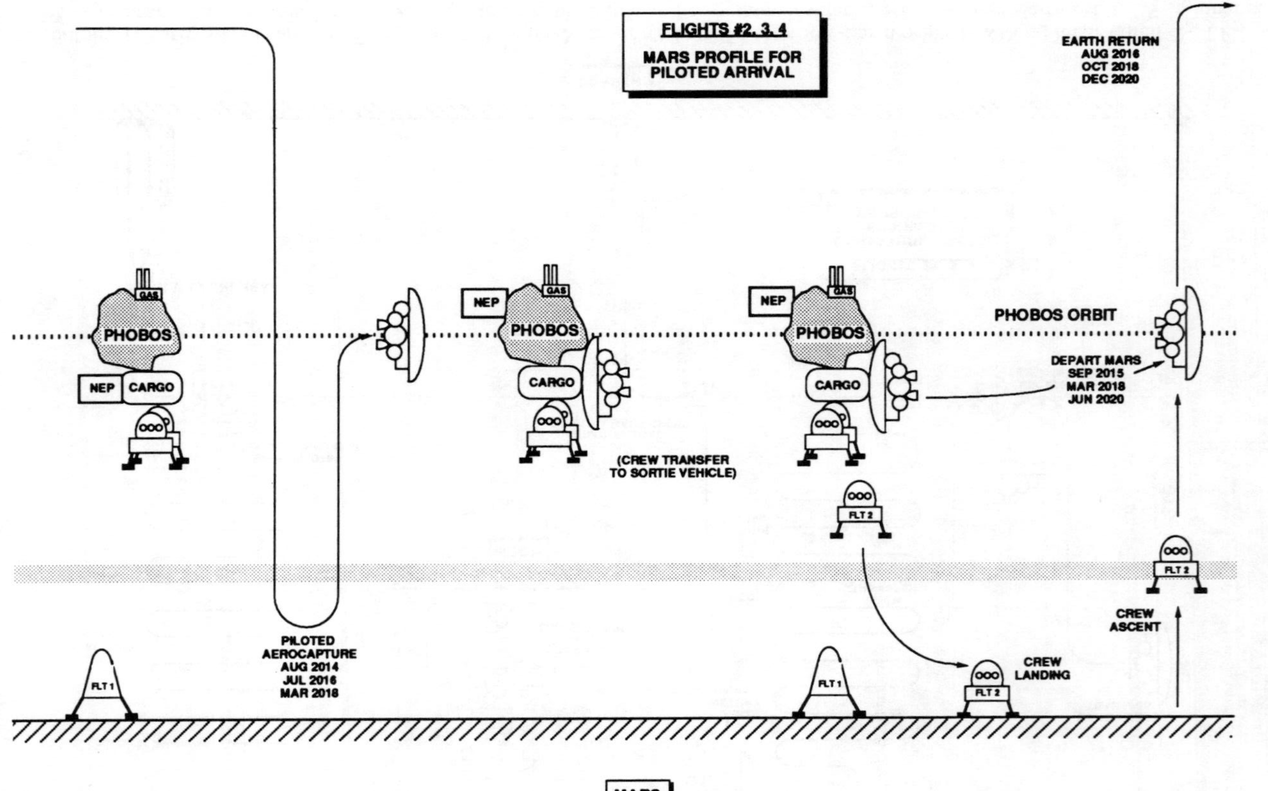

Fig. 1-10. Lunar outpost to early Mars evolution: Flights no. 2, 3, and 4 Mars-orbital operations (NEP = nuclear electric propulsion).

sion. The first crew is transported to the piloted Mars transfer vehicle, and after systems check, they begin their journey to Mars, arriving 32 wk later. The nominal plan is for the crew to stay at Mars approximately 1 yr, and return to Earth after a total mission time of nearly 3 yr. Options exist for the crew to perform a flyby abort mission (e.g., if a problem occurs in-flight to Mars, the mission can return to Earth without landing on Mars, after a total trip time of about 500 d), or to limit their stay at Mars to up to 60 d. A third option exists for a 2-yr stay at Mars.

Piloted excursions to Mars, similar to the first described above, are anticipated at the ensuing launch opportunities. Further cargo flights would be necessary over the duration of the Mars base buildup.

Lunar/Mars Evolutionary Study Results

The evolutionary case study places major demands on LEO operational facilities to assemble, refuel, maintain, and service interplanetary vehicles. The facilities will also be used to transfer substantial quantities of propellants arriving from Earth, and will serve as a transfer facility for crews going to and returning from planetary missions.

This case study also requires orbital staging and refueling operations in low lunar orbit, as well as in the vicinity of Phobos. The electric cargo vehicle serves as a mobile node for operations outside LEO. Therefore, nuclear electric propulsion system technology is a major requirement. Systems and techniques for aerocapture at Mars and Earth are also needed.

Life sciences precursor missions and studies must resolve the issue of zero gravity vs. artificial gravity for extended voyages to Mars. In this case study, that research would be performed in the one-sixth gravity of the lunar surface and at the LEO node. For permanent lunar and Mars bases, maximum possible closure of life support systems must be provided, and significant improvements over Space Station Freedom life support systems desirable for Mars transfer vehicles.

The self-sufficiency embodied in this case levies requirements for development of technology for mining, processing, and storing local resources. Significant power levels to operate surface habitats, systems, and rovers are required. Of necessity, nuclear power sources like that of the SP-100 program are the most promising candidate technologies (French, 1985). However, extensions to the multi-megawatt range must be made.

ECONOMIC VIABILITY

All of the case studies above have one common element. Do the required resources justify the benefits? This is an ever-prevalent question in space development and is particularly difficult to answer because most benefits are hard to quantify. Another way of asking this question is: "Are the benefits realizable without an unacceptable burden to the U.S. economy, both in the short term and the long term?" In other words, "Are these case studies economically viable?'' It is in pursuing the answers to this question that leads many NASA mission planners to feel that the evolutionary scenario is in the best interests of the country. Expedition scenarios could allow the USA significant firsts, but may put the space development community in the same position as after Apollo; a premier, but eroding capability developed at great expense without the continued authority to use it. The potential benefit of a Lunar Observatory is large, but is not clear that is contains enough benefits during development to sustain sufficient enthusiasm, since we have already been to the Moon. A carefully planned and executed evolutionary approach, on the other hand, has the potential of maintaining interest, yet guaranteeing American leadership in space and technology development in the years to come. Essentially, this amounts to learning to live in space permanently, with eventually little to no support required from the Earth. For this to be possible, a number of issues need to be addressed.

EVOLUTIONARY APPROACH

As stated earlier, the evolutionary approach itself is a major factor in demonstrating economic viability of the permanent habitation of space. The phenomenon of growth in any context is essentially the same, continued expansion based on previously developed infrastructure and experience. Space will need to be developed in the same fashion. Common technologies and systems need to be defined and supported, then reused whenever feasible. In this respect, the LEO Space Station common module may be the initial habitat on the Moon, and perhaps on Mars. Every good program manager knows that new untried technology adds significant risk to the program. Many times this is unavoidable, but the parameters and options need to be carefully developed and understood.

HUMAN/MACHINE PRODUCTIVITY

The controversy on the role of humans in space vs. machines has raged on since the beginning of the space program. Some suggest that humans should be involved in every aspect of space development. Others claim that humans have no business in space at all; that machines are just as capable, cheaper, and more dependable. Machines are effective doing routine tasks in well-known environments. Humans are good in situations calling for innovative solutions to unforeseen circumstances. Space Shuttle operations are rife with situations in which the outcome could have been disastrous, had humans not been there to work out solutions. And this is in a well-known environment, with procedures that have been practices hundreds of times. When we find ourselves in the relatively unknown environment beyond LEO for extended periods of time, performing new procedures, the human ability to do problem solving with only sketchy information will be extremely valuable. This is not to say that machines will not have a significant role.

Many activities can be performed without human intervention or with perhaps only consultative responsibilities. This balance between humans and machines needs to be carefully considered to obtain the optimum evolving synergy, to improve the economic viability of permanent settlements.

LIFE SUPPORT

Since the trip from the Earth to LEO is so expensive, any trips that can be eliminated will greatly enhance space operations. One method of doing this is to recycle consumables as much as possible, limiting the number of resupply flights that are needed. The first step in accomplishing this may be the recycling of air and water with physical-chemical regenerative life support systems. This may evolve to the capability for growing food, to the point where almost no additional consumables are needed for life support. This technology is called the controlled ecological life support system (CELSS) (MacElroy et al.; Averner, see Chapter 11 in this book; Prince & Knott, see Chapter 12 in this book; MacElroy et al., see Chapter 13 in this book; Henninger, see Chapter 14 in this book), where air and water are recycled and all food is produced in a relatively small closed or tightly controlled environment, using probably some combination of physical/chemical and biological systems. The task may be simplified somewhat by using locally available resources to augment the resources imported from Earth. This, of course, is the primary subject of this book, to discuss development of lunar or martian-derived soils to serve as a substrate for plant growth. It may also be viable to isolate other products from the lunar regolith, particularly H, C, N, and O (e.g., Gibson & Knudsen, 1985; Williams, 1985; Cutler, 1985; Tucker et al., 1985; White & Hirsch, 1985; Blanford et al., 1985). An important question that remains however, is the level of activity required to justify the added initial expense of developing and deploying the systems needed to do this.

IN SITU RESOURCES UTILIZATION

This is the final and perhaps the most important issue that will be discussed here. The location and concentrations of the all materials on the Moon and Mars are largely unknown. It appears that the terrestrial planets all proceeded down similar evolutionary paths until some unknown factor caused them to diverge. Therefore, it is conceivable that all the materials needed for Earth-like human activities exist on the Moon or Mars. Plus, there is some indication that materials may exist that are extremely rare or nonexistent on the Earth. Once again, the expense of developing and deploying the critical mass of equipment for research and operations needs to be weighed against both near-term and long-term potential.

OFFICE OF EXPLORATION FUTURE ACTIVITIES

The Office of Exploration (OEXP) will continue definition of these case studies with added depth in areas such as propellant production from lunar, Mars, or Phobos materials. In addition, attention is being paid to potential pathways for the execution of these scenarios; and the evaluation criteria are being further defined. The new 1989 Bush administration is sure to have an impact on future development. In fact, the next administration will probably be in a position to define NASA's course for coming years, as no administration has been able to since John F. Kennedy.

REFERENCES

Blanford, G.E., P. Borgensen, M. Maurette, W. Möller, and B. Monart. 1985. Hydrogen and water desorption on the Moon: Approximate, on-line simulations. p. 603–609. *In* W.W. Mendell (ed.) Lunar bases and space activities of the 21st century. Lunar and Planetary Inst., Houston.

Cutler, A.H., and P. Krag. 1985. A carbothermal scheme for lunar oxygen production. p. 559–581. *In* W.W. Mendell (ed.) Lunar bases and space activities of the 21st century. Lunar and Planetary Inst., Houston.

French, J.R. 1985. Nuclear powerplants for lunar bases. p. 99–106. *In* W.W. Mendell (ed.) Lunar bases and space activities of the 21st centruy. Lunar and Planetary Inst., Houston.

Gibson, M.A., and C.W. Knudsen. 1985. Lunar oxygen production from ilmenite. p. 543–550. *In* W.W. Mendell (ed.) Lunar bases and space activities of the 21st century. Lunar and Planetary Inst., Houston.

MacElroy, R.D., H.P. Klein, and M.M. Averner. 1985. The evolution of CELSS for lunar bases. p. 623–633. *In* W.W. Mendell (ed.) Lunar bases and space activities of the 21st century. Lunar and Planetary Inst., Houston.

NASA Office of Exploration. 1988a. Office of Exploration-Exploration Studies Tech. Rep.—FY 1988 Status. Vol. I. NASA TM 4750. NASA, Washington, DC.

NASA Office of Exploration. 1988b. Office of Exploration-Exploration Studies Tech. Rep.—FY 1988 Status. Vol. II. NASA TM 4750. NASA, Washington, DC.

NASA Office of Exploration. 1988c. Office of Exploration-Exploration Studies Tech. Rep.—FY 1988 Status. Vol. III. NASA TM 4750. NASA, Washington, DC.

National Commission on Space. 1986. Pioneering the space frontier. Bantam Books, New York.

Ride, S.K. 1987. Leadership and America's future in space—A report to the administrator. NASA, Gov. Print. Office, Washington, DC.

Tucker, D.S., D.T. Vaniman, J.L. Anderson, F.W. Clinard, Jr., R.C. Feber, J.r, H.M. Frost, T.T. Meek, and T.C. Wallace. 1985. Hydrogen recovery from extraterrestrial materials using microwave energy. p. 583–590. *In* W.W. Mendell (ed.) Lunar bases and space activities of the 21st century. Lunar and Planetary Inst., Houston.

White, D.C., and P. Hirsch. 1985. Microbial extraction of hydrogen from lunar dust. p. 591–602. *In* W.W. Mendell (ed.) Lunar bases and space activities of the 21st century. Lunar and Planetary Inst., Houston.

Williams, R.J. 1985. Oxygen extraction from lunar materials: An experimental test of an ilmenite reduction process. p. 551–558. *In* W.W. Mendell (ed.) Lunar bases and space activities of the 21st century. Lunar and Planetary Inst., Houston.

2 Strategies for a Permanent Lunar Base[1]

Michael B. Duke, Wendell W. Mendell, and Barney B. Roberts

NASA Johnson Space Center
Houston, Texas

Planned activities at a manned lunar base can be categorized as supporting one or more of three possible objectives: scientific research, exploitation of lunar resources for use in building a space infrastructure, or attainment of self-sufficiency in the lunar environment as a first step in planetary habitation. Scenarios constructed around each of the three goals have many common elements, particularly in the early phases. The cost and the complexity of the base, as well as the structure of the Space Transportation System, are functions of the chosen long-term strategy. A real lunar base will manifest some combination of characteristics from these idealized end members.

A MOON IN AMERICA'S FUTURE

The Earth is unique in the solar system, not only for harboring life, but also for its relatively massive satellite. It is speculative that the two attributes are somehow related, but certainly the Earth's companion has left cultural and biological imprints on humanity. As cumulative application of the scientific method has increased our understanding and awareness of the physical universe, fascination with the habitability of the Moon has blossomed. As late as the last century, newspaper stories reported telescopic observations of the daily lives of lunar creatures. The manned landings of the last 10 yr have dispelled such romanticism forever, but in turn have provided the technology and the information necessary to fulfill a greater dream—transport of civilization beyond the confines of Earth.

Cultural expansion is a recurring theme in human affairs. Motivations for exploration or conquest vary from resource limitations (Mongul invasions) to religion (Turkish probings of medieval Europe) to commerce (global

---

[1] Reproduced after slight alterations with permission from Lunar Bases and Space Activities of the 21st Century, W.W. Mendell (ed.), Lunar and Planetary Institute, Houston, 1985, p. 57–68. Copyright © 1985, Lunar and Planetary Institute, Houston, TX 77058-4399, USA.

circumnavigations of the 16th and 17th centuries). American history especially is permeated by the doctrine of manifest destiny. The concept of the frontier has come to symbolize for Americans the exercise of individual freedom, which in collective expression leads to social renewal. Contemporary popular writings cater to this mythos by describing for an overpopulated and confused world the "high frontier" of space. So far, the promise of space has been a reality for a few and only a vicarious experience for most. However, humanity, and the USA particularly, stands today at the threshold of a truly new world—the Moon.

The promise of the Moon is not immediately evident from examination of the current American space program. However, the space shuttle and the proposed space station can be viewed as building blocks in a general purpose space transportation infrastructure (Fig. 2–1). To service geosynchronous orbit, an upper stage is needed in addition to the shuttle. If that upper stage is provided in the form of a reusable orbit-to-orbit transfer vehicle docked at the space station, the transportation system can be multipurpose. In particular, a rudimentary lunar transportation system then will exist because the propulsion requirements for attaining geosynchronous orbit and lunar orbit are essentially identical. A lunar landing vehicle is required to place payloads on the lunar surface, but its design can be a straight forward adaptation of the orbital transfer vehicle (OTV). The space station and the reusable OTV constitute a natural evolutionary path that, when achieved, will make accessible all near-Earth space including the Moon. This "enabling

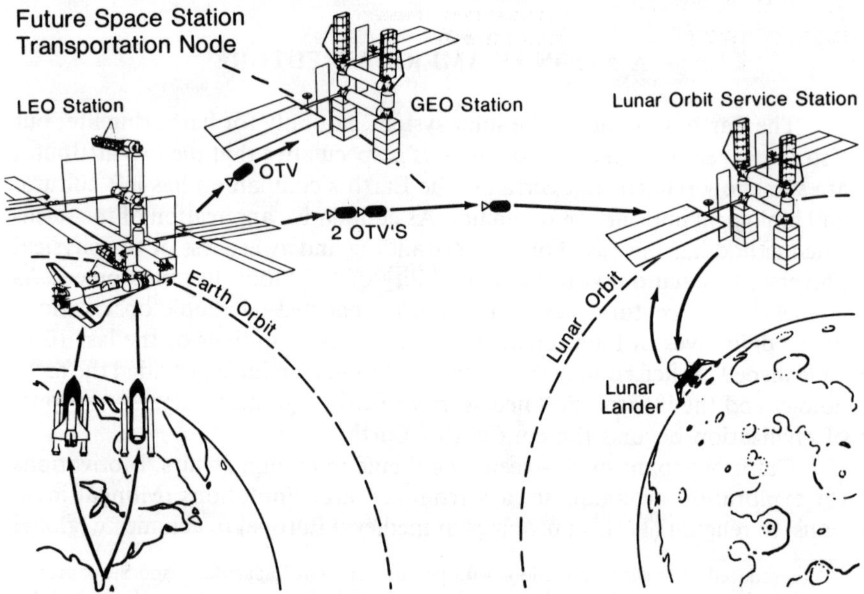

Fig. 2–1. The Space Transportation System of the future may service a station in geosynchronous orbit as well as a lunar base via a station in lunar space. The lift capacity of the Shuttle fleet may be augmented by a unmanned heavy lift vehicle, designed to ship fuel and consumables to space.

technology" is a National Aeronautics and Space Administration (NASA) target for the mid-1990s.

When the requisite technology exists, the American political process inevitably will include lunar surface activities as a major space objective. In fact, some sort of declaration may well precede the actual establishment of the space station. It is, therefore, prudent to consider the nature of a permanent manned presence on the Moon and its potential impact on the evolution of the Space Transportation System (STS).

Although the lunar base program is one in which the USA can assert its leadership in space, it is inherently international in scope and should involve as much participation as possible from other countries. Opportunities for international cooperation exist in the planning stages, in the science and technology development, and in operations at the lunar base. A legal framework will be needed to guarantee that potentially profit-making ventures adequately consider the concerns of the international community.

USES OF THE MOON

A manned lunar base can be discussed in terms of three distinct functions. The first involves the scientific investigation of the Moon and its environment and the application of special properties of the Moon to research problems. The second produces the capability to use the materials of the Moon for beneficial purposes throughout the Earth-Moon system. The last, and perhaps the most intriguing, is to conduct research and development leading to a self-sufficient and self-supporting lunar base, the first extraterrestrial human colony. Although these activities take place on the Moon, the developed technology and the established capability will benefit society on Earth as well as growing industrialization of near-Earth space.

Scientific Research

A lunar base will create new opportunities for investigating the Moon and its environment and for using the Moon as a platform for scientific investigations. Analogous to the function of McMurdo Base in Antarctica, the lunar base will provide logistical and supporting laboratory capability to rapidly expand knowledge of lunar geology, geophysics, environmental science, and resource potential through wide-ranging field investigations, sampling, and placement of instrumentation. Access to large, free-vacuum volumes may enable new experimental facilities such as macroparticle accelerators. The firm, fixed platform will enable new astronomical interferometric measurements to be obtained (Fig. 2–2). The challenge of long-term, self-sufficient operations on the Moon can spur scientific and technological advances in materials science, bioprocessing, physics and chemistry based on lunar materials, and reprocessing systems.

Fig. 2–2. A radio telescope located on the farside of the Moon would be shielded from background noise generated by terrestrial sources. Although depicted here as a parabolic dish in a convenient crater, an initial lunar instrument may well be a phased array of dipole antennas.

Exploitation of Lunar Resources

It has been argued that major industrialization of space cannot occur without access to the resources of the Moon. Studies of immense projects such as solar power satellites have demonstrated that at a sufficiently large scale, it is reasonable to develop the resource potential of the Moon to offset the high Earth-to-orbit transportation costs (Hearth, 1976). The lower gravitational field of the Moon and the absence of an atmosphere that retards objects accelerated from the surface provides a potential 20- to 30-fold advantage for launching from the Moon instead of the Earth. For example, at liftoff, about 1.5% of the space shuttle's mass is payload. Most of the mass is propellant. From the Moon, approximately 50% of the mass can be payload.

The commodity currently envisioned to be most in demand in Earth-Moon space over the next 30 yr is liquid O_2, which makes up six-sevenths of the mass of propellant used by cryogenic (H_2/O_2) rockets, such as the Centaur or postulated OTV's. Although it would appear unlikely that an atmosphereless body is a source for O_2, it is actually an abundant element on the Moon (Arnold & Duke, 1978). It must be extracted, however, from silicate and oxide minerals into its liquid form for use as a propellant. Several processes have been suggested (Criswell, 1980) for accomplishing this, including reduction of raw soil by fluorine (which is recovered) or reduction of iron-titanium oxide (ilmenite) by H (also recovered). Preliminary laboratory studies have verified the concepts behind some of these processes.

Systems studies (e.g., Carroll et al., 1983) show that O_2 production on the Moon could benefit STS in the early years of the next century, even if

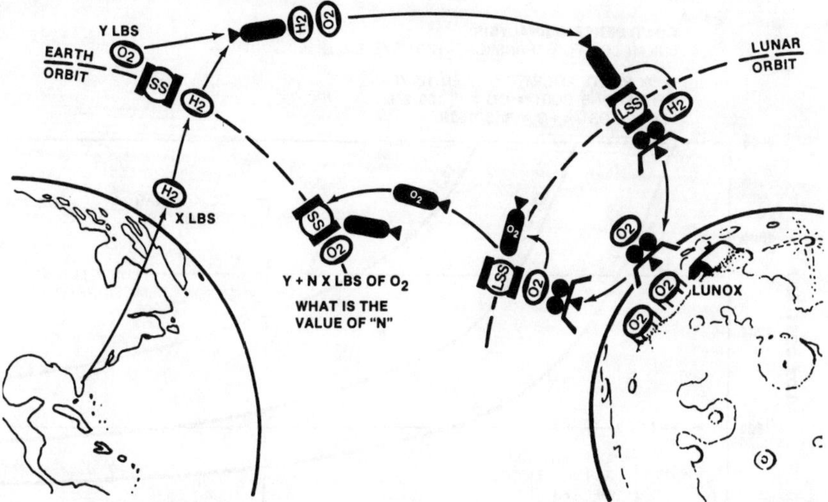

Fig. 2–3. Liquid O_2 fuel (LOX), manufactured on the Moon and delivered to low Earth orbit (LEO) may become a profitable export for a lunar base. A critical parameter in the analyses of the system is the mass payback ratio, defined as the ratio of the excess lunar LOX in LEO to the liquid H fuel delivered from Earth LEO.

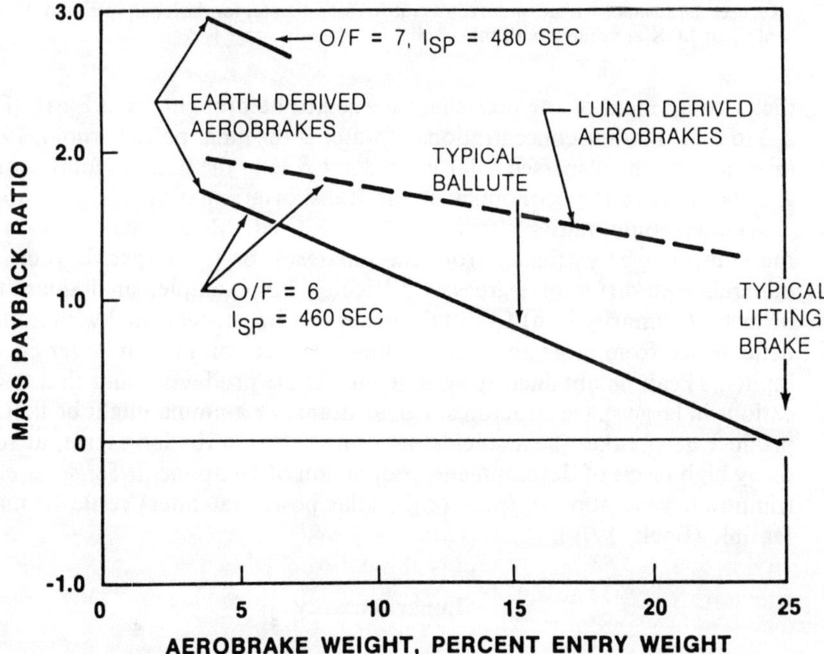

Fig. 2–4. The mass payback ratio for lunar LOX delivered to LEO is sensitive to the design characteristics of the orbital transfer vehicle (OTV) used as a lunar freighter. The fractional mass of the OTV aerobrake and the oxidizer to fuel ratio are key parameters. Manufacture of aerobrakes on the Moon would enhance system performance.

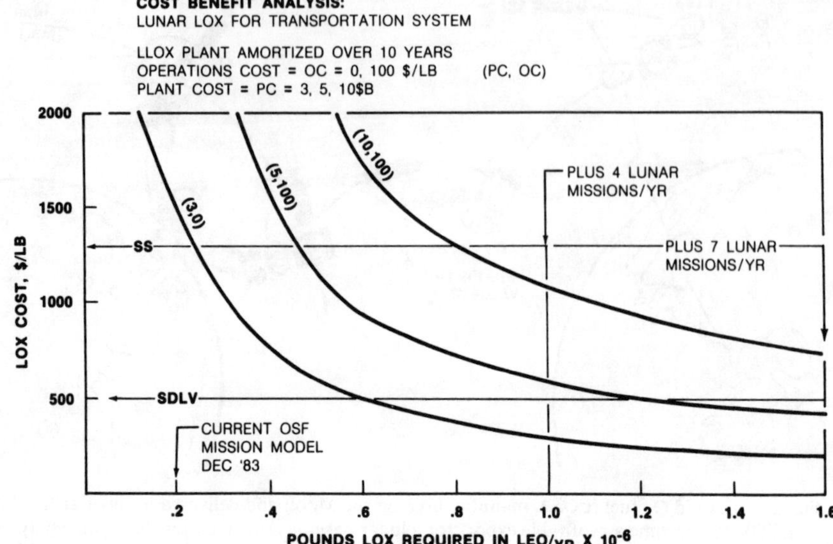

Fig. 2–5. A simple cost-benefit analysis assumes that a lunar O_2 production facility has its capital costs amortized solely by "profits" on delivery of LOX to LEO. While lunar O_2 is competitive with Shuttle delivery in all cases, introduction of a cost-efficient heavy lift vehicle reduces the advantage under more conservative cost estimates for the lunar operation. If costs of lunar LOX are shared with other activities, the advantage is restored.

the H component of the propellant needed to be brought from Earth (Fig. 2–3 to 2–5). Finding concentrations of water at the lunar poles (Arnold, 1979) or extracting the dispersed solar wind-derived H in the lunar regolith would greatly improve the economics of the transportation system.

Other commodities also could be produced. Metals, such as iron or titanium, can be extracted from the lunar soil or from specific rocks or minerals with differing degrees of difficulty. For example, small quantities of metal (primarily iron) from meteorites can be concentrated with a magnetic device from large amounts of lunar soil, or, with much larger energy inputs, Ti can be obtained from ilmenite. These products could find applications in large space structures. Lunar titania or alumina might be used to produce aerobrakes (heat shields) used in OTVs. In the long term, at relatively high levels of development, production of components for solar electric power generation in space (e.g., solar power satellites) could be made feasible (Bock, 1979).

Lunar Autarky

A self-sufficient lunar base is a possible long-term objective that creates new challenges in planning and development. In the near term, emplacement of a controlled environment capsule on the Moon involves known technology. The initial concept for a lunar habitat module is simply an extension of the

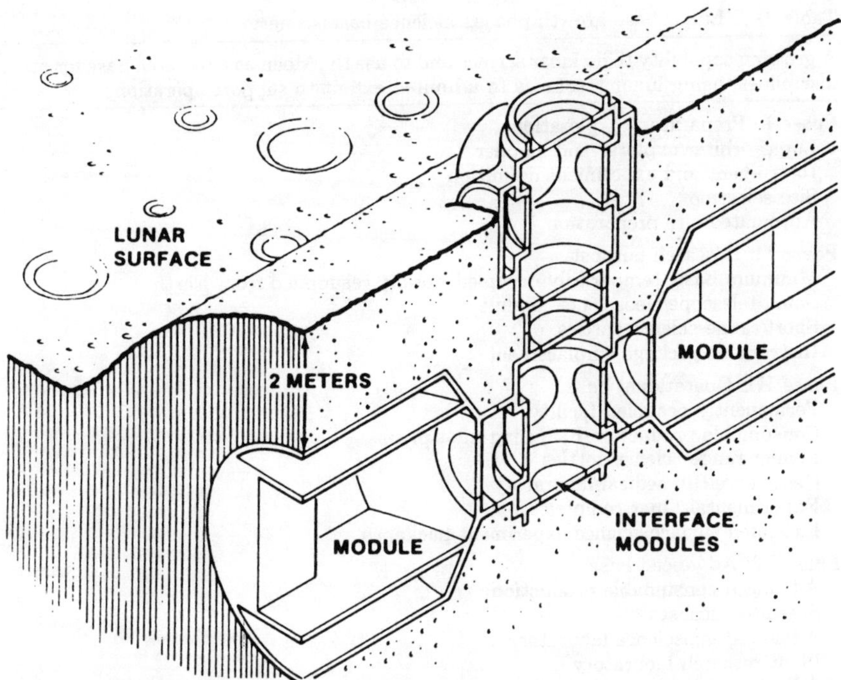

Fig. 2-6. The first lunar base habitats and laboratories could be space station modules, buried in the lunar regolith for protection from solar flare radiation. Interface modules not only interconnect the buried structures but also can be stacked to create exits to the surface.

design experience from Apollo, Skylab, the space shuttle, and space station (Fig. 2-6). A different perspective is required to plan systems that can use the Moon's native materials and energy sources to produce a self-sufficient capability.

Most of the generic technologies for an advanced system are similar to those employed in general space operations (life support, power, thermal control, communications, logistics, and transportation), but they must be modified to use lunar materials for growth and extension. Ultimately, the desire to minimize or eliminate the resupply link from Earth requires a host of applications, new to the space program, carried to new levels of system reliability. Exploration of technologies such as lunar metallurgy, ceramics, manufacturing processes, power systems, and others, will reveal whether autarky is a realistic objective and can prepare the way for achieving it at an operational base. Perhaps this is the most compelling rationale for a lunar base program, as it promises eventual self-sufficiency elsewhere in the solar system.

PHASED EVOLUTION OF A LUNAR BASE

We loosely define three scenarios, each based on one of the long-term rationales described above: scientific research, production, and self-

Table 2-1. Lunar base growth phases: Science base scenario.

A growing capability to do lunar science and to use the Moon as a research base for other disciplines, using lunar resources to a limited extent to support operations.

Phase I: Preparatory exploration
  Lunar orbiter explorer and mapper
  Instrument and experiment definition
  Site selection
  Automated site preparation

Phase II: Research outpost
  Minimum base, temporarily occupied, totally resupplied from Earth
  Small telescope/Geoscience module
  Short-range science sorties
  Instrument package emplacement

Phase III: Operational base
  Permanently occupied facility
  Consumable production/Recycling pilot plant
  Longer range science sorties
  Geoscience/Biomedical laboratory
  Experimental lunar radiotelescope
  Extended surface science experiment packages

Phase IV: Advanced base
  Advanced consumable production
  Satellite outposts
  Advanced geoscience laboratory
  Plant research laboratory
  Advanced astronomical observatory
  Long-range surface exploration

Table 2-2. Lunar base growth phases: Production base scenario.

A lunar base that is intended to develop one or more products for commercial use. Manned activity may be continuous, but a high degree of automation is expected.

Phase I: Preparatory exploration
  Lunar orbiter explorer and mapper
  Lunar pilot plant definition
  Site selection
  Automated site preparation

Phase II: Research outpost
  Minimum base, temporarily occupied, totally resupplied from Earth
  Surface mining pilot operation
  Lunar O_2 pilot plant
  Lunar materials utilization research module

Phase III: Operational base
  Permanently occupied facility
  Expanded mining facility
  Consumables supplied locally
  Oxygen production plant
  Lunar materials processing pilot plant(s)

Phase IV: Advanced base
  Large-scale O_2 production
  Ceramics/Metals production facility
  Locally derived consumables for industrial use
  Industrial research facility

Table 2–3. Lunar growth phases: Lunar self-sufficiency research base scenario.

A lunar base that grows in its capacity to support itself and expand its capabilities using the indigenous resources of the Moon, with the ultimate objective of becoming independent of Earth.

Phase I: Preparatory exploration
 Lunar orbiter explorer and mapper
 Process definition
 Site selection
 Automated site preparation

Phase II: Research outpost
 Minimum base, temporarily occupied, totally resupplied from Earth
 Surface mining pilot operation
 Lunar O_2 production pilot plant
 Closed systems research module

Phase III: Operational base
 Permanently occupied facility
 Expanded mining facility
 Lunar agriculture research laboratory
 Lunar materials processing pilot plant(s)

Phase IV: Advanced base
 Lunar ecology research laboratory
 Lunar power station—90% lunar materials-derived
 Agricultural production pilot plant
 Lunar manufacturing facility
 Oxygen production plant
 Lunar volatile extraction pilot plant

Phase V: Self-sufficient colony
 Full-scale production of exportable O_2
 Volatile production for agriculture, Moon-orbit transportation
 Closed ecological life support system
 Lunar manufacturing facility: tools, containment systems, fabricated assemblies, etc.
 Lunar power station—100% lunar materials-derived
 Expanding population base

sufficiency (Tables 2–1 to 2–3). Each scenario passes through several phases, some of which are common to the other scenarios. The distinction among the three views lies with the culminating phase of each.

Precursor Exploration

Because the scientific data base is incomplete, particularly in the polar regions, the first step in Phase I is global mapping of the Moon, both with relatively high resolution imagery and with remote-sensing measurements to determine the chemical variability. This task can be accomplished with an unmanned satellite, a Lunar Geochemical Orbiter (LGO) (Minear et al., 1977), which is a proposed mission in NASA's planetary program and could be flown in the 1990–1992 time frame. The LGO is in the Planetary Observer mission class, a low-cost approach to planetary exploration recommended by the report of the Solar System Exploration Committee (1983). Secondly, Phase I should include research on technologies necessary to exploit lunar resources. Technology development in resource problems on Earth is typically a long lead time process. At the conclusion of Phase I, the initial site for a base will have been defined and planned activities understood in some detail. Concurrently with this preliminary phase in the lunar program, development of a space station and on OTV capable of supporting a lunar base would be carried out in NASA's STS program.

Research Outpost

At Phase II, an initial surface facility would establish limited research capability for science, materials processing, or lunar surface operations. Depending on the long-term objectives of the lunar base program, the detailed studies and the experimental plans start to diverge at this phase for the different scenarios. A focus on lunar science and astronomy would result in local geological exploration, the establishment of a small astronomical observatory, and emplacement of automated instruments. If production were to be the focus, a pilot plant for lunar O_2 extraction could be set up instead, and study of the fabrication of aerobrakes from lunar material could be initiated. If the program goal pointed to achieving self-sufficiency, the emphasis at this stage could be on agricultural experiments using lunar soil as substrate and recycling water, O_2, and CO_2.

To accomplish Phase II in any of the scenarios, the STS must have the capability of landing and taking off from the Moon, transporting manned capsules (about 10 000 kg) to and from the lunar surface, and delivering payloads of about 20 000 kg to the lunar surface. This involves delivering approximately 40 000 kg into lunar orbit using OTV's. The requirement for storage of the return vehicle on the Moon for extended periods (2–12 wk) may require new high-performance, storable propellant systems at this phase of development.

Permanent Occupancy

At Phase III, permanent occupancy is the objective. The surface infrastructure would include greater access to power, better mobility in and away from the base, and more diversified research capability. Still, depending on the long-term objectives, the nature of the base can vary. A science base might emphasize long-range traverses for planetological studies or extension of observational capability with larger telescopes. A production base will incorporate highly automated systems to produce and transfer liquid O_2 for use in the transportation system. Advanced research for a self-sufficient base would be making the first extensions of the base using indigenous materials. The production and the self-sufficiency scenarios require a small cousin to the Earth-orbit space station in lunar space (lunar orbit or an Earth-Moon libration point) to provide for transfer, refueling, and maintenance of the lunar lander and the OTVs.

Advanced Base

The advanced base, Phase IV, is even more specialized. Depending on the long-term plan, it produces more types or a greater range of scientific investigations, adds products to the growing lunar industrial base, or enters a phase of significant expansion of capabilities using lunar materials as the majority of the feedstock. This is the terminal phase for science and production scenarios. Growth may occur by enlarging the number of experiments or products produced on the Moon, but a self-sustaining capability is not included. The production base might develop toward a highly automated state where permanent occupancy was unnecessary. For the production and independence scenarios, the base should begin paying its own operational costs. In the self-sufficiency scenario, research and development of pilot plants aimed at a broad range of indigenous lunar technologies would be pursued. The final phase of the self-sufficient scenario is a truly autarkic settlement, a lunar colony, in which the link to Earth can be discretionary.

EVOLUTION OF THE PROGRAM

Figure 2–7 ties the possible development of a lunar base to the growth of lunar resource support of the transportation system. Initially, the base is totally dependent on terrestrial supply where 7 kg in low Earth orbit (LEO) is required to place 1 kg on the lunar surface. With the introduction of lunar O_2 first into near-Moon operations and then into the return leg of the transportation system, the slope of the curve changes from 7:1 to 3.5:1. As the lunar manufacturing capability increases to the point where aerobrakes can be manufactured, the slope decreases to something slightly greater than 1:1. Further growth of lunar capability allows expansion of the base mass to be more or less independent of the quantity of imported terrestrial mass. At the point of self-sufficiency, only trace minerals and crew changeout are chargeable weights to lunar operations; the slope of the curve in Fig. 2–7 is essentially flat.

Another consideration in the growth of lunar activities is the economic "balance of trade" between Earth orbit and the lunar surface. The value of lunar products may support lunar operations before a true mass balance is achieved. It is difficult to calculate the economic value of lunar O_2 and other products in LEO. However, these lunar credits are shown qualitatively in Fig. 2–7 at the point where a closed ecological life support system (CELSS) and a significant manufacturing capability are available. The slope of the credits line will be a function of many things, such as the amount of O_2 required to support nonlunar activities, the value and quantity of lunar resources required in LEO, and the more intangible value of science and research enabled by the lunar base. Finally, the dashed line of constant slope indicates the continued total dependency that would exist if these technologies are not pursued on the Moon, that is, if a self-sufficiency element is not included in the lunar base program.

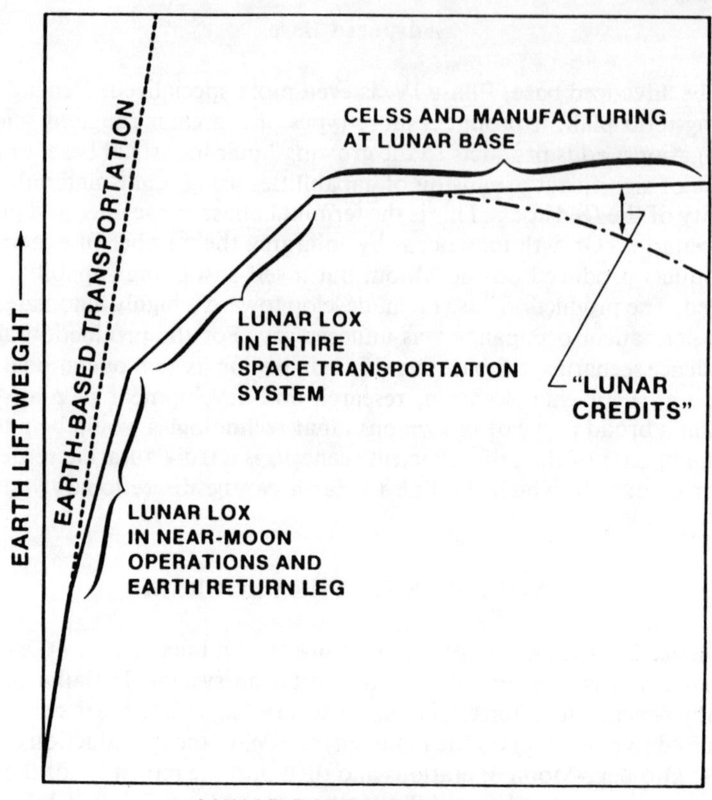

Fig. 2–7. Initially, almost 7 kg must be lifted into LEO for every kilogram landed on the Moon. As lunar O_2 is introduced into the transportation system, the ratio improves as a unit mass goes from Earth to Moon with only little overhead in the system. In a Phases IV advanced base, the growth of lunar-surface infrastructure becomes only weakly dependent on imports from Earth. A favorable balance of trade is ultimately conceivable.

The real lunar base will evolve as some combination of the above scenarios. Determination of the right mix requires research, development, and debate. Even if a program is started now, several years should be devoted to the study of the detailed lunar base scenario. The time is available because the development of the space transportation infrastructure and the completion of the orbital science survey will take 7 to 10 yr. Proper preparation will make it possible to decide on a specific lunar base design in the mid-1990s. That time frame is consistent with the development of the infrastructure that will enable the lunar base program to be carried out to its full potential. The first manned landings could occur early in the first decade of the next century; permanent occupancy could be achieved by the year 2007, the 50th anniversary of the Space Age.

There are potential technological problems that may slow lunar base development. At each phase, serious questions will also arise about whether to proceed and how and when to proceed. A commitment need not be made now to the whole plan. Nevertheless, the long-term objective is one of immense significance in history and should not be casually discarded. It is inevitable that people will settle the Moon and other bodies in the Solar System. We live in a generation that has already taken significant steps along that path. With careful planning, we can nurture the capability to move from the planet, to provide benefits to the Earth, and to satisfy humanity's spirit of adventure.

REFERENCES

Arnold, J.R. 1979. Ice in the lunar polar regions. J. Geophys. Res. 84:5659–5668.

Arnold, J.R., and M.B. Duke. 1978. Summer workshop on near-Earth resources. NASA CP-2031. NASA, Washington, DC.

Bick, E.H. 1979. Lunar resources utilization for space construction. Final report for contract NAS9-15560. General Dynamics Convair Div., Advanced Space Programs, San Diego, CA.

Carroll, W.F., W.H. Steurer, R.H. Frisbee, and R.M. Jones. 1983. Extraterrestrial materials— Their role in future space operations. Astronaut. Aeronaut. 21:80–85.

Criswell, D.R. 1980. Extraterrestrial materials processing and construction. Final Rep. Contr. NR/SR 09-051-001. Lunar and Planetary Inst., Houston, TX.

Hearth, D.P. 1976. Outlook for space. NASA SP-386. NASA, Washington, DC.

Minear, J.W., N. Hubbard, T.V. Johnson, and V.C. Clarke, Jr. 1977. Mission summary for lunar polar orbiter. JPL Doc. 660-41, Rev. A, Jet Propulsion Lab. California Inst. of Technology, Pasadena, CA.

Solar System Exploration Committee. 1983. Planetary exploration through the year 2000: A core program. U.S. Gov. Print. Office, Washington, DC.

3   The Environment at the Lunar Surface

G. Jeffrey Taylor
University of New Mexico
Albuquerque, New Mexico

The Moon's geologic environment is dramatically different from Earth's and presents fascinating challenges to engineers designing facilities on the lunar surface. This chapter summarizes the geologic nature of the stark lunar surface and its tenuous atmosphere.

GENERAL CHARACTERISTICS

The strength of the Moon's gravitational field is about one sixth that at Earth's surface; the surface gravity is 1.62 m s^{-2} and the escape velocity is 2.37 km s^{-1}. The lower gravity allows use of materials of lower strength than on Earth for structures of equivalent size. Alternatively, much larger structures can be built on the Moon. The Moon has a slow sidereal (the time it takes to complete one revolution) rotation period of 27.3 Earth days, so days and nights each last almost 2 wk. Consequently, solar energy systems require some way to store energy and plants will need artificial light during the long lunar night. Finally, because of the Moon's smaller radius, its surface has a larger curvature than does the Earth's surface. For example, a chord along a 10-km baseline would have a bulge along it of 7.2 m; a 60-km baseline would have a bulge of 260 m.

STABLE PLATFORM

The Moon provides a stable platform on which to build lunar structures. Seismic properties are summarized in Table 3–1, which is adapted from Goins et al. (1981). There are two main categories of lunar seismic signals, based on the depth at which they originate. Almost all occur deep within the Moon at depths of 700 to 1100 km; on the average, about 500 deep events were recorded during the 8 yr that the Apollo network operated. These deep moonquakes are related to tidal forces inside the Moon.

Moonquakes also occur at much shallower depths (<200 km), but apparently below the crust (Nakamura et al., 1979). They occur much less frequently than do deep moonquakes, only about five per year. Shallow moonquakes do not appear to be related to tidal flexing of the Moon or to surface features. For comparison, most earthquakes occur at depths of 50 to 200 km.

Lunar seismic activity is drastically less than terrestrial seismicity (Table 3-1). Lunar seismographs detected only 500 quakes per year. In contrast, 10 000 detectable quakes occur each year on Earth. Note that the magnitudes of detectable quakes is different on Earth and the Moon, due mostly to greater seismic noise on Earth. In fact, most moonquakes are in the magnitude 1 to 2 range on the Richter scale, which is in the Earth's seismic background.

The total seismic energy released in the Moon is about 10^8 times less than in Earth. The magnitudes of the largest events on the Moon are also much less than the largest events on Earth (Table 3-1). The most energetic lunar events are the shallow ones, the largest recorded one being only 4.8 magnitude, corresponding to an energy of 2×10^{10} J. The largest recorded earthquake measured 9.5 magnitude, corresponding to an energy of 10^{19} J.

Seismic waves are intensely scattered near the lunar surface. This causes the energy of the waves arriving at a given point to be spread out, so the damaging effects of a moonquake would be less than those of an earthquake of the same magnitude. (In fact, values of seismic energy and magnitudes that Goins et al. (1981) reported for the Moon are greater than those that Lammlein et al (1974) reported because the latter authors had not accounted for scattering of seismic waves near the lunar surface or for some instrument effects.) Consequently, it appears that the lunar surface is far more stable than any place on Earth.

The scattering of seismic waves in the Moon is significant down to a depth of 25 km, but is most intense in the upper few hundred meters. This implies a lack of coherent layering in this region.

Tidal forces raise and lower the lunar surface about as much as on Earth, where body tides deflect the ground about 10 to 20 cm twice each day. Because the Moon is locked into a synchronous orbit, the main tidal bulge on the Moon is a permanent feature. Nevertheless, small tidal deflections stemming

Table 3-1. Comparison of moonquake and earthquake intensities. From Goins et al. (1981).

	Moon	Earth
No. of events/yr	5 shallow (m $>$ 2.2)† 500 deep (m $>$ 1.6)†	10^4 (m $>$ 4)
Energy release of largest event	2×10^{10} J (shallow) 1×10^6 J (deep)	10^{19} J
Magnitude of largest event	4.8 (shallow) 3.0 (deep)	9.5
Seismic energy release/yr	2×10^{10} J yr^{-1} (shallow) 8×10^6 J yr^{-1} (deep)	10^{18} J yr^{-1}

† m = magnitude.

from librations do occur, but have much longer periods than on Earth. The tidal flexing of the lunar surface in both horizontal and vertical directions is about 2 mm along the length of a 10-km baseline (J. Williams, 1986, personal communication). The precise amount of motion depends on the position of the Moon. Tidal motions must be taken into account when designing, for example, long transportation systems or telescope arrays that require accurate alignment.

ATMOSPHERE

The lunar atmosphere is a collisionless gas. The total nighttime concentration is only 3×10^{-13} mol m^{-3} (Hoffman et al., 1973). Its total mass is only 10^4 kg, about the mass of air in a movie theater on Earth at 1 bar. This flimsy atmosphere will eliminate engineering problems associated with wind (Johnson, 1988), but might add others, such as difficulty in lubricating moving parts.

The composition of the lunar atmosphere appears in Table 3–2. The gases are derived from the solar wind, except for ^{40}Ar, which is produced by the decay of ^{40}K inside the Moon and then diffuses out. No daytime measurements of gas concentrations were made due to instrument limitations, but Hodges (1976) calculates that gases of carbon compounds, specifically CO_2, CO, and CH_4, probably dominate. They are absent at night because they condense out of the atmosphere onto soil particles.

The tenuous lunar atmosphere can be altered significantly by large-scale operations on the Moon. Vondark (1974) has calculated that if the density of the lunar atmosphere is increased, a point is reached where the rate at which gas is lost is slowed dramatically. This could compromise a number of scientific experiments requiring a hard vacuum, such as observations of the solar wind. Considering that each Apollo mission contributed 10^4 kg of gas (Johnson, 1971), temporarily doubling the atmosphere's nighttime mass, it would appear easy to contaminate the Moon's fragile atmosphere when regular flights to and from the lunar surface begin. The atmosphere must be monitored carefully when a lunar base is established. Studying the evolution of the Moon's atmosphere will, in fact, be an interesting research project in itself.

Table 3–2. Composition of the lunar atmosphere at night. From Hoffman et al. (1973).

Gas	Concentration, mol m^{-3}
H_2	1.1×10^{-13}
^{4}He	6.6×10^{-14}
^{20}Ne	1.3×10^{-13}
^{36}Ar	5.0×10^{-15}
^{40}Ar	1.2×10^{-14}
O_2	$<3 \times 10^{-16}$
CO_2†	$<5 \times 10^{-15}$

† Carbon gases (CO_2, CO, CH_4) probably dominate the daytime lunar atmosphere (Hodges, 1976).

SURFACE TEMPERATURES

Surface temperatures change drastically from high noon to dawn on the Moon, presenting a challenge to those designing lunar structures subject to thermal expansion and contraction. At Apollo 17, for example, the temperature ranged from 384 to 102 K during the month-long lunar day (Keihm & Langseth, 1973). Furthermore, the temperature decreases rapidly as sunset approaches, falling about 5 K h^{-1}. These data apply to equatorial regions only. In polar regions, the predawn temperature is about 80 K (Mendell & Low, 1970). The temperature in permanently shadowed areas at the poles could be lower. The cold night temperature will permit cooling of many systems without use of cryogenics.

The temperature variation is damped out rapidly at depth in the lunar soil (Keihm & Langseth, 1973). At a depth of 30 cm, the temperature is about 250 K and varies only 2 to 4 K from noon to dawn. This steady temperature might be useful for some purposes, but not as a heat sink because the lunar soil has a sluggish thermal conductivity (see below).

MAGNETIC FIELD

No magnetic field is now being generated inside the Moon, although there was a source of magnetism several billion years ago. It is not known whether this was generated by a dynamo in a metallic core, as on Earth, or by local, transient events such as meteorite impacts. Whatever its source, the lunar magnetic field is much weaker than is Earth's (Dyal et al., 1974). On the surface, the lunar magnetic field strength ranges from 3×10^{-9} to 3.3×10^{-7} t. For comparison, Earth's field at the equator is 3.0×10^{-5} t. Also, the lunar field varies locally on the Moon. For example, at the Apollo 16 landing site, the field varied from 1.13×10^{-7} to 3.27×10^{-7} t.

There is also a field external to the Moon, derived from the solar wind. This ranges from 5×10^{-9} t in the free-streaming solar wind to about 10×10^{-9} t in Earth's geomagnetic tail, in which the Moon resides 4 d during each lunation.

RADIATION ENVIRONMENT

Because of the Moon's small magnetic field and nearly absent atmosphere, solar and galactic nuclear particles hit its surface unimpeded. The Sun's spectrum peaks in the visible, at about 0.5 μm, but a significant amount of it, 7%, is in the ultraviolet (UV), between 0.28 and 0.40 μm (Robinson, 1966). Since the solar constant is 1393 W m^{-2} at the Earth-Moon distance from the Sun (Coulson, 1975), the total UV flux is 95 W m^{-2}.

There are three sources of radiation with different energies and fluxes (see Taylor [1975] for a summary):

1. High energy (1–10 Gev/nucleon) galactic cosmic rays, with fluxes of about 1 cm^{-2} s^{-1} and penetration depths of up to a few meters.
2. Solar flare particles with energies of 1 to 100 Mev/nucleon, fluxes up to 100 cm^{-2} s^{-1}, and penetration depths up to 1 cm.
3. Solar wind particles, which have much lower energies of about 1000 ev, tiny penetration depths (10^{-8} cm), but high fluxes (10^8 cm^{-2} s^{-1}).

These penetration depths refer to the primary particles only. Reactions between them and lunar material cause a cascade of radiation that penetrates much deeper (Silberberg et al., 1985), up to several meters. The combination of high flux and energy make solar flare particles the most dangerous to people working on the lunar surface, to electronic devices deployed directly on the surface, and perhaps to plants shielded only by transparent glass.

MICROMETEORITE FLUX

The lack of a significant atmosphere on the Moon allows even the tiniest particles to impact with their full cosmic velocities, 10 to several tens of kilometers per second. This rain of minute impactors could damage some structures and instruments on the lunar surface. Almost all lunar rock samples contain numerous microcraters, commonly called *zap pits,* on surfaces that were exposed to space while on the lunar surface. Studies of lunar rocks (Fechtig et al., 1974) have revealed the average flux of projectiles over the past several hundred million years. Table 3–3 shows the number of craters of a given size (or larger) expected per square meter per year. It is obvious from these data that microcraters in the 1- to 10-μm range will be common on surfaces exposed at the lunar surface. Even 100-μm craters will not be uncommon, with one produced on each square meter of surface every other year or so. It appears that sensitive surfaces, such as mirrors on optical telescopes, will have to be protected. Furthermore, evidence from the Surveyor 3 TV camera shroud and from Apollo windows (Cour-Palais, 1974) suggest that the present flux of particles capable of making craters up to about 10 μm across is 10 times greater than indicated from studies of rocks.

Table 3–3. Micrometeorite fluxes on the moon. Calculated from data given by Fechtig et al. (1974).

Crater diam.	Craters
μm	m^{-2} yr^{-1}
≥ 0.1	30 000
≥ 1.0	1 200
≥ 10	300
≥ 100	0.6
≥ 1000	0.001

REGOLITH

The lunar regolith, also called the lunar *soil,* is a global veneer of debris generated from underlying bedrock by meteorite impacts. It contains rock and mineral fragments and glasses formed by melting of soil, rock, and minerals (see Chapter 4 in this book). It also contains highly porous particles called *agglutinates,* glass-bonded aggregates of rock and mineral fragments. Agglutinates are produced by micrometeorite impacts into the lunar regolith.

Regolith depth ranges from 2 to 30 m, with most between 5 to 10 m. Impacts by micrometeorites have reduced much of the regolith material to a powder. Its grain size ranges from 40 to 268 μm and varies in a highly complex fashion with depth (Heiken, 1975). The chemical composition of the regolith reflects the composition of the underlying bedrock, modified by admixture of material excavated from beneath or thrown in by distant impacts.

Mitchell et al. (1972) measured the mechanical properties of lunar regolith samples. The bulk density of the regolith is low, 0.8 to 1.0 Mg m^{-3}, in the upper few millimeters but increases to 1.5 to 1.8 Mg m^{-3} at depths of 10 to 20 cm. Its porosity is 35 to 45% in the upper 15 cm, accounting in part for the low density. Except for the uppermost few millimeters, the lunar regolith is more cohesive, 0.1 to 1.0 kN m^{-2}, than most terrestrial soils and has an angle of internal friction of 30 to 50°. Agglutinates and shock-damaged rock fragments are weak and break under loads, leading to an increase in soil density (Carrier et al., 1973).

The lunar regolity is an excellent insulator. Its thermal diffusivity at depths of 30 cm is 0.7 to 1.0 $\times$ 10^{-8} m^2 s^{-1} and its thermal conductivity is 0.9 to 1.3 $\times$ 10^{-2} W m^{-1} K^{-1} (Langseth et al., 1976). This is not surprising considering the high porosity and lack of air. At depths <30 cm, thermal diffusivity is lower.

A small amount of lunar dust might be transported by charge differences built up by photoconductivity effects. Criswell (1972) described a bright glow photographed by Surveyor 7 and explained the phenomena as leviation of dust grains about 6 μm in radius. The grains were lifted only 3 to 30 cm above the local horizon, and had a column density of five grains cm^{-2}. This does not appear to be a significant transport mechanism on the lunar surface.

UPPER FEW HUNDRED METERS OF THE MOON

The upper few hundred meters of the Moon have been intensely fragmented by meteorite impacts. In the heavily cratered highlands and regions underlying mare basalt flows, the fragmental region extends for at least a few kilometers. Consequently, it might be difficult to find extensive areas of intact bedrock.

Active seismic experiments (Cooper et al., 1974) indicate that the velocity of compressional waves is about 100 m s^{-1} at depths of $<$ 10 m, which is in the regolith, and about 300 m s^{-1} at depths between 10 and 300 m.

These velocities are too slow to correspond to coherent rock, implying that the upper few hundred meters of the lunar surface is rubble (Cooper et al., 1974). Rocks returned from the highlands confirm the fragmental nature of the upper lunar crust. Most are complicated mixtures of other rocks, and many are weakly consolidated. Furthermore, the rims of all craters are by their nature weakly or unconsolidated materials and, therefore, not able to withstand tensional stresses.

A few localities might have bedrock, however. Many mare basalt flows, for example, form visible layers in crater walls or, as at the Apollo 15 landing site, in the walls of sinuous rilles (river-like depressions). Also, extensive sheets of impact-generated melt rocks occur on the floors of many large craters, such as Copernicus, which is 95 km in diameter.

CRATER MORPHOLOGIES

Fresh lunar craters up to 15 km in diameter have a consistent diameter/depth ratio of 5 (Pike, 1974). More specifically, craters < 15 km across follow the relation $R_i = 0.196 \, D_r^{1.010}$; craters > 15 km follow the relation $R_i = 1.044 \, D_r^{0.301}$ where R_i is the crater depth and D_r is the diameter as measured from rim crest to rim crest (Pike, 1974). Large craters are much shallower for their diameters than are smaller ones. Crater morphology changes as a crater is eroded by meteorite bombardment, during which a crater becomes wider and shallower, thereby increasing the diameter-to-depth ratio. Thus, even the smoothest areas on the lunar surface are undulating plains, so building horizontal transportations systems might require cut-and-fill operations. Finally, as noted above, rim materials consist of weak, unconsolidated rock. This could present problems for construction if certain facilities had to be built on crater rims.

ACKNOWLEDGMENT

I thank Dr. James Williams of the Propulsion Laboratory for providing information about tidal flexing. Preparation of this chapter was supported in part by NASA Grant NAG 9-30 (K. Keil).

REFERENCES

Carrier, W.D. III, L.G. Bromwell, and R.T. Martin. 1973. Behavior of returned lunar soil in vacuum. J. Soil Mech. Found. Div., ASCE 99:979–996.

Cooper, M.R., R.L. Kovach, and J.S. Watkins. 1974. Lunar near-surface structure. Rev. Geophys. Space Phys. 12:291–308.

Coulson, K.L. 1975. Solar and terrestrial radiations. Academic Press, New York.

Cour-Palais, B.G. 1974. The current micrometeoroid flux at the moon for masses $< 10^{-7}$ g from the Apollo window and Surveyor 3 TV camera results. Proc. Lunar Sci. Conf. 5th, 1974 3:2451–2462.

Criswell, D. 1972. Lunar dust motion. Proc. Lunar Sci. Conf. 3rd, 1972 3:2671–2680.

Dyal, P., C.W. Parkin, and W.D. Daly. 1974. Magnetism and the interior of the Moon. Rev. Geophys. Space Phys. 12:568–591.

Fechtig, H., J.B. Hartung, K. Nagel, and G. Neukum. 1974. Lunar microcrater studies, derived meteoroid fluxes, and comparison with satellite-borne experiments. Proc. Lunar Sci. Conf. 5th, 1974 3:2463–2474.

Goins, N.R., A.M. Dainty, and M.N. Toksoz. 1981. Seismic energy release of the Moon. J. Geophys. Res. 86:378–388.

Heikien, G. 1975. Petrology of lunar soils. Rev. Geophys. Space Phys. 13:567–587.

Hodges, R.R., Jr. 1976. The escape of solar-wind carbon from the moon. Proc. Lunar Sci. Conf. 7th, 1976 1:493–500.

Hoffman, J.H., R.R. Hodges, and F.S. Johnson. 1973. Lunar atmospheric composition results from Apollo 17. Proc. Lunar Sci. Conf. 4th, 1973 3:2865–2875.

Johnson, F.S. 1971. Lunar atmosphere. Rev. Geophys. Space Phys. 9:813–823.

Johnson, S.W. 1988. Engineering for a 21st century lunar observatory. J. Aerospace Eng. 1:35–51.

Keihm, S.J., and M.G. Langseth. 1973. Surface brightness temperatures at the Apollo 17 heat flow site: Thermal conductivity of the upper 15 cm of regolith. Proc. Lunar Sci. Conf. 4th, 1973 3:2503–2513.

Lammlein, D.R., G.V. Lathan, J. Dorman, Y. Nakamura, and M. Ewing. 1974. Lunar seismicity, structure, and tectonics. Rev. Geophys. Space Phys. 12:1–21.

Langseth, M.G., S.J. Keihm, and K. Peters. 1976. Revised lunar heat-flow values. Proc. Lunar Sci. Conf. 7th, 1976 3:3143–3171.

Mendell, W.W., and F.J. Low. 1970. Low-resolution differential drift scan of the moon at 22 microns. J. Geophys. Res. 75:3319–3324.

Mitchell, J.D., W.N. Houston, R.F. Scott, N.C. Costes, W.D. Carrier, III, and L.G. Bromwell. 1972. Mechanical properties of lunar soil: Density, porosity, cohesion, and angle of internal friction. Proc. Lunar Sci. Conf. 3rd, 1972 3:3235–3253.

Nakamura, Y., G.V. Latham, H.J. Dorman, A.-B.K. Ibrahim, J. Koyama, and P. Horvath. 1979. Shallow moomquakes: Depth, distribution and implications as to the present state of the lunar interior. Proc. Lunar Planet. Sci. Conf. 10th, 1979 3:2299–2309.

Pike, R.J. 1974. Depth/diameter relations of fresh lunar craters: revision from space-craft data. Geophys. Res. Lett. 1:291–294.

Robinson,N. 1966. Solar radiation. Elsevier Publ. Co., New York.

Silberberg, R., C.H. Tsao, and J.A. Adams, Jr. 1985. Radiation transport of cosmic ray nuclei in lunar material and radiation doses. p. 663–669. *In* W.W. Mendell (ed.) Lunar bases and space activities of the 21st century. The Lunar and Planetary Inst., Houston, TX.

Taylor, S.R. 1975. Lunar science: A post-Appolo view. Pergamon Press, New York.

Vondrak, R.R. 1974. Creation of an artificial lunar atmosphere. Nature (London) 248:657–659.

4 Mineralogical and Chemical Properties of the Lunar Regolith

David S. McKay and Douglas W. Ming

NASA Johnson Space Center
Houston, Texas

Lunar regolith is the layer of loose, incoherent rock material that forms the majority of the Moon's surface. The finer-grained (< 1 mm) part of the lunar regolith has been loosely called *soil* and, in most locations, is the result of numerous and repeated bombardment by mostly small meteorites. Unlike terrestrial soils, the regolith lacks the influence of organic matter and the Moon is thought to be completely lacking in water. The regolith holds the most promise as a source of raw materials for use at lunar bases as well as materials for use in building large infrastructures in space. Lunar regolith may also provide rocket propellants and life support consumables (including O_2, H, and manufactured water).

No doubt, the regolith will play an important role in the development of lunar-base agriculture. The regolith may (i) act as a soil and provide a solid support substrate for plant growth, (ii) be a source for essential, plant-growth elements, and (iii) provide a source of O_2 and H, which may be used to manufacture water—possibly the most important component of an agricultural system.

It has been 20 yr since the first Apollo mission returned samples from the Moon in 1969. Since that time, lunar materials have been thoroughly studied and a wealth of information has been published on their physical, chemical, and mineralogical properties. The study of nonterrestrial surface materials and how they may react as a soil and/or a source of plant nutrients opens up a whole new frontier for researchers in the agricultural community. Because the Moon may be the first stepping stone for the presence of people away from the Earth, the following chapter has been prepared to acquaint agricultural scientists with the mineralogical and chemical properties of the lunar regolith.

REGOLITH SOIL-FORMING PROCESSES

The surface geology of the Moon is dominated by two primary units—light-colored highlands and smooth dark maria. The lunar highlands, which have developed on mostly anorthositic bedrock, are the oldest surface features of the Moon and are heavily cratered and brecciated. The highlands and the giant lunar basins (e.g., Imbrium and Orientale) were formed during the time interval from about 4.5 Gyr ago to about 3.9 Gyr ago. The maria are younger than the highlands and contain fewer craters. The maria were formed by basaltic lava flows, mostly after the primary flux of meteorites that bombarded the highlands (although some old mare basalt samples have also been found). Some of the large-ring basins (e.g., Imbrium) were filled by the flooding of lava, which gives rise to the smooth dark morphological characteristics of the mare. The flooding or extensive lava flows lasted from about 3.8 to about 3.0 Gyr ago. In addition to the extensive lava flows that were formed during this period of lunar history, volcanic ash or pyroclasts were produced and have been found in lunar regolith samples from all of the Apollo sites. These ash deposits are most abundant at the Apollo 17 site where they make up the "orange soil" reported by the astronauts. The heavy bombardment by meteorites during the evolution of the Moon has resulted in the pulverization of bedrock at the surface and, hence, a fine-grained material or regolith has formed at the lunar surface.

Although the Moon is geologically dead, there are important soil-forming processes occurring at the surface; the two most prominent processes are (i) meteorite bombardment and (ii) implantation of solar winds. Of these two processes, the most important weathering element in the lunar environment is micrometeorite bombardment. As discussed in a later section, micrometeorites are acting as both destructional and constructional forces in the regolith. Solar-wind species and other more energetic solar flare particles are slowly weathering the surfaces of regolith materials.

The thickness of the regolith ranges from a few meters in some areas overlying mare basalt flows to tens of meters over much of the older highland terrain. Core samples taken on the Apollo and Soviet Luna missions have shown that the lunar regolith varies considerably with depth in many properties related to maturity and sometimes also in chemical properties. However, the variation is more random than systematic; that is the deepest samples are not always the least mature and freshest. This is because occasional larger impacts may randomly overturn whole sections of regolith so that mature soils are often found near the bottom of the cores.

LUNAR MINERALS

The origin of minerals on the lunar surface is somewhat complex because of the extensive reworking of the regolith by meteorites. Some of the minerals, however, have survived the bombardment and are probably of plutonic or crustal origin. Recrystallization of primary minerals and melting and growth of new minerals is widespread on the lunar surface as rocks and other regolith materials are melted by meteorite bombardment. Lunar minerals and materials have formed in an extremely reduced environment; hence, they lack ferric iron. Since the lunar environment is void of water, lunar minerals do not have any water of crystallization.

Table 4-1. Lunar minerals (Williams & Jadwick, 1980).

Major minerals[†]	Minor minerals[‡]
Olivine $(Mg,Fe)_2SiO_4$ Pyroxene $(Ca,Mg,Fe)SiO_3$ Plagioclase feldspar $(Ca,Na)(Al,Si)_4O_8$	Spinels $(Fe,Mg, Al,Cr,Ti)_3O_4$ Armalcolite (Fe_2TiO_5) Silica (quartz, tridymite, cristobalite) SiO_2 Iron Fe (variable amounts of Ni and Co) Troilite FeS Ilmenite $FeTiO_3$

Trace minerals[§]

Phosphates	Oxides
Apatite $Ca_5(PO_4)_3(F,Cl)$¶ Whitlockite $Ca_9(Mg,Fe)(PO_4)_7(F,Cl)$	Rutile TiO_2 Corrundum (?)# Al_2O_3 Hematite (?) Fe_2O_3 Magnetite Fe_3O_4 Geothite (?) FeO(OH)
Zr minerals	
Zircon $ZrSiO_4$¶ Baddeleyite ZrO_4	
	Metals
Silicates	Copper (?) Cu Brass (?) Tin (?) Sn
Pyroxferroite $(Fe,Mg,Ca)SiO_3$ Amphibole $(Ca,Mg,Fe)(Si,Al)_8O_{22}F$ Garnet (?) Tranquilletyite $Fe_8Zr_2Ti_3Si_3O_4$¶	**Zr-rich minerals**
	Zirkilite or zirconolite $CuZrTi_2O_7$¶
Sulfides	**Meteoritic minerals**
Mackinawite $(Fc,Ni)_9S_8$ Pentlandite $(Fe, Ni)_9S_8$ Chalcopyrite $CuFeS_2$ Cubanite $CuFe_2S_3$ Sphalerite $(Zn,Fe)S$	Schreibernite $(Fe,Ni)_3P$ Cohenite $(Fe,Ni,Co)C$ Niningerite $(Mg,Fe,Mn)S$ Lawrencite (?) $(Fe,Ni)Cl_2$

[†] Major minerals may occur in concentrations up to 100%.
[‡] Minor minerals generally occur at <2%.
[§] Trace minerals never exceed a few tenths of a percent.
[¶] These minerals are known to exhibit complete substitution, particularly of elements like Y, Nb, Hf, U, and the rare-earth elements that are concentrated in these minerals.
[#] (?) Controversial with respect to indigenous lunar origin.

The mineralogy at the lunar surface is dominated by three mineral groups—plagioclase feldspar, pyroxene, and olivine (Table 4–1). Plagioclase feldspar is the most predominant group of minerals at the lunar surface (Table 4–2). Plagioclase feldspars are primarily associated with anorthositic rocks, therefore, regolith that has formed on the highlands contains more plagioclase

Table 4-2. The abundance of plagioclase feldspar, pyroxene, olivine, and ilmenite in lunar materials (Williams & Jadwick, 1980).

Lunar material	Plagioclase feldspar	Pyroxene	Olivine	Ilmenite
		vol. %		
Mare basalts	15–35	40–65	0–35	0–25
Anorthositic rocks	40–98	0–40	0–40	trace
Crystalline breccias	50–75	--	1–5	1–2
Fragmental breccias	--	5–30	0–5	2–12
Vitric breccias	15–50	--	--	--
Soil	10–60	5–20	0–4	0.5–5

Table 4–3. Typical chemical compositions of common lunar minerals (Williams & Jadwick, 1980).

Oxide	Plagioclase feldspar		Pyroxene		Olivine		Ilmenite	
	Mare†	Highlands‡	Mare§	Highlands‡	Mare§	Highlands‡	Mare§	Highlands‡
					wt. %			
SiO_2	46.06	46.67	47.84	53.53	37.36	37.66	0.01	0.21
TiO_2	0.15	0.02	3.46	0.90	0.11	0.09	53.58	54.16
Cr_2O_3	--	--	0.80	0.50	0.20	0.15	1.08	0.44
Al_2O_3	33.71	33.51	4.90	0.99	<0.01	0.02	0.07	<0.01
FeO	0.68	0.25	8.97	15.42	27.00	26.24	44.88	37.38
MnO	0.01	--	0.25	0.19	0.22	0.32	0.40	0.46
MgO	0.31	0.09	14.88	26.36	35.80	35.76	2.04	6.56
CaO	18.07	17.78	18.56	2.43	0.27	0.16	--	--
BaO	--	<0.01	--	--	--	--	--	--
Na_2O	0.67	1.51	0.07	0.06	--	--	--	--
K_2O	0.04	0.13	--	--	--	--	--	--
ZrO	--	--	--	--	--	--	0.08	0.01
V_2O_2	--	--	--	--	--	--	0.01	<0.01
Nb_2O_5	--	--	--	--	--	--	<0.01	0.13
Total	99.70	99.97	99.73	100.39	100.97	100.40	102.16	99.37

† Lunar sample 12021 (Weill et al., 1971).
‡ Lunar sample 72395 (Dymek et al., 1976).
§ Lunar sample 74255 (Dymek et al., 1975).

feldspar than regolith that has formed on the mare. Most lunar feldspars have anorthite ($CaAl_2Si_2O_8$) contents greater than the 80 mol%. Lunar plagioclase, however, may have other elements in addition to Ca, Al, Si, and minor amounts of Na present in their structures, including Ti, Fe, Mn, Mg, K, and Ba (Table 4-3).

Olivine exists as a solid solution of forsterite (Mg_2SiO_4) and fayalite (Fe_2SiO_4). Most of the lunar olivine compositions are between 50 to 75 mol% forsterite; the chemical composition for a typical forsterite (fa_{50}-fa_{75}) can be found in Table 4-3. Olivine is not as common as plagioclase feldspar on the lunar surface, but it is fairly common in mare basalts (Table 4-2).

Lunar pyroxenes are mixtures of enstatite ($MgSiO_3$), wollastonite ($CaSiO_3$), and ferrosilite ($FeSiO_3$). Pyroxenes have several structural types on Earth. This is no different on the Moon, where three structural forms are found—orthopyroxene, pigeonite (low-Ca clinopyroxene) and augite (high-Ca clinopyroxene). Typical compositions of highland and mare pyroxenes are given in Table 4-3; however, it is difficult to define an average or representative chemical composition for lunar pyroxenes. Pyroxenes are the predominant mineral phases found in mare basalts (Table 4-2).

Ilmenite and spinel may also occur in fairly large quantities at the lunar surface. Lunar ilmenites are mixtures of $FeTiO_3$ with small amounts of $MgTiO_3$ and they are the most abundant of the lunar oxide minerals. Ilmenite is most common in high-Ti basalts in the mare regions, where it may account for 15 to 25% of the total volume. Ilmenite is an ideal lunar mineral from which to extract O_2 for use in propulsion and life support systems. Oxygen can be produced by the reduction of ilmenite. Spinels are complex mixtures of ulvöspinel (Fe_2TiO_4), chromite ($FeCr_2O_4$), hercynite ($FeAl_2O_4$), picrochromite ($MgCr_2O_4$), spinel ($MgAl_2O_4$), and magnesium-titanite (Mg_2TiO_4). The major sulfide is the mineral troilite (FeS) which contains nearly all of the lunar S present in minerals. Native S has not been found in lunar samples.

Minerals in the lunar regolith reflect the mineralogy and chemistry of the underlying bedrock, which has been pulverized by meteorite bombardment. In many cases, the mineral has been fused to other lunar materials (e.g., glass and agglutinates) during impact (vide infra). Several minerals found in the regolith in trace amounts (e.g., zirkilite) may not be indigenous to the lunar environment; instead, these minerals may have been deposited in the regolith by meteorites. Other minerals that occur only in trace amounts at the lunar surface are listed in Table 4-1 and several of these minerals that may have unique contributions to lunar base agriculture (e.g., apatite) are discussed in more detail in the next section. Additional information on the mineralogy of surface materials may be found in Smith (1974) and Smith and Steele (1976).

REGOLITH MATERIALS

The regolith is the continuous layer of fine-grained debris that covers the lunar surface and the finer component (< 1 mm) is often referred to as

Table 4-4. Grain-size distribution (<1 mm) for Apollo 11 soil 10084,853 (Williams & Jadwick, 1980). This soil has formed on mare material.

Grain size	Weight	Cumulative wt.
mm	%	
10–4	1.67	1.67
4–2	2.39	4.06
2–1	3.20	7.26
1–0.5	4.01	11.27
0.5–0.25	7.72	18.99
0.25–0.15	8.23	27.22
0.15–0.090	11.51	38.72
0.090–0.075	4.01	42.73
0.075–0.045	12.40	55.14
0.045–0.020	18.02	73.15
<0.020	26.85	100.00

soil. Regolith materials include mineral and glass phases and will be discussed in more detail below.

The bulk density of the regolith ranges from 0.9 to 1.1 Mg m^{-3} in the top few centimeters and increases with depth to as high as 1.9 Mg m^{-3}. The average bulk densities for 0- to 30-cm and 30- to 60-cm regolith depths are 1.58 and 1.74 Mg m^{-3}, respectively (Houston et al., 1974). The surface material is loose because of reworking by micrometeorites, and apparently material below about 20 cm is compacted by the shaking and compression caused by the micrometeorite impacts. Generally, the regolith color does not vary to any great degree from site to site with most materials being gray. Most of the regolith ranges from dark gray (10YR3/1) to white (10YR8). However, in some cases, volcanic glasses may give the soil various colors. For example, pale-green regolith material was collected at the Apollo 15 landing site, and orange-brown material was collected at the Apollo 17 site. The colors of these two materials have been attributed to the presence of abundant pyroclastic green and orange glasses, respectively.

The particle-size distribution for an Apollo 11 mare soil is given in Table 4-4. Mean grain sizes for the regolith vary between 40 to 802 μm (see Fig. 4-1), depending on the degree of impact. However, the mean grain size of most regolith materials fall between 45 and 100 μm (Williams & Jadwick, 1980). As a whole, lunar regolith material is poorly to very poorly sorted and has no soil structure such as clear layering, graded beds, or weathering zones. The coarsest samples are generally the most poorly sorted and vice versa for the finest samples.

The regolith consists of lithic and mineral debris as well as glass particles formed by impact melting. Regolith materials have a variety of morphological characteristics and can be divided into several major categories (i) agglutinates; (ii) crystalline, igneous rock fragments; (iii) breccia fragments; (iv) mineral fragments; and (v) glass.

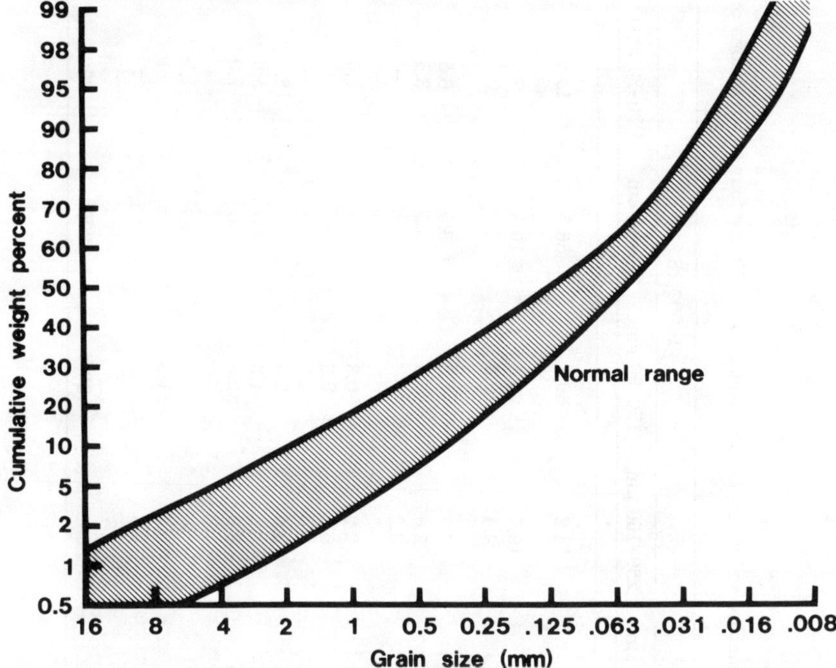

Fig. 4-1. Cumulative grain-size distribution for a large number of lunar soils (Williams & Jadwick, 1980).

Agglutinates

A major phase in many lunar regolith samples is a particle type that is so far unique to the Moon—agglutinates. Agglutinates are constructional particles made of lithic, mineral, and glass fragments welded together by a glassy matrix containing extremely fine-grained metallic iron and formed by micrometeorite impacts at the lunar surface. Chemical compositions of agglutinates, therefore, tend to represent the bulk chemistry of the soil that they have formed in (Gibbons et al., 1976), although minor but systematic differences may occur. Agglutinates are a major regolith component, making up to 60% by volume of some of the returned lunar soils (McKay et al., 1974) (see Table 4-5). They have absolutely no equivalent on Earth or, with possibly a few rare exceptions, in meteorites. Agglutinates are also unlikely to occur on Mars because of the thin atmosphere that will slow down or burn up micrometeorites, but may be present on the planet Mercury which lacks any significant atmosphere.

Lunar agglutinates dominate the optical and morphological properties of many regolith samples. The formation of agglutinates may be the most active soil-forming process presently occurring at the lunar surface and thus, most of the glass now being produced on the Moon is in the form of agglutinates. Micrometeorites are continually bombarding the regolith surface

Table 4-5. Comparison of a mature soil (smooth terrain, 14003,28) to an immature soil (Cone Crater ejecta, 14141,30) for four size fractions of Apollo 14 samples (McKay et al., 1972b).

	14141,30				14003,28			
Components	250–150 μm	150–90 μm	75–60 μm	30–20 μm	250–150 μm	150–90 μm	75–60 μm	30–20 μm
	%							
Agglutinates	5.3	5.2	6.5	12.5	54.2	60.3	56.5	43.5
Microbreccias								
Recrystallized	57.5	47.2	36.5	15.5	19.2	20.5	16.5	7.0
Vitric	6.9	6.8	16.5	5.5	4.4	3.0	1.0	0.5
Angular glass fragments								
Brown	2.3	6.2	4.0	7.0	7.2	4.3	8.0	6.5
Colorless	0.3	2.6	1.5	5.0	2.0	3.0	3.0	3.5
Pale green	0.3	--	--	--	0.2	--	--	--
Glass droplets								
Brown	0.8	2.0	0.5	2.0	2.2	0.3	2.0	7.0
Colorless	--	0.2	--	--	0.2	0.3	--	--
Ropy glasses	--	--	--	--	--	1.0	0.5	--
Clinopyroxene	4.1	8.0	15.5	18.5	3.0	2.3	5.5	18.5
Orthopyroxene	1.2	2.0	2.5	9.5	1.2	1.3	4.5	5.0
Plagioclase	5.6	6.6	11.5	21.5	3.6	2.3	1.5	7.0
Olivine	0.5	0.4	1.5	0.5	1.0	--	--	--
Opaque minerals	--	--	--	--	--	--	--	1.5
Basalt	5.9	4.2	0.5	--	0.8	1.3	--	--
Anorthosite	0.3	1.2	--	--	0.2	--	--	--
Tachylite	8.9	7.0	3.0	2.5	--	--	1.0	--
No. of grains counted	400	500	200	200	500	300	200	200

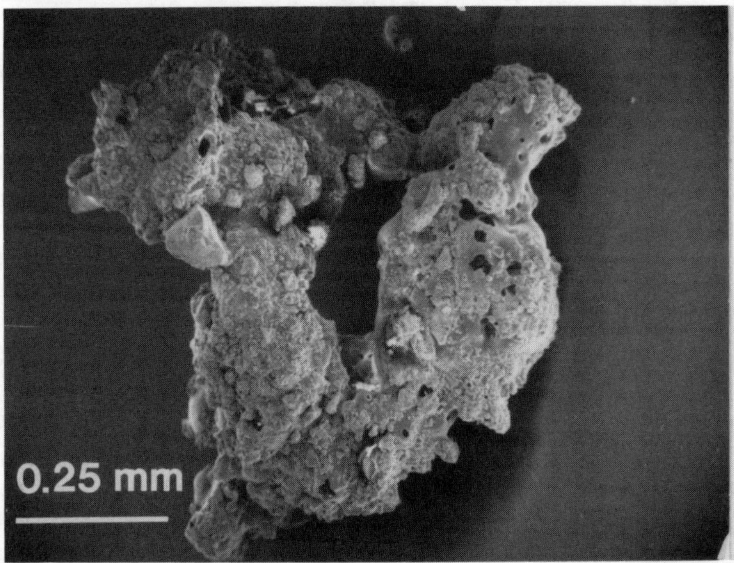

Fig. 4–2. Scanning electron microscope (SEM) view of a convoluted branching agglutinate. While some relatively clean glassy surfaces can be seen, most of the agglutinate is covered with welded fine-grained fragments from the regolith. NASA Photo S-71-24575.

and result in the construction of agglutinates. Glass produced by micrometeorite impacts weld together lithic and mineral fragments, glass, and fine-grained metallic iron. Agglutinates may dominate the regolith matrix, depending on the degree of micrometeorite impact and maturity of the soil (vide infra). Morphologies of agglutinates may vary from simple irregular grains to complex branch forms. A typical branching shape is shown in Fig. 4-2.

Crystalline, Igneous Rock Fragments

This category includes basaltic fragments that are made of intergrown plagioclase feldspar, pyroxene, ilmenite, and sometimes olivine or cristobalite. Crystalline basalt fragments are the remains of igneous rocks derived from the interior of the Moon as liquids by well-known igneous processes. Constituent mineral grain size is usually < 1 mm, and a wide variety of igneous textures have been described in these fragments. Mare basalts can be divided into two categories—high-Ti basalts and low-Ti basalts. Hi-Ti basalts have TiO_2 contents usually >9.0 wt. % and low-Ti basalts have TiO_2 contents usually <5.0 wt. %. Other kinds of crystalline fragments include nonmare basalts rich in K, Rare-Earth Elements, P (KREEP), anorthosites (plutonic rocks composed mostly of plagioclase, generally found in the highlands), microgabbros (very fine-grained plutonic rocks), and granular-textured lithics including norites (orthopyroxene gabbros) and troctolites (plagioclase/olivine gabbros). The KREEP basalts may be a source of K and P to use as plant growth nutrients.

Breccias

Breccias are basically fragmental rocks made from crystalline lithic fragments, regolith components, or other breccias that have been lithified (turned to rock) during complex impact-driven processes (e.g., large meteorite impacts). Regolith breccias contain some identifiable regolith components such as glass spheres or agglutinates; however, breccias mostly consist of rock and mineral phases. The matrix of breccias can be made of either finer fragmental material or of impact-melted and crystallized material. A wide variety of breccias are present in lunar regolith samples. In fact, breccias are by a wide margin, the most volumetrically important rock type on the Moon and in the lunar regolith. Some of the more common types of breccias include (Williams & Jadwick, 1980; Stöffler et al., 1980):

1. **Vitric-matrix breccias.** These breccias consist of a mixture of mineral, glass and rock fragments bonded together by grain-to-grain sintering and are cemented by smaller glass fragments that form during impact.
2. **Fragmental-matrix breccias.** These breccias are similar to vitric-matrix breccias except they lack glass fragments and are cemented together by grain-to-grain sintering. The matrix consists entirely of fine-grained fragments.
3. **Cataclastic Anorthosites.** These very friable breccias are crushed rocks consisting of 50 to 99% plagioclase feldspar.
4. **Crystalline-matrix breccias.** These breccias consist of interlocking crystals of plagioclase feldspar, pyroxene, olivine, and ilmenite.

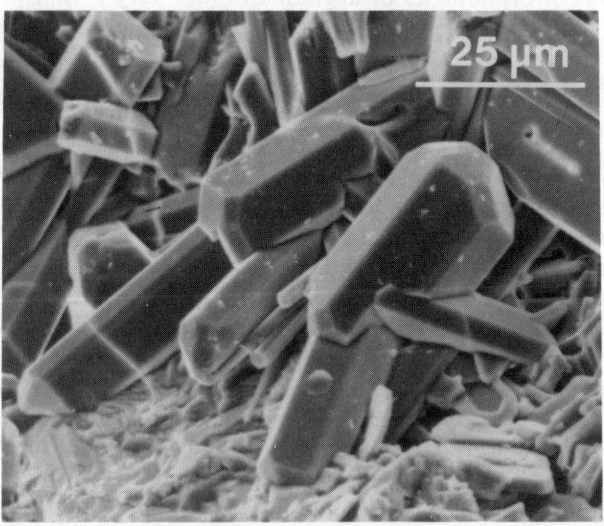

Fig. 4–3. An SEM photograph of a group of apatite (calcium phosphate) crystals in a vug in an Apollo 14 breccia. These crystals grew from a hot vapor phase generated by the large impact that formed the breccia. NASA Photo S-72-30019.

Table 4-6. Chemistry of breccias (Williams & Jadwick, 1980; McGee et al., 1979; Ryder & Norman, 1980).

| | Mare | | | | Highland | | | | | |
| | Vitric matrix | | | | Fragmental | Cataclastic | Crystalline matrix | | | Granulitic |
Chemical	Low-Ti 12034	High-Ti 10060	KREEP 14047	Anorthositic gabbro 60255	matrix 14063	Anorthosite 60025	KREEP 14305	KREEP 76015	Anorthositic gabbro 68415	matrix 79215
Oxide					wt. %					
SiO_2	47.8	40.0	47.2	45.2	45.5	45.3	48.3	46.2	45.3	43.8
TiO_2	2.3	8.5	1.7	0.69	1.3	0.02	1.5	1.5	0.3	0.3
Al_2O_3	15.5	11.3	18.2	26.1	23.0	34.2	16.2	17.2	28.7	27.7
Cr_2O_3	0.27	0.3	0.1	0.1	0.16	0.003	0.2	0.2	0.1	0.2
FeO	12.4	17.7	10.5	5.9	5.8	0.5	10.4	9.8	4.1	4.6
MnO	0.2	0.2	0.1	0.06	0.1	0.008	0.1	0.1	0.05	0.06
MgO	8.3	7.7	8.9	6.4	9.6	0.2	10.3	13.0	4.3	6.3
CaO	10.8	14.5	11.5	15.1	13.0	19.8	9.9	10.8	16.2	15.9
Na_2O	0.7	0.5	0.7	0.5	0.7	0.5	0.8	0.7	0.5	0.5
K_2O	0.5	0.2	0.5	0.1	0.1	0.1	0.6	0.3	0.09	0.1
P_2O_5	0.5	0.1	0.5	0.1	--	0.003	0.6	0.3	0.06	0.4
S	0.09	0.15	0.08	0.04	--	0.024	--	0.09	0.04	--
Total	99.36	101.15	99.98	100.29	99.26	100.66	98.90	100.19	99.74	99.86
Elements					mg kg^{-1}					
Zr	630	580	780	--	325	0.48	--	507	72	--
Ni	--	70	--	391	--	1.1	200	135	184	255
Zn	--	25	20	21	5.3	3.25	2.1	2.8	4.8	2.3
Cd	--	0.300	0.102	0.0608	0.018	0.005	--	0.0032	0.00275	0.00098
Cu	--	--	--	--	--	8.4	--	--	12	--
Co	0.035	0.029	--	0.035	0.019	0.073	--	--	11	0.0188

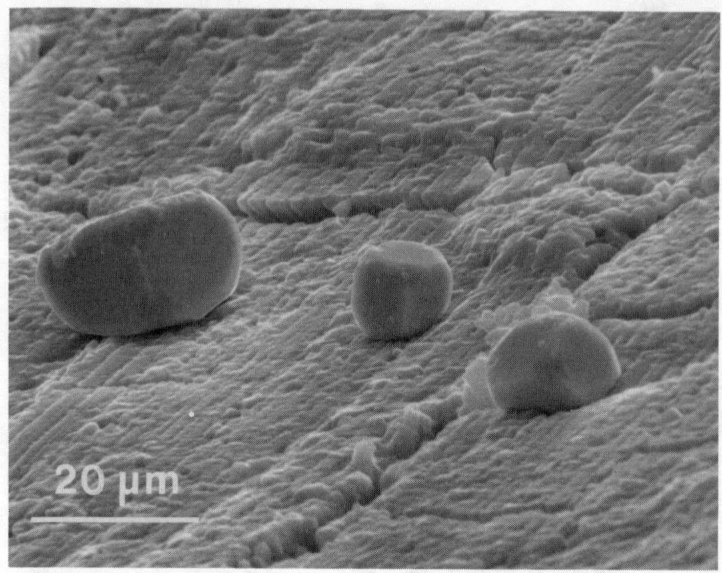

Fig. 4–4. An SEM photograph of a trio of metallic iron crystals resting on a pyroxene substrate. These iron crystals condensed from a hot vapor generated by the impact that formed the breccia fragment in which they were found. NASA Photo S-72-53357.

5. **Granulitic-matrix breccias.** These hard breccias are metamorphosed rocks that consist mostly of crystalline matrix. They are similar to crystalline-matrix breccias except that they have an obvious granulitic metamorphic texture with 120° grain boundaries common.

The chemical composition of breccias reflects the chemistry of the lunar materials struck by the meteorite and, to a much less extent, the chemistry of the meteorite. The chemistry of some of the more common breccias are given in Table 4–6.

During the complex impact processes that create breccias, volatile phases may be mobilized and redistributed. Figure 4–3 shows an intergrowth of the mineral apatite in a vug in an Apollo 14 breccia. These doubly terminated hexagonal apatite crystals grew from a hot vapor created in the impact that formed the breccia (McKay et al., 1972a). No doubt, apatite and possibly whitlockite will be an important source for fertilizer P. A similar process formed the iron metal crystals shown in Fig. 4–4 (Clanton et al., 1973).

Mineral Fragments

During meteorite impact, lithic fragments are broken, which results in the separation of mineral fragments and subsequent deposition of minerals in the lunar regolith. The primary mineral fragments found in the lunar regolith are plagioclase feldspars and pyroxene; however, olivine, ilmenite, and other trace minerals may be present in the lunar regolith. We have previously discussed the variety of lunar minerals found in the regolith.

The relative abundance of lunar minerals varies considerably from site to site. For example, at the most highland-like site, Apollo 16, the feldspar/pyroxene ratio in some samples is on the order of 10:1 (Heiken et al., 1973); whereas, at the Apollo 12 site (a mare site) this ratio is approximately 1:2 in favor of pyroxene (McKay et al., 1971).

Glass

A wide variety of glasses are found in the regolith; however, there are only two primary sources for regolith glass—impact and volcanic glasses. Glasses may be classified into several different groups based on their shape, color, or chemistry. The chemistry of several of the more common glass types are given in Table 4–7. Note that KREEP glasses have high K and P_2O_5 contents; these glasses may release high amounts of K and P into soil solution.

Regolith glasses have a variety of shapes. For example, ropy glasses, thought to form in relatively large impacts, display ropy textures and coatings of welded fragments (Fig. 4–5). Glass spheres are common in the lunar regolith and are formed both by impact (Fig. 4–6) and volcanic activity (Fig. 4–7). The volcanic spheres often have thin adhering coatings of complex S-rich sublimates (Fig. 4–8) and also occasionally potassium and sodium chloride crystals (Fig. 4–9). These sublimates apparently were formed during the original volcanic eruptions. These submicron sublimates may rapidly release S, K, and Na into solution; hence, the study of how these materials will react in aqueous environments will be necessary to evaluate whether regolith-containing volcanic glass spheres can be used for plant growth.

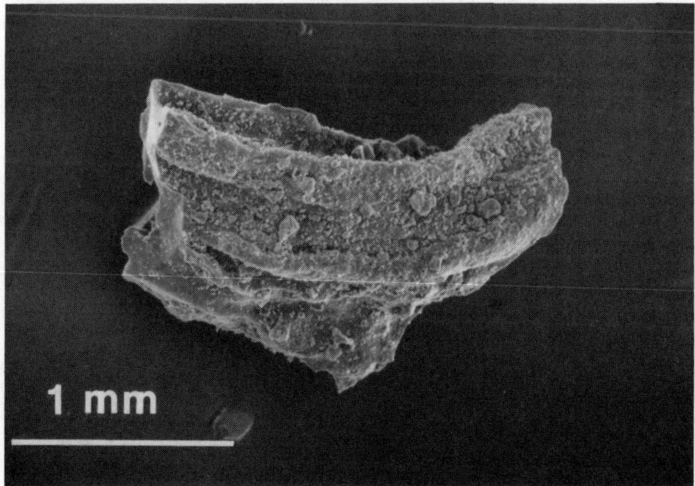

Fig. 4–5. A ropy glass from Apollo 12 regolith. This ropy-textured glass is covered with welded fragmental material but is not like an agglutinate. It is interpreted to be splash glass from a large impact, possibly Corpernicus. NASA Photo S-71-17751.

Tabel 4–7. Chemistry of lunar glasses.

Oxide	Anorthosite (plagioclase) glass†	Highland basalt glass†	Green glass†	Orange glass‡	High-Ti mare glass‡	Low-Ti mare glass‡	KREEP glass§	Granite glass¶
				wt. %				
SiO_2	44.89	45.35	44.14	39.2	41.7	46.7	48.2	73.13
TiO_2	0.04	0.43	0.37	8.9	8.6	0.72	2.0	0.50
Al_2O_3	35.76	27.63	7.81	5.9	10.8	10.4	15.5	12.37
Cr_2O_3	--	0.07	0.33	0.67	0.30	0.60	--	0.35
FeO	0.18	4.62	21.05	22.4	17.4	18.4	10.6	3.49
MnO	--	--	--	0.28	0.22	0.24	--	--
MgO	0.12	6.13	16.72	14.6	9.2	12.7	7.7	0.13
CaO	19.03	15.64	8.41	7.2	10.9	9.7	10.9	1.27
Na_2O	0.55	0.42	0.13	0.36	0.31	0.18	0.6	0.64
K_2O	0.03	0.05	0.03	0.09	0.07	0.03	1.1	5.97
P_2O_5	--	--	--	0.04	0.05	0.03	0.8	--

† Glass composition in soil from Apollo 16 site (Ridley et al., 1973).
‡ Glass composition in soil from Apollo 17 site (Warner et al., 1979).
§ Glass composition in soil from Apollo 12 site (Meyer et al., 1971).
¶ Glass composition in soil from Apollo 15 site (Reid et al., 1972).

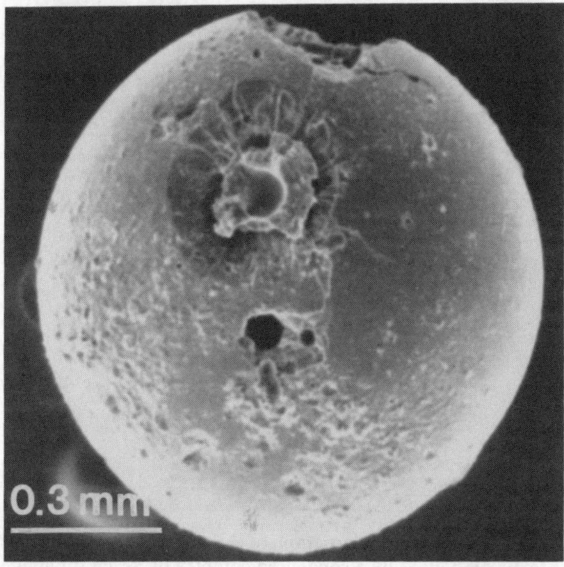

Fig. 4–6. An impact-produced glass sphere from Apollo 11 regolith. This sphere is partially covered with splash glass and has also been impacted by several micrometeorites that have made hypervelocity craters in its surface. NASA Photo S-71-41321.

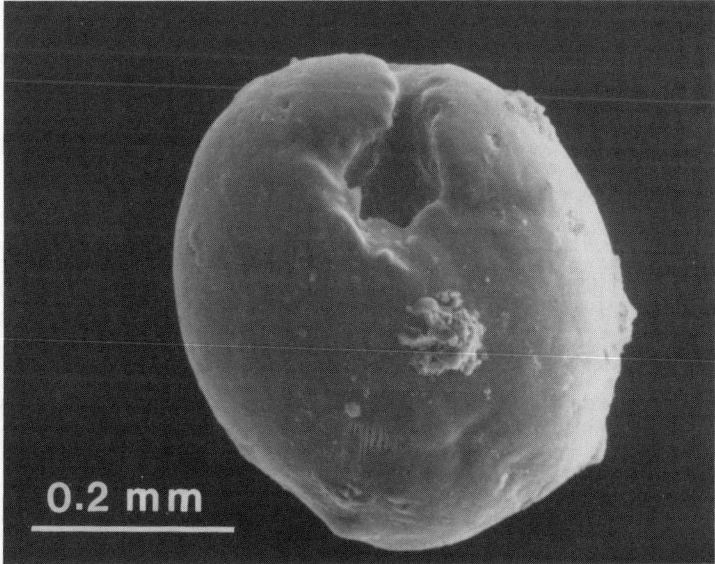

Fig. 4–7. Volcanic glass sphere from the Apollo 17 regolith. This sphere is part of the orange glass sample, and is undoubtedly a volcanic pyroclastic product. It contains one large vesicle that is partly open to the exterior. NASA Photo S-73-35102.

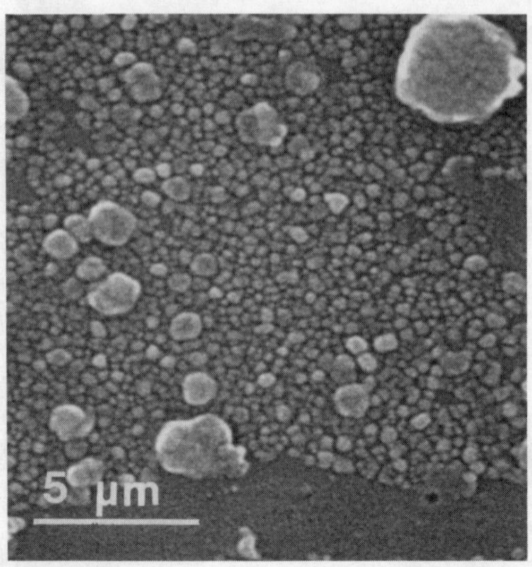

Fig. 4–8. Higher magnification SEM micrograph of the surface of a volcanic glass sphere from Apollo 17. The texture is caused by a condensed complex of S-rich sublimates on the glassy surface. These sublimates may have eventual resource potential for lunar base. NASA Photo S-78-28000.

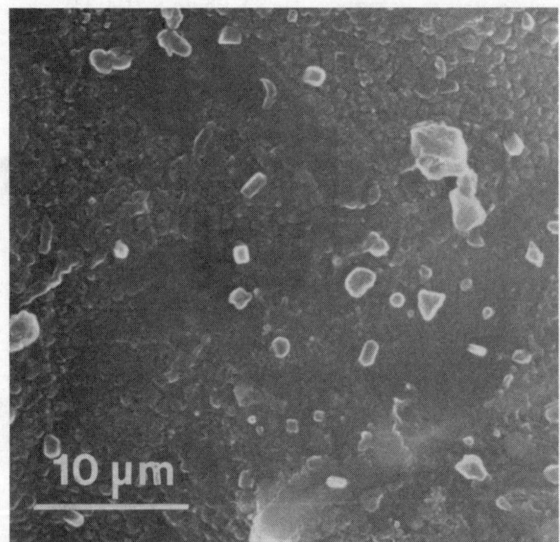

Fig. 4–9. Scanning electron microscope photo of small chloride crystals on the surface of another Apollo 14 pyroclastic droplet. Both K and Na chloride crystals are present. NASA Photo S-78-28002.

CHEMISTRY OF THE LUNAR REGOLITH

The chemistry of the lunar regolith varies from site to site. These chemical differences generally reflect the chemistry of the underlying bedrock. During regolith formation, however, debris from meteorite impacts may travel considerable distances before being deposited; hence, mixtures of underlying bedrock and debris materials from impacts are not uncommon. Bulk chemistry from representative surface samples for the Apollo sites is given in Table 4-8.

No doubt, dissolution studies will have to be conducted on lunar regolith materials to determine the potential to release plant-nutrient and plant-toxic elements in plant growth systems. Glass is a major component of many lunar soils and will probably be the most reactive regolith material in aqueous environments. Most of the time, the bulk chemistry of the regolith is reflected in the chemistry of glass phases, particularly agglutinates. Unlike most terrestrial soils, lunar regolith has high concentrations of Fe, Mn, Mg, Ca, S, Cr, and Ni; however, the regolith is fairly low in Zn and K (Henninger et al., 1989). How these elements will behave in lunar soils subjected to water is still a major question (see Chapter 16 in this book). For example, high rates of release for toxic elements (e.g., Cr, Ni) may be a serious limitation for using in situ lunar regolith as soils (see Chapter 6 in this book).

VOLATILES IMPLANTED IN LUNAR MATERIALS

Three types of radiation interact with lunar materials; (i) solar wind, (ii) solar flares, and (iii) galactic cosmic rays. The solar wind species are the lowest energy radiation to strike the lunar surface; whereas, galactic cosmic rays are the highest energy radiation that strikes the surface. Solar wind species will penetrate lunar surface materials to depths of more than 100 nm; solar flare particles interact with lunar rocks to produce nuclear reaction products and may etch lunar materials to a depth of about 1 cm; and the penetration of galactic cosmic rays produces tracks and nuclear reaction products at depths up to several meters.

Solar-implanted volatiles are probably of most interest to agriculturists. Most of the H, C, and N found in lunar regolith is derived from solar winds. Hydrogen may be the most important solar-implanted gas for lunar base agriculture; because combined with the abundant O_2 present in the regolith, H becomes an important component for the production of water. Solar-implanted H is not abundant in the regolith; in fact, the average H concentration of lunar soils is around 55 mg kg^{-1}. This is equivalent to about 500 mg kg^{-1} water, if O_2 is combined with this H. Typical concentrations for solar-wind H as well as C and N imbedded in lunar fines (< 1 mm) are given in Table 4-9.

Carbon is more abundant in lunar soils than in lunar rocks; the range for C in lunar rocks is from 4 to 280 mg kg^{-1} and the average composition for all lunar soils tested from the six Apollo missions is 115 mg kg^{-1} (Gibson, 1975). Most of the C found in lunar materials has been derived from the solar winds (Moore et al., 1970, 1971; Kaplan et al., 1970), although some C is also contributed by meteorites.

Nitrogen, like C, is almost entirely implanted in lunar materials by solar winds; the average composition for lunar soils is around 85 mg kg^{-1} with

Table 4–8. Comparative chemistry of surface soils for the Apollo sites.

Chemical	Apollo						
	11†	12‡	14§	15¶	16#	17††	17‡‡
Oxide				wt. %			
SiO$_2$	41.3	46.0	48.08	46.62	45.07	39.82	45.17
TiO$_2$	7.5	2.8	1.77	1.50	0.64	9.52	1.55
Al$_2$O$_3$	13.7	12.5	17.59	16.57	26.39	11.13	20.63
FeO	15.8	17.2	10.45	12.14	6.08	17.41	8.74
MgO	8.0	10.4	9.27	10.66	6.14	9.51	9.87
CaO	12.5	10.9	11.12	11.24	15.29	10.85	12.84
Na$_2$O	0.41	0.48	0.65	0.41	0.38	0.32	0.46
K$_2$O	0.14	0.26	0.54	0.196	0.12	0.07	0.17
MnO	0.213	0.220	0.14	0.157	0.08	0.25	0.13
Cr$_2$O$_3$	0.290	0.41	0.26	0.344	0.12	0.46	0.23
P$_2$O$_5$	0.12	0.25	0.58	0.222	0.15	0.06	0.15
S	0.1	<0.1	0.071	0.084	0.09	0.12	0.06
Element				mg kg^{-1}			
Pb	1.4	--	10	2.8	2.9	--	--
Zn	23	--	28	21	22	33	21
Cu	10	7.2	16	9	7.9	--	--
Co	28.0	43	38	41	21	--	31
Ni	200	310	430	220	330	131	231
Ba	170	430	1000	300	120	--	190
Sr	160	140	135	130	145	157	153
B	--	--	18.0	--	--	--	--
Zr	320	--	790	390	130	214	293

† Apollo 11 soil 10084 (Laul & Papike, 1980; Taylor, 1982; Frondel et al., 1970).
‡ Apollo 12 soil 12001 (Laul & Papike, 1980; Taylor, 1982; Frondel et al., 1971).
§ Apollo 14 soil 14003 (Rose et al., 1972; Gibson & Moore, 1973).
¶ Apollo 15 soil 15271 (Duncan et al., 1975; Taylor, 1982).
# Apollo 16 soil 66041 (Apollo 16 PET, 1972; Rose et al., 1973).
†† Apollo 17 soil 71501 (Rhodes et al., 1974).
‡‡ Apollo 17 soil 72501 (Rhodes et al., 1974; Laul et al., 1974).

Table 4-9. Abundances of volatile elements in lunar soils from the six Apollo missions (Gibson, 1975).

Mission	Hydrogen[†]	Carbon	Nitrogen	Sulfur
		mg kg^{-1}		
Apollo 11	50	140–240	100–150	1000–1400
Apollo 12	80	20–180	40–130	700–1200
Apollo 14	70	40–190	80–165	700–1000
Apollo 15	13–120	20–190	25–115	400–1000
Apollo 16	10–79	30–280	30–155	300–1000
Apollo 17	41–211	5–200	5–130	500–1400

† From Williams and Jadwick (1980).

a range of 7 to 164 mg kg^{-1} (Gibson, 1975). Sulfur is abundant in the lunar regolith; however, most of the S is thought to be indigenous to the lunar environment and only a small portion of the total S is thought to be implanted in lunar materials by solar winds. The major S phase found in the lunar regolith is the mineral troilite (FeS). The average S content of lunar soils is 848 mg kg^{-1} with a range of 290 to 1400 mg kg^{-1} (Gibson, 1975). The total S abundances for lunar samples from the six Apollo missions are given in Table 4-9.

MATURITY OF LUNAR REGOLITH

Maturity of lunar soil is a measure of its integrated exposure time to surface process that tend to change or weather fresh material. Unlike ter-

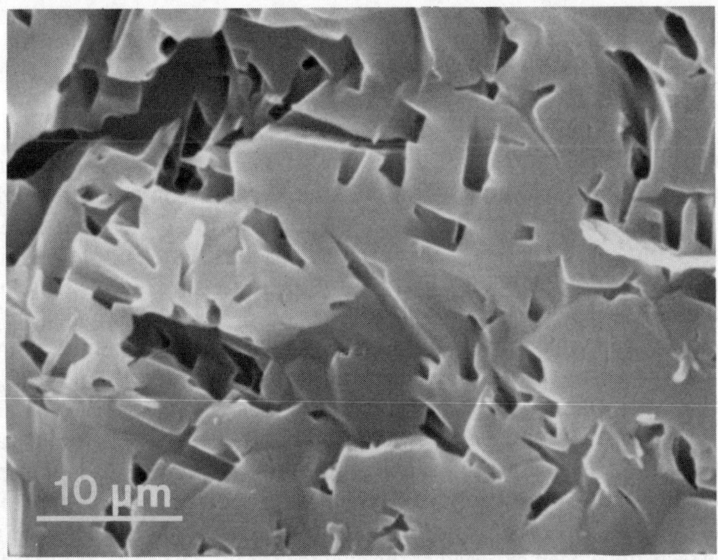

Fig. 4-10. Solar flare particle and cosmic ray particle tracks in a plagioclase crystal. The plagioclase crystal structure has been severely damaged where these high-energy particles have penetrated into the crystal. Etching of the crystal with NaOH has preferentially removed the damaged material leaving elongated,rectangular holes or tracks. Such track damage is common on many regolith crystals. NASA Photo S-72-55171.

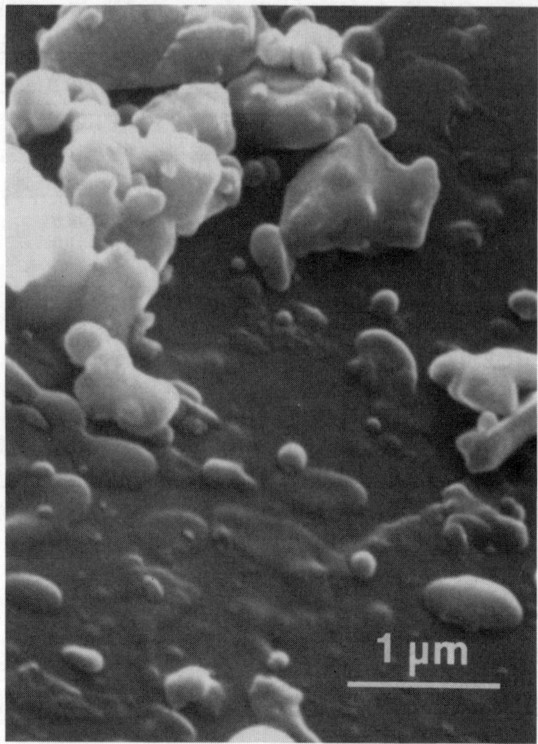

Fig. 4–11. High magnification view of the surface of a small mineral grain from the Apollo 17 regolith. This surface is covered with small attached fragments and glass droplets and pancake splashes. Ion etching and vapor deposition is also likely present. The original mineral grain surface has been completely modified and obscured by this weathering process. NASA Photo S-73-34927.

restrial soils where water and wind are primarily responsible for surface weathering, lunar weathering is caused mainly by micrometeorite bombardment, but also include implantation of solar wind species and radiation damage from more energetic solar flare particles. An example of mineral weathering due to solar flare particles is illustrated in Fig. 4–10, where a plagioclase feldspar crystal (Apollo 14) has been severely damaged by solar flare iron group nuclei and subsequently etched to remove the damaged material. The net result of this exposure at the lunar surface is that the regolith becomes finer grained, the agglutinate abundance increases, the solar wind gases increase in abundance, the radiation damage increases, and a variety of fine-scale weathering processes including glass splashing (Fig. 4–11), sputter erosion, impact vaporization, and resultant vapor deposition increase in intensity and importance. Soils that have the highest concentration of all of these effects have been exposed at the surface for long periods of time; therefore, they are the most mature and contrast most strongly with fresh soils derived from bedrock. Several methods are used to estimate the maturity of lunar soils; including (i) mean grain size, (ii) agglutinate content, (iii) ferromagnetic resonance maturity index, and (iv) solar wind concentration.

Mean Grain Size

Grain-size parameters have been widely used as an index to determine the relative surface exposure of a lunar soil (Mendell & McKay, 1975). Regolith exposed at the surface for a long time increases its chance for meteorite pulverization. The most mature soils are the finest-grained materials and have mean grain size of around 45 μm. For example, McKay et al. (1974) found that Apollo 17 soils could be grouped into three broad categories— immature, submature, and mature. Typical immature (71061, 1), submature (75081,36), and mature (72141,1) Apollo 17 soils have mean grain sizes of 114.2, 83.6, and 49.0 μm, respectively.

Agglutinates

The percentages of agglutinates determined petrographically in the 90 to 150 μm fraction have been used as an index to determine the exposure time of a surface (McKay et al., 1972b). A correlation exists between grain size and the number of agglutinates in which the finest samples have the highest agglutinate content (McKay et al., 1972b, 1974). For example, Apollo 17 immature soils had an average agglutinate content of 15.3%, whereas, mature soils had an average agglutinate content of 46.9% (Fig. 4–12). In another example, an Apollo 14 mature soil (14003,28) has an agglutinate (150–90 μm) content of 60.3%; whereas, an Apollo 14 immature soil (14141,30) has an agglutinate (150–90 μm) content of 5.2% (Table 4–5). Reworking by micrometeorites at the surface increases the agglutinate content of the regolith. Regolith formation may reach a steady-state when the con-

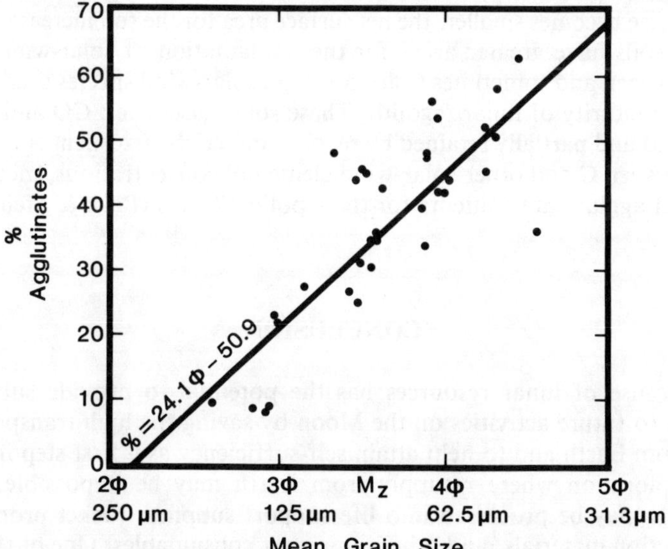

Fig. 4–12. Mean grain size vs. the percentage of agglutinates in the 90 to 150 μm fraction for Apollo 17 soils (M_z = graphic mean grain size) (McKay et al., 1974).

structional forces (agglutinate formation) are equal to the destructional forces (pulverization of large particles) and replenishment with fresh particles from bedrock.

Ferromagnetic Resonance Maturity Index (I_s/FeO)

The intensity of the characteristic ferromagnetic resonance (FMR) of a lunar soil normalized to its total Fe content (I_s/FeO) is an index to the relative surface exposure age (Housley et al., 1975; Morris, 1976). The characteristic resonance results from fine-grained metal particles that form from the reduction of Fe^{2+} in response to micrometeorite impacts. Regolith exposed to the surface for a long time increases its chances for micrometeorite impacts, which in turn increase the chances for the reduction of Fe^{2+} to fine-grained metal. Because the concentration of fine-grained metal is proportional to both the exposure time to micrometeorites and the FeO content of the soil, the concentration of fine-grained metal normalized to the FeO content of the soil is proportional to the length of surface exposure. A linear correlation is observed between I_s/FeO and the N content from implanted solar wind (Morris, 1976). Agglutinate abundances also correlate with I_s/FeO.

Solar Wind

Mature soils have the longest exposure time at the surface and, therefore, are more likely to be bombarded by solar winds. The net result of this exposure is an increase in implanted solar-wind gases in regolith materials. As discussed previously, mature soils have smaller mean grain sizes. As the particle size becomes smaller, the net surface area for the soil increases; hence, mature soils have greater areas for the implantation of solar-wind gases.

Nitrogen and sometimes C are common solar-wind species used to indicate the maturity of lunar regolith. These solar gases (i.e., CO and N_2) are implanted and partially retained by regolith materials. Excellent correlations exist between C and other solar-wind elemental concentrations, mean grain size, and agglutinate contents for the Apollo 17 soils (Basu & Meinschein, 1976).

CONCLUSION

The use of lunar resources has the potential to provide substantial benefits to future activities on the Moon by saving the high transportation costs from Earth and to help attain self-sufficiency as a first step in planetary exploration where resupply from Earth may be impossible. Lunar materials may be processed into life support supplies, rocket propellants, construction materials, and other important consumables. One of the long-term technologies necessary for developing a permanent, human-occupied lunar base is food production, particularly the production of edible plants.

Lunar materials will be able to supply most of the elements essential for the growth of plants; these materials may (i) act as a soil and provide a solid-support substrate for plant growth and/or (ii) be a source for the extraction of essential, plant-growth elements, which will be used as fertilizers.

A wealth of information has been published on the mineralogical, chemical, and physical properties of lunar materials. This information should aid agriculturalists in the planning of highly productive plant growth systems at lunar bases. The lunar samples returned during the Apollo missions are a national treasure and there is not enough material available on Earth to adequately test the potential of plant growth in lunar materials. However, the published information that is available on the mineralogical and chemical properties of lunar materials is such that terrestrial analogs may be designed and tested in plant growth systems to prepare us for the day when we permanently return to the Moon.

REFERENCES

Apollo 16 Preliminary Examination Team (PET). 1972. Preliminary examination of lunar samples. Part A: A petrographic and chemical description of samples from the lunar highlands. p. 7-1-7-24. *In* Apollo 16 preliminary science report. NASA SP-315. NASA, Washington, DC.

Basu, A., and W.G. Meinschein. 1976. Agglutinates and carbon accumulation in Apollo 17 lunar soils. Proc. Lunar Sci. Conf. 7th, 1976 1:337-349.

Claton, U.S., D.S. McKay, and R.B. Laughon. 1973. Iron crystals in lunar breccias. Proc. Lunar Sci. Conf. 4th, 1973 1:925-931.

Dymek, R.F., A.L. Albee, and A.A. Chodos. 1975. Comparative mineralogy and petrology of Apollo 17 basalts: Samples 70215, 71055, 74255, and 75055. Proc. Lunar Sci. Conf. 6th, 1975 1:49-77.

Dymek, R.F., A.L. Albee, and A.A. Chodos. 1976. Petrology and origin of Boulder #2 and #3, Apollo 17, Station 2. Proc. Lunar Sci. Conf. 7th, 1976 2:2335-2378.

Duncan, A.R., M.K. Sher, Y.C. Abraham, A.J. Erlank, J.P. Willis, and L.H. Ahrens. 1975. Interpretation of the compositional variability of Apollo 15 soils. Proc. Lunar Sci. Conf. 6th, 1975 2:2309-2320.

Frondel, C., C. Klein, Jr., J. Ito, and J.C. Drake. 1970. Mineralogical and chemical studies of Apollo 11 lunar fines and selected rocks. Proc. Apollo 11 Sci. Conf., 1:445-474.

Frondel, C., C. Klein, Jr., and J. Ito. 1971. Mineralogical and chemical data on Apollo 12 lunar fines. Proc. Lunar Sci. Conf. 2nd, 1971 1:719-726.

Gibbons, R.V., F. Hörz, and R.B. Schaal. 1976. The chemistry of some individual lunar soil agglutinates. Proc. Lunar Sci. Conf. 7th, 1976 1:405-422.

Gibson, E.K., Jr. 1975. Distribution, movement, and evolution of the volatile elements in the lunar regolith. The Moon 13:321-326.

Gibson, E.K., Jr., and G.W. Moore. 1973. Carbon and sulfur distributions and abundances in lunar fines. Proc. Lunar Sci. Conf. 4th, 1973 2:1577-1586.

Heiken, G.H., D.S. McKay, and R.M. Fruland. 1973. Apollo 16 soils: Grain size analyses and petrography. Proc. Lunar Sci. Conf. 4th, 1973 1:251-265.

Henninger, D.L., D.W. Ming, and C.W. Lagle. 1989. A case for the development of lunar agricultural soils. *In* W.W. Mendell (ed.) Second symposium on lunar bases and space activities of the 21st century. Lunar and Planetary Inst., Houston. (In review.)

Housley, R.M., E.H. Cirlin, I.B. Goldberg, H. Crowe, R.A. Weeks, and R. Perhac. 1975. Ferromagnetic resonance as a method of studying the micrometeorite bombardment history of the lunar surface. Proc. Lunar Sci. Conf. 6th, 1975 3:3173-3186.

Houston, W.N., J.K. Mitchell, and W.D. Carrier, III. 1974. Lunar soil density and porosity. Proc. Lunar Sci. Conf. 5th, 1974 3:2361-2364.

Kaplan, I.R., J.W. Smith, and E. Ruth. 1970. Carbon and sulfur concentration and isotopic composition in Apollo 11 samples. Proc. Apollo 11 Lunar Sci. Conf., 1970 2:1317-1329.

Laul, J.C., and J.J. Papike. 1980. The lunar regolith: Comparative chemistry of the Apollo sites. Proc. Lunar Planet. Sci. Conf. 11th, 1980 2:1307-1340.

Laul, J.C., D.W. Hill, and R.A. Schmitt. 1974. Chemical studies of Apollo 16 and 17 samples. Proc. Lunar Sci. Conf. 5th, 1974 2:1047-1066.

McGee, P.E., C.H. Simonds, J.L. Warner, and W.C. Phinney. 1979. Introduction to the Apollo collections: Part II. Lunar breccias. Curatorial Branch Publ., NASA Johnson Space Center, Houston.

McKay, D.S., U.S. Clanton, D.A. Morrison, and G.H. Ladle. 1972a. Vapor phase crystallization in Apollo 14 breccia. Proc. Lunar Sci. Conf. 3rd, 1972 1:739-752.

McKay, D.S., G.H. Heiken, R.M. Taylor, U.S. Clanton, D.A. Morrison, and G.H. Ladle. 1972b. Apollo 14 soils: Size distribution and particle types. Proc. Lunar Sci. Conf. 3rd, 1972 1:983-994.

McKay, D.S., R.M. Fruland, and G.H. Heiken. 1974. Grain size and the evolution of lunar soils. Proc. Lunar Sci. Conf. 5th, 1974 1:887-906.

McKay, D.S., D.A. Morrison, U.S. Clanton, G.H. Ladle, and J.F. Lindsay. 1971. Apollo 12 soil and breccia. Proc. Lunar Sci. Conf. 2nd, 1971 1:755-773.

Mendell, W.W., and D.S. McKay. 1975. A lunar soil evolution model. The Moon 13:285-292.

Meyer, C., Jr., R. Brett, N.J. Hubbard, D.A. Morrison, D.S. McKay, F.K. Aitken, H. Takeda,and E. Schonfeld. 1971. Mineralogy, chemistry and origin of the KREEP component in soil samples from the ocean of storms. Proc. Lunar Sci. Conf. 2nd, 1971 1:393-411.

Moore, C.B., E.K. Gibson, Jr., J.W. Larimer, C.F. Lewis, and W. Nichiporuk. 1970. Total carbon and nitrogen abundances in Apollo 11 lunar samples and selected achondrites and basalts. Proc. Apollo 11 Lunar Sci. Conf., 1970 2:1375-1382.

Moore, C.B., C.F. Lewis, J.W. Larimer, F.M. Delles, R.C. Gooley, W. Nichiporuk, and E.K. Gibson, Jr. 1971. Total carbon and nitrogen abundances in Apollo 12 lunar samples. Proc. Lunar Sci. Conf. 2nd, 1971 2:1343-1350.

Morris, R.V. 1976. Surface exposure indices of lunar soils: A comparative FMR study. Proc. Lunar Sci. Conf. 7th, 1976 1:315-335.

Reid, A.M., J. Warner, W.I. Ridley, and R.W. Brown. 1972. Major element composition of glasses in three Apollo 15 soils. Meteoritics 7:395-415.

Rhodes, J.M. K.V. Rogers, C. Shih, B.M. Bansal, L.E. Nyquist, H. Wiesmann, and N.J. Hubbard. 1974. The relationships between geology and soil chemistry at the Apollo 17 landing site. Proc. Lunar Sci. Conf. 5th, 1974 2:1097-1117.

Ridley, W.I., A.M. Reid, J.L. Warner, R.W. Brown, R. Gooley, and C. Donaldson. 1973. Glass compositions in Apollo 16 soils 60501 and 61221. Proc. Lunar Sci. Conf. 4th, 1973 1:309-321.

Rose, H.J., Jr., F. Cuttitta, C.S. Annell, M.K. Carron, R.P. Christian, E.J. Dwornik, L.P. Greenland, and D.T. Ligon, Jr. 1972. Compositional data for twenty-one Fra Mauro lunar materials. Proc. Lunar Sci. Conf. 3rd, 1972 2:1215-1229.

Rose, H.J., Jr., F. Cuttitta, S. Berman, M.K. Carron, R.P. Christian, E.J. Dwornik, L.P. Greenland, and D.T. Ligon, Jr. 1973. Compositional data for twenty-two Apollo 16 samples. Proc. Lunar Sci. Conf. 4th, 1973 2:1149-1158.

Ryder, G., and M.D. Norman. 1980. Catalog of Apollo 16 rocks. Curatorial Branch Publ. 52, NASA Johnson Space Center, Houston.

Smith, J.V. 1974. Lunar mineralogy: A heavenly detective story. Part I. Am. Mineral. 59:231-243.

Smith, J.V., and I.M. Steele. 1976. Lunar mineralogy: A heavenly detective story. Part II. Am. Mineral. 61:1059-1116.

Stöffler, D., H.D. Knöll, U.B. Marvin, C.H. Simonds, and P.H. Warren. 1980. Recommended classification and nomenclature of lunar highland rocks—a committee report. p. 51-70. In J.J. Papike and R.B. Merill (ed.) Proceedings of the conference on the lunar highlands crust. Geochim. Cosmochim. Acta, Suppl. 12. Pergamon Press, New York.

Taylor, S.R. 1982. Planetary science: A lunar perspective. Lunar and Planetary Inst., Houston.

Warner, R.D., G.J. Taylor, and K. Keil. 1979. Composition of glasses in Apollo 17 samples and their relation to known lunar rock types. Proc. Lunar Planet. Conf. 10th, 1979 2:1437-1456.

Weill, D.F., R.A. Grieve, I.S. McCallum, and Y. Bottinga. 1971. Mineralogy-petrology of lunar samples. Microprobe studies of samples 12021 and 12022; viscosity of melts of selected lunar compositions. Proc. Lunar Sci. Conf. 2nd, 1971 1:413-430.

Williams, R.J., and J.J. Jadwick. 1980. Handbook of lunar materials. NASA Ref. Publ. 1057. Natl. Tech. Info. Serv., Springfield, VA.

5   Pedology, Pedogenesis, and the Lunar Surface

L. R. Drees and L. P. Wilding

Texas A&M University
College Station, Texas

In the waning years of the 20th century and into the 21st century, soil scientists will be faced with new and unique opportunities and challenges. Not all of these challenges will be focused on solving problems in the terrestrial environment. Policy decisions by the U.S. Government point towards increased activity in space exploration, including a manned lunar base. From an agronomic standpoint, lunar materials are not well understood. They need to be investigated, not only to satisfy the scientists basic inquisitiveness, but to explore the feasibility of using in situ or modified lunar materials as a plant growth medium.

Food crops grown in a lunar *soil* would be an integral component of a bioregenerative life support system for extended lunar habitation. Lunar materials may provide favorable attributes as a medium for growing plants in a controlled environment: physical support, nutrient reserves, buffering capacity, low maintenance, medium for recycling waste by-products, and nutrient recycling. Live plants also provide a psychological and aesthetic quality of an Earth-like environment.

To better understand how the lunar regolith may be converted or changed to a viable soil material under a human life support environment, the processes and reactions that may occur in an Earth-like environment need to be explored. Although large-scale plot experiments are not feasible for lunar soil research, it is possible to examine analogous processes of soil formation in a terrestrial environment. Our long-term experience with crop production and soil-plant-environmental relationships enables us to better understand processes of soil formation and environmental responses. This basic understanding can then be extrapolated to the processes that would yield a viable lunar soil. In this context, soil concepts will be discussed from a pedological viewpoint.

CLASSICAL CONCEPTS AND DEFINITIONS OF SOIL

Through the ages, people have had different and varied concepts of soil, most being nonpedogenic concepts (Simonson, 1968; Arnold, 1983). Early humans were hunters and gatherers, relying on nature to provide food, shelter and clothing, and with minimal interaction with the natural environment. They likely recognized certain areas as yielding more fruit or game, but cultivated neither. The initiation of soil tillage may be considered as marking the birth of civilization. Aristotle, around 350 B.C., developed a philosophy that the Earth was one of the four basic elements of the universe (earth, wind, fire, and water). These four elements opposed each other, but in combination produced the qualities of hot-cold, hard-soft, light-heavy, and dry-wet.

As early as 5000 B.C., the Chinese noted differences in soil productivity for rice (*Oryza sativa* L.), thus recognizing soil as a medium for plant growth and differences due to spatial variability. The western world recognized this quality much later, but slowly came to realize that soil fertility could be maintained or improved by land drainage, crop rotations, farm manures, and natural fertilizers such as ground marl. Only in the 18th and 19th century did the concept develop that the soil was the storehouse of essential plant nutrients. Quantitative field and laboratory research is a product of the 19th and 20th century. Painstaking research has resulted in various cropping, tillage, and management systems that aim to optimize crop yields while sustaining soil fertility and minimizing soil loss. These systems are adjusted and modified to account for differences in soils, climate, and environmental conditions.

As scientific disciplines evolved, each developed a different definition of soil, often related to use-oriented concepts. This has delayed the progress in developing a unified definition applicable to all users. In general, soil to the geologist is the Earth's rind or regolith, a product of rock weathering. To an engineer who uses soil as foundations for structures, soil is any material on the Earth's surface that is unconsolidated or can be ripped. To an agronomist, soil is a medium for plant growth that differs spatially physically and chemically. Soil has been called the interface between the living and the dead—where plants combine solar energy and CO_2 of the atmosphere with nutrients and water from the soil to form living tissue (Foth, 1984). A pedologist's concept of soil is a natural three-dimensional body at the Earth's surface capable of supporting plant growth. An accepted soil definition of the USDA-SCS (Soil Survey Staff, 1975) is a "collection of natural bodies on the Earth's surface, in places modified or even made by man of earthy materials, containing living matter and supporting or capable of supporting plants out-of-doors."

PEDOGENIC IMPLICATIONS OF SOIL CONCEPTS

This definition, and current pedological concepts, indicate that soils have four basic constituents: air, water, organic matter, and minerals (Fig. 5–1). Air and water occupy the pore space and any change in air content will result in a corresponding change in water content, and vice versa. Likewise, an increase in organic matter will result in a proportional decrease in mineral matter.

The Soil Environment

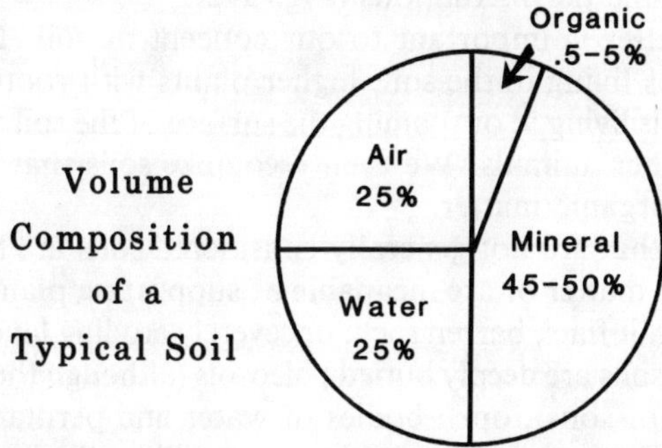

Fig. 5-1. Volume composition of a silt loam surface soil. The proportion of each component phase determines suitability for plant growth.

The pedological concepts and definition also imply that the soil is a dynamic medium, not static, and is continually perturbed by internal and external forces. It is a product of its environment of which it is an integral component. In a multi-component dynamic soil system, individual components will change or adjust in response to long- and short-term perturbations in its environment (Fig. 5-2). Loss of soluble salts may represent a rapid change with time while, translocation of clays and the formation of an argillic horizon (zone of clay accumulation) may represent a slow adjustment to the environment. Other properties may show periodic changes in response

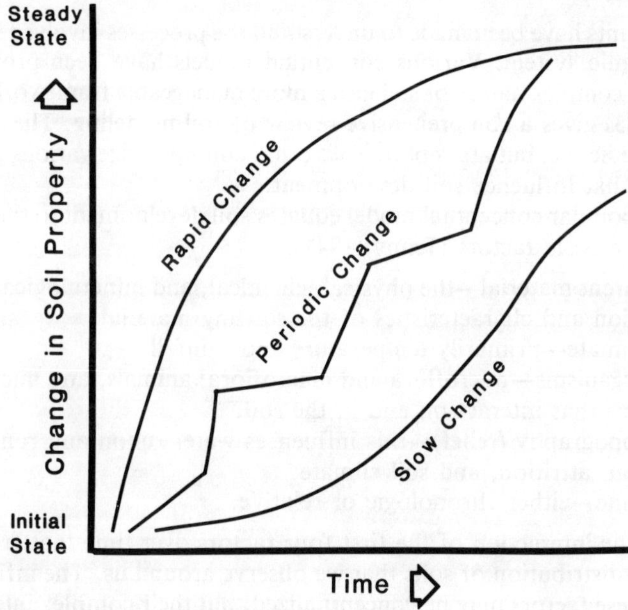

Fig. 5-2. Schematic diagram of rate change of soil properties in approaching steady-state conditions.

to environmental changes such as wetting and drying cycles, seasonal changes or long-term climatic fluctuations (Fig. 5–2).

Living matter is important to our concept of soil. Examples include microorganisms living in the soil, higher plants with roots anchored in the soil, and animals living in or roaming the surface of the soil consuming plants, residues, or other animals. We even recognize soils that are composed of almost 100% organic matter.

Materials that are not generally considered soils are those that are devoid of organic matter or are incapable of supporting plants in their natural setting such as salt flats, barren rock, or severely eroding land. Other materials not considered soils are deeply buried paleosols (although they may have many characteristics of soils), open bodies of water and permanently submerged materials. Such limitations may not be compatible with alternate or nonpedogenic concepts of soil.

Living organisms and their residues are deemed significant in soil development. Humus affects soil structure and consistency while organic acids accelerate weathering and mineral decomposition. Some pedologists however, suggest that soil-forming processes occur under abiotic conditions (Ugolini & Edmonds, 1983). Although chemical weathering may occur under abiotic conditions (e.g., reddening of fine-grained material on Mars, Arvidson et al., 1978), rates of soil formation are considerably faster in the presence of biota and free water.

SOIL MODELS

Attempts have been made to understand the processes involved in a complex dynamic system. Various conceptual models have been proposed to reduce the complex nature of soil into a more manageable framework. Smeck et al. (1983) gives a comprehensive review of soil modeling. These models are not predictive, but attempt to isolate and conceptualize various processes or forces that influence soil development.

One popular conceptual model equates soil development to the interaction of five basic factors (Jenny, 1941):

1. Parent material—the physical, chemical, and mineralogical composition and characteristics of the starting material.
2. Climate—primarily temperature and rainfall.
3. Organisms—microflora and macroflora, animals, and microorganisms that interact on and in the soil.
4. Topography (relief)—this influences water runon and runoff, erosion, attrition, and soil climate.
5. Time—either chronologic or relative.

It is the interaction of the first four factors over time that creates the landscape distribution of soils that we observe around us. The influence of each of these factors may be conceptualized, but their complex interactions preclude holding all but one factor constant so it may be evaluated in terms of sole pedogenic contributions.

The Generalized Process Model (Simonson, 1959) suggests that soil development and horizonation results from the vector equilibrium of the processes of gains, losses, transformations, and translocations (Fig. 5–3). Losses are due to erosion and leaching of soluble or dissolved weathering products. Losses may also be due to nutrient removal by plants. Additions include organic matter from plant residues or applied manures, soil amendments and fertilizers, sediments of aeolian or fluvial sources, salt precipitation, water, etc. In a weathering regime, various chemical reactions promote the alteration of primary minerals and the formation of secondary minerals (clays, carbonates, and salts) and hydrous oxides. Such reactions do not represent losses, but the rearrangement of atoms into new structures. It is the weathering processes over time that have transformed a relatively few primary minerals into a large variety of secondary minerals common to most soils (Allen & Fanning, 1983). Many of the weathering products may be physically translocated within the profile by gravity or moving water (eluviation), or as ions in solution to be precipitated elsewhere in the profile. Translocation of soluble products may be upward or downward depending on the net direction of water movement. In some loci, ants, termites, or burrowing animals play a significant role in the vertical translocation of soil material.

A more recent model (Energy Model of Runge, 1973) equates soil formation as the product of two intensity factors representing energy relationships in soils. These two intensity factors (water available for leaching and organic matter production) act together through time. Water available for leaching, equated with gravitational energy, is a determinant-driving energy flux in soils. The intensity of this factor is conditioned by intensity and duration of rainfall, runon and runoff, and soil permeability. This factor strongly affects a multitude of soil chemical processes and reaction kinetics. On the lunar surface, the lower gravitational constant could mediate the inten-

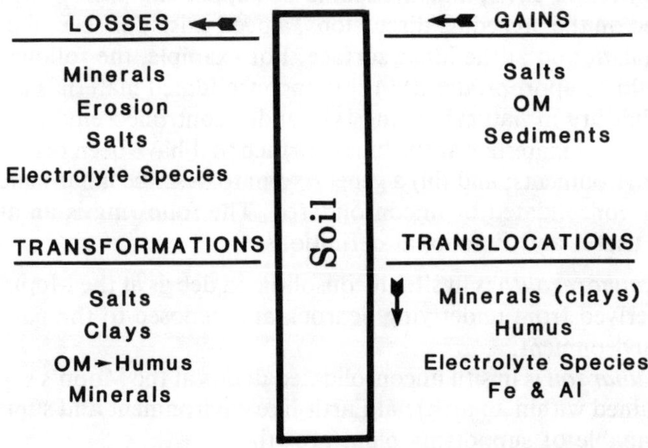

Fig. 5–3. Schematic representation of Generalized Process Model for soil development (Simonson, 1959).

sity of this factor. In a modified lunar environment, however, water movement could be monitored and controlled. Organic matter production over time is a function of the radiant energy available for plant growth. The intensity of this factor is conditioned by nutrient availability, soil-air-water relationships and plant characteristics.

In considering the definition of soils and the various models of soil processes, it is evident that water plays a critical role in soil formation. Water is the driving force behind most chemical reactions on the Earth or a modified Earth-like environment on an extraterrestrial body. These chemical reactions are largely responsible for changing a sterile regolith into a viable soil. Water is also largely responsible for the translocation of constituents and controls the intensity of biological processes and resultant biochemical reactions within the soil profile. Without organic matter and the capability of growing plants we have a sterile soil environment, although some chemical reactions may occur. Thus, water and organic matter are essential ingredients in a pedological concept of soil and controlling reaction rates.

PROPOSED DEFINITIONS OF LUNAR MATERIALS

In support of a manned lunar base, plants will not likely be grown on the natural lunar environment, but rather will be placed in specialized sealed enclosures that maintain favorable temperature, moisture, lighting, pressure, etc. Such enclosures would be akin to greenhouses on the Earth. Crops grown in greenhouses seldom use soil in situ, but rather use a soil sample removed from its natural setting. Once a soil unit is removed from its natural habitat, it ceases to be soil. It is called a soil sample. Alternatively, lunar materials may be left in their *natural* setting and enclosed within a structure to provide a controlled environment capable of supporting plant growth.

Based on the preceding discussions, a need exists for several definitions of materials found at the lunar surface. For example, the following definitions would be appropriate: (i) in situ unconsolidated materials at the lunar surface that are in natural settings vs. under-controlled environments; (ii) unconsolidated materials at the lunar surface that have been displaced from natural environments; and (iii) a general term to describe lunar materials that are either consolidated or unconsolidated. The following is an attempt to arrive at useful and functional definitions:

1. *Lunar regolith* is in situ unconsolidated debris at the Moon's surface derived from underlying bedrock and exposed to the natural lunar environment.
2. *Lunar soil* is in situ unconsolidated debris at the Moon's surface contained within an artificial Earth-like environment and supporting or capable of supporting plant growth.
3. *Lunar soil sample* is unconsolidated debris at the Moon's surface transported from an in situ environment and placed in an artificial Earth-like environment for manipulation and utilization such as supporting plant growth.

4. *Lunar material* is a general term that does not differentiate lunar rocks, lunar regolith, lunar soil, and lunar soil sample.

It is realized that the above definitions will not satisfy all users of lunar materials. These definitions however, should provide a more rational meaning to the term *soil*. Two items in the definitions should be noted. First, each definition has the prefix "lunar" to restrict the definition to the Moon so it will not be confused with comparable terrestrial definitions. Secondly, the term soil is used only where the lunar material is in a controlled environment conducive to the biotic or abiotic weathering of lunar materials and the formation of secondary weathering products. The above lunar prefixes could be modified and extended to encompass other extraterrestrial bodies.

LUNAR ENVIRONMENT

The first step in understanding how lunar materials may respond in a controlled Earth-like environment is to examine the characteristics of the lunar regolith, the unconsolidated parent material for lunar soil or lunar soil samples. The active processes on the Moon are cosmic, not terrestrial. They are mechanical, not chemical. The lunar surface has been bombarded by objects ranging from several micrometers to 100 km. The lunar surface has been tortured by a rain of projectiles until it resembles a World War I battlefield after intensive artillery bombardment. This bombardment has had the beneficial effect of producing a fairly continuous and somewhat homogeneous regolith 5- to 10-m thick (Taylor, 1975). The lunar rock has been physically fragmented, pulverized, and mixed, with evidence of shock-induced metamorphism and brecciation.

On a smaller scale, cosmic rays and high-velocity nuclei from the solar wind have produced mineral damage on an atomic scale. The surface material is implanted with noble gases from the solar wind, which also accounts for trace amounts of C, N, and H found in the lunar regolith (Williams & Jadwick, 1980).

From a pedological viewpoint, the lunar regolith is static and sterile, not dynamic and viable. Except for some physical movement, mixing and physical alterations due to meteorite impact, the lunar regolith will remain static and sterile. It does not have the capability of becoming the terrestrial equivalent of soil in its current environment.

The abundance of mineral species on the Moon is limited. Whereas there are more than 2200-known terrestrial minerals, only about 100 have been identified on the Moon (Mason, 1971; Frondel, 1975); most occur in minor or trace quantities. The difference is that the Earth contains many secondary minerals as weathering products. Dominant minerals on the lunar surface are olivine [$(Mg,Fe)_2SiO_4$], pyroxene [$(Ca,Mg,Fe)SiO_3$], anorthite [$CaAl_2Si_2O_8$], and minor amounts of ilmenite [$FeTiO_3$]. These minerals are similar to those found in terrestrial basalts and anorthosites, but not in the same proportions (Wood, 1970). Also common, but with no terrestrial ana-

Table 5-1. Percentage range in abundance (%) of common lunar minerals in mare basalts
and lunar regolith. Data from Williams and Jadwick (1980).

Mineral	Mare basalt	Lunar regolith
Olivine	<5–35	0–4
Pyroxene	35–70	5–30
Anorthite	15–35	10–60
Ilmenite	0–25	0.5–5
Agglutinates	10–17	30–50
Silica (Quartz, Crist., Trid)	<10	

logue, are agglutinates that are composed of the above minerals and glass
fragments bonded by glass droplets due to high-velocity meteorite impact
(Clanton et al., 1974; Williams & Jadwick, 1980). The suite of minerals is
variable depending on location (mare vs. highland), but not as variable as
terrestrial soils (Williams & Jadwick, 1980). Mineralogical differences also
exist between the lunar bedrock and lunar regolith (Table 5–1). Lunar materials can be characterized as being Fe and Ti rich and alkali (K and Na) poor,
which is unusual among most terrestrial soils and sediments.

If broad generalizations are made concerning the mineralogical composition of terrestrial soils and lunar materials, there are notable differences.
Minerals that tend to be common or abundant in terrestrial soils are rare
or absent on the lunar surface (Table 5–2). Likewise, abundant or common
minerals on the lunar surface are often rare or absent in terrestrial soils. Exceptions to these generalities do exist. Because the Moon lacks free O_2, many
elements (Fe, Mn, Ti, and other metallic elements) are in a chemically reduced
state, unlike the elements in minerals of terrestrial soils. Because the Moon
lacks water, it does not have amphiboles, serpentines, micas, and clay minerals
(phyllosilicates) containing bound water in the crystal structure that are so
plentiful in terrestrial soils (Frondel, 1975). The Moon also lacks carbonates
that are abundant in terrestrial soils and rocks (Table 5–2).

Compared to terrestrial soils, the lunar regolith has a well-graded particle-
size distribution. About 80% of lunar regolith is between 0.5 and 0.02 mm
(medium sands through coarse silts). Most of the remainder is <0.02 mm

Table 5-2. Mineral compositions of lunar and terrestrial soils.

Mineral	Lunar surface	Terrestrial soils
Quartz	Trace	Abundant
K and Na-Feldspars	Trace	Common
Micas	Absent	Common
Carbonates	Absent	Common
Ca-Feldspars	Abundant	Trace
Olivines	Abundant	Trace
Pyroxenes	Abundant	Trace
Ilmenite	Common	Trace
Agglutinates	Abundant	Absent

(Williams & Jadwick, 1980). Clay minerals, responsible for many physical and chemical properties in terrestrial soils, are notably absent from the lunar surface.

Terrestrially, a knowledge of soil spatial variability provides a means to quantify pedogenic processes and better understand causal factors for soil-distribution patterns (Wilding & Drees, 1983). The lunar surface has been sampled in only a few isolated locations. Based on the sparse data available, it is not yet possible to document the systematic and random spatial variability with respect to various lithogenic properties (mineralogy, bulk density, texture, depth to bedrock, etc.). Strongly disturbed soil material (e.g., meteorite-mixed lunar regolith) would be expected to have a larger component of random rather than systematic variability (Upchurch et al., 1987). However, until statistically valid sampling schemes are implemented on the lunar surface, it will be almost impossible to identify the magnitude of spatial variability.

WEATHERING OF LUNAR REGOLITH

Lunar minerals would be susceptible to chemical weathering if free water and O_2 were available. In an Earth-like environment, chemical weathering would be expected to occur because lunar minerals are not currently in equilibrium with an Earth-like environment. Minerals common to the lunar surface are among the least stable in terrestrial soils and sediments (Allen & Fanning, 1983) (Fig. 5–4). In a terrestrial-like environment, such minerals

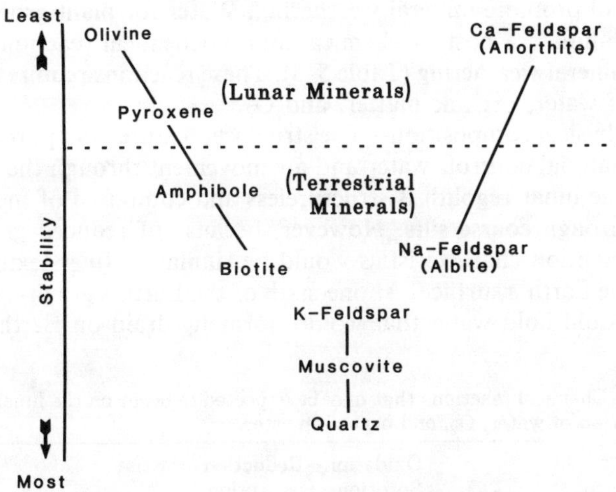

Fig. 5–4. Relative stability of common soil-forming minerals in a terrestrial environment. Minerals above the horizontal dashed line are common on the lunar surface while minerals below the dashed line are common in terrestrial soils.

would be susceptible to chemical weathering, releasing Ca, Mg, Fe, Ti, Al, and Si ions into solution, and forming a variety of secondary minerals. The release of essential plant nutrients has been documented by simulated weathering of Apollo 12 dust samples (Keller & Huang, 1971). The high proportion of glass and agglutinates in the lunar materials will likely influence weathering kinetics and soil solution chemistry. Amorphous constituents tend to be more soluble than crystalline minerals. Solubility is controlled by factors such as temperature, mineralogy, and chemical composition. Initial lunar soil solution chemistry and secondary products may be determined, in large part, by the chemistry of the glasses and agglutinates.

In weathering of lunar soils or soil samples, a dynamic equilibrium will develop between the lunar soil solution, exchangeable cations, and mineral reserves. Ions in solution are readily available to plants. Extraction or removal of ions from solution promotes the release of exchangeable ions which in turn promotes the slow release of ions from mineral reserves (primary lunar minerals). It is the dynamic equilibrium between the three phases (water-soluble nutrients, exchangeable cations, and mineral reserves) that will determine weathering kinetics.

Alteration of the lunar regolith to a viable lunar soil will be dependent on several factors internal and external to the lunar material. Some of these factors may be predictively modeled but others will need to be investigated. Factors affecting the alteration of the lunar surface include:

1. Climate—Temperature and moisture distribution will need to be controlled for effective water utilization and optimum crop yields. However, atmospheric pressure may be different from the Earth's environment, but optimized for human habitation.
2. Water movement—Sufficient water flux will be important to prevent surface salt accumulation, remove or translocate weathering products, and promote mineral weathering. Water for plant growth also will stimulate a host of chemical and biochemical reactions enhancing mineral weathering (Table 5–3). These reactions require the presence of water, organic matter, and O_2.
3. Physical composition—The structure, texture, and porosity of lunar material controls water and air movement through the soil system. The lunar regolith is structureless and composed of medium sands through coarse silts. However, because of reduced gravity, water retention characteristics would be similar to finer-textured soils at the Earth's surface. At one-sixth of the Earth's gravity, larger pores would hold water that would normally drain on Earth.

Table 5–3. Chemical reactions that may be expected to occur on the lunar surface with the addition of water, O_2, and organic matter.

Oxidation—Reduction
Sorption—Desorption
Complexes—Chelation
Diffusion
Hydration and hydrolysis

The generalized process model (Fig. 5–3), modified for a lunar environment yields a more graphic illustration of expected processes (Fig. 5–5).

1. Losses—In a closed ecological system, losses would be minimal. However, some weathering products may have to be removed if they prove to be toxic or detrimental to plant growth.
2. Gains—The following gains would be expected to occur:
   a. Moisture will be added to the system for plant growth and to enhance weathering of mineral reserves.
   b. Organic matter will occur as waste by-products of human metabolism (after pathogen removal) and via plant growth and decomposition. This addition would be beneficial in improving tilth, cation exchange capacity (CEC), moisture retention, buffering capacity, and accumulating nutrients in a sandy growth medium.
   c. Nutrients in the form of fertilizers may be required to compensate for inherent deficiencies.
   d. Amendments, such as zeolites, could be mixed with lunar materials to create a more favorable medium for plant growth (see Chapter 7 in this book). Zeolites have the potential to act as traps for toxic heavy metals, as a carrier for slow-release fertilizers and as soil conditioners (Henninger et al., 1987).
3. Translocations—Chemical weathering of lunar minerals would release soluble weathering products with a corresponding accumulation of residual products. Dissolution reactions proceed more rapidly if common ion-weathering products are translocated or leached from the weathering environment.
   a. Release of ions in solution will determine the base and nutrient status of the soil and the consequent pH.
   b. Secondary minerals will likely be formed and partially translocated downward in suspension. However, their rate of movement and depth of accumultion will be governed by lunar gravity forces.

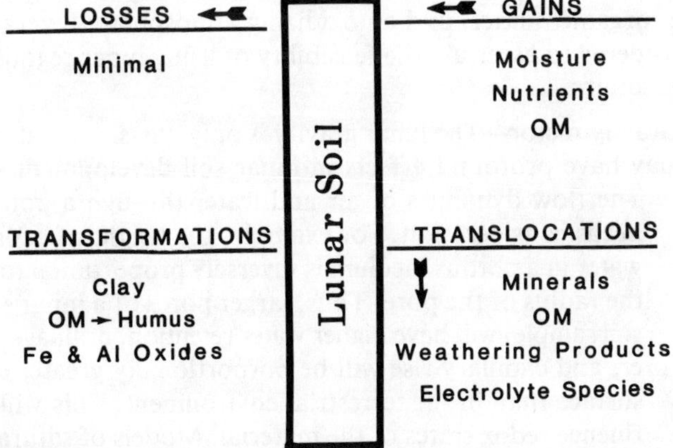

Fig. 5–5. Generalized Process Model for soil development modified to represent a lunar environment.

4. Transformations.

    a. Formation of clay minerals is an important consequence of mineral weathering. Terrestrial basalts (similar to lunar regolith minerals) commonly weather to smectite and/or kaolinite depending on the leaching environment and chemical composition of leachates. Stability diagrams based on solution composition of lunar dusts suggests kaolinite as the principal weathering product (Keller & Huang, 1971). Ionic concentration of equilibrium solutions will control the minerals formed. However, commonly in open soil systems kinetics of reactions and perturbation of chemical equilibria are more important. Considering the composition of lunar materials, there is the potential of forming the clay mineral nontronite that has Fe^{2+} instead of Al^{3+} as the octahedral cation.

    b. Aluminum and Fe oxides and oxyhydroxides are often the weathering products of basalts, but tend to be stable weathering products.

    c. Weathering of basalts in humid tropics often yield thick iron crusts (petroferric layers) that are nutrient poor and detrimental to water movement and plant root growth.

With adequate environmental controls, lunar farmers may be able to alter the direction and intensity of these processes. One must keep in mind that many of the above transformations, and consequent secondary minerals, are long-term events unless they are accelerated by lowering threshold energies biologically, chemically, or thermally (Yaalon, 1983). Hence, from a short-term agronomic perspective, reactions that govern kinetics in the lunar soil may be more important than thermodynamic equilibrium.

ITEMS FOR ADDITIONAL STUDY

Because the lunar surface has not been intensively investigated and lacks moisture, organic matter, and an oxidizing atmosphere, several items need to be considered to ascertain the feasibility of using lunar regolith as a plant growth medium:

1. Gravity factor—The lunar gravity is only one-sixth that of Earth and may have profound effects in lunar soil development.

    a. The flow dynamics of air and water through a porous medium needs consideration. For example, the height of capillary rise of water in a porous medium is inversely proportional to gravity and the radius of the pore. Thus, larger pores of a lunar soil (or lunar soil sample) will have higher water retention, drainage will be slower, and capillary rise will be porportionally greater on the lunar surface than in the terrestrial environment. This will directly influence redox states of the material. Models of saturated and unsaturated flow dynamics will have to be developed to fit a lunar environment.

    b. Lower gravity forces will affect the translocation of suspended and dispersed colloids. Due to lower gravity, larger suspended particles will be translocated, but at slower rates. A 2-μm particle at the Moon's surface will settle in water similar to an 0.8-μm particle on the Earth.

2. Lunar soil-forming factors.

    a. Due to the mineralogical nature of the lunar surface, essential plant nutrients such as K and P may be limiting while other elements such as Cr, Ni, and Al may occur in toxic levels.

    b. At the present time, the nature of weathering products of lunar materials placed in an Earth-like environment are only speculative. The speciation and concentration of ions in solution and the type of secondary minerals formed have yet to be verified.

    c. Weathering products may remain in place, be translocated within the lunar profile, or be leached out of the system. This will affect porosity, redox state, water retention, and water movement through the system. Lunar scientists may be able to partially control the formation and fate of weathering products by controlling temperature, solution speciation, and moisture movement in the lunar environment.

3. Spatial variability—There is a need to document the magnitude and loci of spatial variability. Statistically valid sampling schemes should permit predictions with a given degree of confidence. Such sampling schemes should be implemented to identify suitable sites for human habitation and the growth of food crops.

CONCLUSION

Although the lunar surface has an inhospitable environment, astronauts have demonstrated their ability to survive for a short time with suitable life support systems. Long-term lunar habitation will necessitate a lunar-derived or lunar-supplemented food supply to achieve a moderate degree of independence and self-sufficiency. The use of lunar soil or lunar soil samples offers some favorable benefits for food production. Soil is a resilient medium. It is slow to change both physically and chemically despite human abuses and environmental perturbations. Soils with clay mineral weathering products and organic matter have high-buffering capacities that act as shock absorbers to external forces. Many of the nutrients (major and minor plant nutrients) are available for slow release by weathering mineral reserves. These mineral reserves are in dynamic equilibria with exchangeable and soluble chemical species. Soil as a support medium has the distinct advantage that nutrient requirements would not need to be transported from Earth, or manufactured on the Moon. Plants, and other organisms, have the ability to recycle nutrients from lower depths and return them to the surface for future use. Soil also supplies the needed physical support for the growth of higher plants, especially tuberous plants and trees.

Many crops have shown a positive response to the incorporation of plant residues and animal manures as organic fertilizer supplements. They are valuable in improving the physical tilth characteristics and moisture retention of the soil for seedbed preparation and plant root media. Human waste byproducts, once detoxified and pathogens removed, could prove to be a valuable recycled commodity. Human waste byproducts and plant residues could provide essential nutrients (C, N, and P) deficient in lunar regolith. This aspect complements the controlled ecological life support system (CELSS) program as envisioned by the National Aeronautics and Space Administration (NASA) (Henninger et al., 1987).

Compared to other procedures for growing foodstuffs on the Moon (e.g., hydroponics and aeroponics), lunar soils (lunar soil samples) used for the growth of food crops would require low maintenance, releasing human energies and equipment for more productive activities. The buffering capacity, moisture retention, and nutrient release in the soil system governed by dynamic equilibria are self-contained within the soil system.

On Earth, soil is our most valuable nonrenewable natural resource. Soils sustain the food, feed and fiber production, serve as a medium to recycle our wastes, and give sustenance to the plants and animals that maintain gaseous equilibrium; these are all components essential to our survival on planet Earth. With a suitable environment, we should expect no less from the lunar soil.

REFERENCES

Allen, B.L., and D.S. Fanning. 1983. Composition and soil genesis. p. 141–192. *In* L.P. Wilding et al. (ed.) Pedogenesis and soil taxonomy. Vol. 1. Concepts and interactions. Elsevier, New York.

Arnold, R.W. 1983. Concepts of soils and pedology. p. 1–21. *In* L.P. Wilding et al. (ed.) Pedogenesis and soil taxonomy. Vol. 1. Concepts and interactions. Elsevier, New York.

Arvidson, R.E., A.B. Binder, and K.L. Jones. 1978. The surface of Mars. Sci. Am. (Mar.) 238:76–89.

Clanton, U.S., D.S. McKay, and G.H. Ladle. 1974. Micromorphology of lunar regolith particles. p. 642–654. *In* G.K. Rutherford (ed.) Soil microscopy. The Limestone Press, Kingston, Ontario.

Foth, H.D. 1984. Fundamentals of soil science. John Wiley and Sons, New York.

Frondel, J.W. 1975. Lunar mineralogy. John Wiley and Sons, New York.

Henninger, D.L., C.W. Lagle, and D.W. Ming. 1987. A lunar derived "soil" for the growth of higher plants. p. 1–20. *In* G.M. Andrus (ed.) Symposium '86: The first U.S. Maglev transportation conference and the first lunar development symposium. S86-35. Lunar Develop. Counc., Pitman, NJ.

Jenny, H. 1941. Factors of soil formation. McGraw-Hill, New York.

Keller, W.D., and W.H. Huang. 1971. Response of Apollo 12 lunar dust to reagents simulative of those in the weathering environment of earth. Proc. 2nd Lunar Sci. Conf. 1:973–981.

Mason, B. 1971. The lunar rocks. Sci. Am. (Oct.) 225:49–58.

Runge, E.C.A. 1973. Soil development sequences and energy models. Soil Sci. 115:183–193.

Simonson, R.W. 1959. Outline of a generalized theory of soil genesis. Soil Sci. Soc. Am. Proc. 23:152–156.

Simonson, R.W. 1968. Concept of soil. Adv. Agron. 20:1–47.

Smeck, N.E., E.C.A. Runge, and E.E. Mackintosh. 1983. Dynamics and genetic modeling of soil systems. p. 51–81. *In* L.P. Wilding et al. (ed.) Pedogenesis and soil taxonomy. Vol. 1. Concepts and interactions. Elsevier, New York.

Soil Survey Staff. 1974. Soil taxonomy: A basic system of soil classification for making and interpreting soil surveys. Agric. USDA-SCS Handb. 436. U.S. Gov. Print. Office, Washington, DC.

Taylor, S.R. 1975. Lunar science: A post-Apollo view. Pergamon Press, New York.

Ugolini, F.C., and R.L. Edmonds. 1983. Soil biology. p. 193–231. *In* L.P. Wilding et al. (ed.) Pedogenesis and soil taxonomy. Vol. 1. Concepts and interactions. Elsevier, New York.

Upchurch, D.R., L.P. Wilding, and J.L. Hatfield. 1988. Methods to evaluate spatial variability. p. 201–229. *In* L.R. Hossner (ed.) Reclamation of surface-mined lands. Vol. II. CRC Press, Boca Raton, FL.

Wilding, L.P., and L.R. Drees. 1983. Spatial variability and pedology. p. 83–116. *In* L.P. Wilding et al. (ed.) Pedogenesis and soil taxonomy. Vol. 1. Concepts and interactions. Elsevier, New York.

Williams, R.J., and J.J. Jadwick. 1980. Handbook of lunar materials. NASA Ref. Publ. 1057. NASA Sci. and Tech. Info. Office, Washington, DC.

Wood, J.A. 1970. The lunar soil. Sci. Am. (Aug.) 223:14–23.

Yaalon, D.H. 1983. Climate, time and soil development. p. 233–251. *In* L.P. Wilding et al. (ed.) Pedogenesis and soil taxonomy. Vol. 1. Concepts and interactions. Elsevier, New York.

6 Nutrient Availability and Element Toxicity in Lunar-Derived Soils

L. R. Hossner and E. R. Allen

Texas A&M University
College Station, Texas

The cultivation of higher plants on the lunar surface will provide a means of supplying food and recycling O_2 for inhabitants of a lunar base. Ideally, the plants will be grown directly in the lunar regolith or "soil"; however, as is common on Earth, soil amendments and fertilizers will be required to prevent deficiencies of essential plant nutrients. Further adjustments may be necessary to reduce the level of toxic elements that may occur. This chapter examines the use of lunar soils as a growth medium for higher plants and focuses on nutrient availability and element toxicity problems that may arise due to the chemical properties of the lunar regolith.

NUTRIENTS REQUIRED BY HIGHER PLANTS

There are 16 elements generally considered to be essential for the growth of higher plants. These nutrients are presented in Table 6-1. The elements C, H, O, N, P, K, Mg, Ca, and S are classified as macronutrients; while Fe, Mn, Zn, Cu, B, Mo, and Cl are classified as micronutrients. It has also been found that four other elements, Na, Si, Co, and V are necessary for or can enhance the growth of some plants as well (Tisdale et al., 1985);

Table 6-1. Nutrients required for the growth and reproduction of higher plants.

Macronutrients	Micronutrients
Carbon (C)	Iron (Fe)
Hydrogen (H)	Manganese (Mn)
Oxygen (O)	Zinc (Zn)
	Copper (Cu)
Nitrogen (N)	Boron (B)
Phosphorus (P)	Molybdenum (Mo)
Magnesium (Mg)	Chlorine (Cl)
Calcium (Ca)	
Sulfur (S)	

however, they are not essential for all plants. The nutrients C, H, and O are obtained by the plants from CO_2 and water. The other 13 nutrients must be supplied by the lunar soil.

NUTRIENT AVAILABILITY AND UPTAKE

A nutrient must be both chemically and positionally available before plant uptake can occur. A chemically available nutrient is one that is in the form of a soluble or exchangeable ion. All of the nutrients present in a soil are, therefore, not chemically available to the plant. The unavailable portion may be part of the structure of a slowly soluble mineral, may be tightly sorbed to a mineral surface, or may be fixed in a nonexchangeable form by organic matter. The nutrients that are chemically available become positionally available when the ions come into direct contact with the plant root surface. Absorption of the ions by the root can then proceed.

The movement of a nutrient ion from the solid phase through the soil solution to the plant root is demonstrated in Fig. 6–1. The total soil element includes both the available and unavailable nutrients. The chemically available nutrients consist of the labile soil fraction and the nutrients present in the soil solution. Reactions are shown to be reversible and the amount of an ion supplied to the plant root is controlled by the rate-limiting reaction. Normally the amount of element in solution is small compared to the labile fraction. Also, the labile fraction is commonly small compared to the total amount of that nutrient in the soil. The positionally available nutrients are those contacting the root surface. Nutrient ions can move to the root or come in contact with a root surface through root interception, ion diffusion, and/or mass flow (Barber, 1962). Once in contact with the root, the ion is either passively or actively absorbed into the plant.

Most information in the literature useful for predicting nutrient availability in lunar soils contains only a measure of the total soil element concentrations. A limited amount of data is also found regarding lunar soil solution concentrations of various nutrients; however, no measurements of nutrient

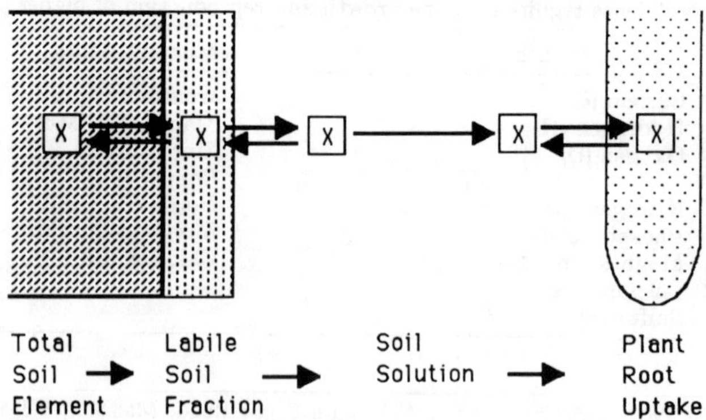

Fig. 6–1. Movement of nutrients from the soil to the plant.

labile fractions have been performed. This lack of data makes it difficult to evaluate the nutrient-supplying capacity of lunar soils. Predictions of nutrient availability must be extrapolated from the total soil element and the known soil solution concentrations with no consideration given to the nutrient labile fractions.

ELEMENT TOXICITY

Micronutrients essential for plant growth and other trace elements in terrestrial soils can be toxic to plants if their concentrations are too high. Bohn et al. (1985) lists 13 elements that have been designated as being toxic. These elements are presented in Table 6-2 along with their total soil and soil solution concentrations.

A high total soil concentration value does not always indicate that an element will be toxic. Factors such as the mineral or ionic form of the element and the plant species being grown, as well as the soil composition, pH, Eh, and moisture status all affect the element's availability to a plant (Mitchell, 1964). The potential for a soil to provide toxic concentrations of an element has been estimated using various types of extractants, such as pH 7 ammonium acetate, pH 2.5 acetic acid, pH 2.5 hydrochloric acid, 0.1 M hydrochloric acid, 0.5 M EDTA, and 0.005 M DTPA (Baker & Amacher, 1982; Thornton & Webb, 1980; Mitchell, 1964). Because no such extractions have been performed on lunar soil samples, predictions of potentially toxic elements in lunar soils must be extrapolated from the known total soil element and soil solution concentrations, as is the case with predictions concerning nutrient availability.

Table 6-2. Natural and apparently safe, soil and plant concentrations of elements that have been designated as being toxic. Adapted from Bohn et al. (1985).

Element	Total soil		Soil solution	Plant range
	Typical value	Range		
	$mg\ kg^{-1}$		$mg\ L^{-1}$	$mg\ kg^{-1}$
Aluminum (Al)	50 000	10 000–200 000	0.1–0.6	--
Arsenic (Ar)	5	1–50	0.1	--
Beryllium (Be)	1	0.2–10	0.001	--
Cadmium (Cd)	0.06	0.01–7	0.001	0.1–0.8
Chromium (Cr)	20	5–1 000	0.001	--
Cobalt (Co)	8	1–40	0.01	0.05–0.5
Copper (Cu)	20	2–100	0.03–0.3	4–15
Lead (Pb)	10	2–200	0.001	0.1–10
Manganese (Mn)	850	100–4 000	0.1–10	15–100
Mercury (Hg)	0.05	0.02–0.2	0.001	--
Nickel (Ni)	40	10–1 000	0.05	~1
Selenium (Se)	0.5	0.1–2.0	0.001–0.01	--
Zinc (Zn)	50	10–300	<0.005	8–15

LUNAR SOILS AND THEIR COMPOSITION

Lunar soils can be divided into two broad categories based on topography: those located in low-lying basins and depressions called *maria*, and those located in the highlands. Both soil types are composed primarily of pyroxenes, olivines, plagioclase feldspars, opaques, and glass spheres and chips (Williams & Jadwick, 1980; Keller & Huang, 1971). Even though the mineralogical compositions of the soils are similar, their chemical compositions differ (see Chapter 4 in this publication). The mare soils for example, have consistently lower Al/Si ratios than the highland soils (Williams & Jadwick, 1980). A comparison of the chemical compositions of mare and highland soils for selected elements is presented in Table 6–3. The average elemental composition for the same elements in a typical terrestrial soil is listed as well for reference. The Ca present in the lunar soils is primarily in the form of olivine and plagioclase, Mg is present primarily as olivine and pyroxene, S is present as various forms of iron sulfides, P is present as apatite and whitlockite, K is present in pyroclastic and granitic glasses and in feldspars, and N is present as entrapped N_2 (Williams & Jadwick, 1980). These mineral forms will, in part, govern the portion of each nutrient that is available to the plant.

An estimate of the variability in elemental composition of lunar soils was made using data from the analyses of samples collected during the six

Table 6–3. Average elemental composition of lunar and terrestrial soils.

Major components	Soil-type		
	Highland[†]	Mare[†]	Terrestrial[‡]
		mg kg^{-1}	
Aluminum	127 010	78 850	71 000
Silicon	210 270	212 230	320 000
Calcium	113 640	84 330	13 700
Iron	45 860	109 600	38 000
Magnesium	45 230	55 480	5 000
Sulfur[§]	300–1400	290–600	700
Titanium	3 600	23 380	4 000
Sodium	4 450	4 450	6 300

[†] Turkevich (1973).   [‡] Lindsay (1979).   [§] Compositional range (Gibson, 1975).

Table 6–4. Mean, standard deviation, and range of the total concentration of selected elements in lunar soil samples collected during the six Apollo missions. Adapted from Morris et al. (1983).

Element	No. of samples	Mean	Standard deviation	Range
			mg kg^{-1}	
Phosphorus	94	800	600	200–3500
Potassium	119	1700	1400	500–7500
Sulfur	50	900	300	300–1900
Chromium	98	2233	1154	479–6705
Nickel	80	291	132	55–720

Apollo missions (Morris et al., 1983). The results for five elements, P, K, S, Cr, and Ni, are presented in Table 6–4. Mare and highland soils were not differentiated in the calculations. Coefficients of variation ranged from 33 to 82% for the five elements indicating that a high degree of variability occurs in the chemical composition of lunar soils. The large range in values for the various elements indicate that nutrient availability and element toxicity problems will differ depending on which lunar soils are used for growing higher plants. Some selection will be necessary so that the best possible media can be provided for plant growth.

NUTRIENT AVAILABILITY IN LUNAR SOILS

Some interpretations can be made concerning the nutrient-supplying potential of lunar soils based on their composition. Both the mare and highland soils contain high levels of the macronutrients Ca, Mg, and S when compared to terrestrial soils. When exposed to a moist, aerobic, Earth-like environment, dissolution of the easily weathered olivines and pyroxenes (Buol et al., 1980) should result in adequate levels of plant available Ca and Mg. Oxidation of the sulfide minerals should provide adequate levels of S.

The relatively low K concentrations and absence of plant-available N in lunar soils will require additions of N and to a lesser extent K to ensure good plant growth. Fixation of N_2 using leguminous plants is also a possible means of supplying N.

Phosphorus is present in lunar soils primarily in the form of apatite. The decreasing solubility of this mineral with increasing pH (Lindsay, 1979) in combination with the near-neutral pH of the lunar soil (pH = 7.38 in deionized water, pH = 6.32 in CO_2-charged water; Keller & Huang, 1971) will result in slow dissolution of the apatite and limited availability of P to plants. According to Tisdale et al. (1985), apatite is effective in supplying P to plants only if it is finely ground and the soil pH is 6 or less. Modifications of the existing phosphate minerals in the lunar soil or additions of P fertilizers may be necessary to provide this nutrient in adequate amounts.

A better indication of the nutrient-supplying ability of lunar soils could be found by measuring the plant-available fraction with extractants common to plant and soil analyses (e.g., 1 M NH_4OAc, 0.5 M $NaHCO_3$, 1 M NH_4HCO_3, 0.005 M DTPA, 0.02 M EDTA; Page et al., 1982; Cox & Kamprath, 1972). Unfortunately, no such analyses have been performed on the samples of lunar soil collected by the Apollo missions. One study, however, (Keller & Huang, 1971) examined the dissolution of lunar dust (Sample 12070,128) in deionized water, CO_2-charged water, 0.01 M acetic acid (a weakly complexing solution), and 0.01 M salicylic acid (a strongly complexing solution). Simultaneous dissolution of pulverized basalt from Washington was also performed for comparison with the lunar dust. Even though the extracting solutions that Keller and Huang used do not give the best correlation with plant availability, some information can be extrapolated from their results.

Table 6–5. The ratio of the quantity of ions dissolved from lunar dust relative to the quantity of ions dissolved from terrestrial basalt, after 81 d. Source: Keller and Huang (1971).

Element	Solvent		
	CO_2–H_2O	0.01 M HAc†	0.01 M H_2Sal†
	ratio		
Silicon	0.56	1.9	2.1
Aluminum	1.2	0.78	2.7
Iron	0.6	3.8	0.86
Magnesium	0.86	4.1	0.52
Calcium	0.79	1.9	3.3
Sodium	1.0	1.9	2.0
Potassium	0.92	0.92	4.1
Titanium	--	--	1.1
Manganese	--	--	2.2
pH Basalt	6.6		
pH Lunar dust	6.32		

† HAc = acetic acid; H_2Sal = salicylic acid.

The ratio of the quantity of ions dissolved from lunar dust relative to the quantity of ions dissolved from basalt are presented in Table 6–5 for CO_2-charged water, 0.01 M acetic acid, and 0.01 M salicylic acid. The measurements were taken after gently shaking the samples for 81 d. The concentrations of these ions were not reported by the authors. The plant nutrients in most cases dissolved to a greater degree from lunar dust than from basalt when an organic acid was used as the solvent. The dissolution of Fe and Mg in acetic acid and Ca and K in salicylic acid was 3 to 4 times greater from lunar dust than from basalt.

The dissolution of seven other selected elements from lunar dust and basalt in 0.01 M salicylic acid after 81 d is presented in Table 6–6. Lower levels of the micronutrients, Cu, Zn, and Mn, had dissolved from lunar dust than from basalt; however, according to Keller and Huang (1971), the amounts dissolved are relatively proportional to the amounts present in the original samples. These levels are also equal to or greater than the soil solution concentration values normally found in soils (Table 6–2), indicating that the lunar dust can provide micronutrients to plants. Whether or not deficiencies occur will depend to a large degree on the particular requirements of the type of plants grown and on the soil pH.

Table 6–6. Dissolution of selected metals from lunar dust (12070,128) and basalt in 0.01 M salicylic acid after 81 d. Source: Keller and Huang (1971).

Element	Lunar dust	Basalt
	mg L^{-1}	
Titanium	3.0	0.95
Chromium	0.8	0.03
Strontium	0.26	0.09
Copper	0.06	0.13
Zinc	0.04	0.18
Manganese	0.72	0.83
Barium	0.16	0.15

ELEMENT TOXICITY IN LUNAR SOILS

Soil composition, pH, and Eh play major roles in determining whether element toxicity problems will occur. These same factors will be considered in this examination of potential toxicity problems in lunar soils. Because data are lacking with regard to the plant-available fractions of the various toxic elements present in lunar soils, the assessment of the potential toxicity problems must be extrapolated from the known total composition of the lunar soil, the dissolution study by Keller and Huang (1971), and the occurrence of toxicities in terrestrial soils.

Of the potentially toxic elements listed in Table 6–2, only Cr and Ni are present in the lunar soils at levels unusually high for terrestrial soils. Typical terrestrial soil concentration values for Cr and Ni are 20 and 40 mg kg^{-1}, respectively (Table 6–2). The mean total soil concentration values for Cr and Ni in the samples collected during the Apollo missions were 2233 and 291 mg kg^{-1}, respectively (Table 6–4). The ranges were 479 to 6705 mg kg^{-1} for Cr and 55 to 720 for Ni. These high values make Cr and Ni likely candidates to cause toxicity in plants grown in lunar soils.

Within the range of pH and Eh normally found in soils, Cr has the capability of existing in two trivalent states, the Cr^{3+} cation and the CrO^{2-} anion, and two hexavalent states, the $Cr_2O_7^{2-}$ and CrO_4^{2-} anions (Bartlett & Kimble, 1976). The hexavalent state appears to be more toxic than the trivalent state (Ross et al., 1981). Trivalent Cr is more common in soils than Cr(VI) because of the ease with which Cr is reduced; however, Bartlett and James (1979) have shown that oxidation of Cr(III) to Cr(VI) does take place in aerobic nonacid soils. According to Bohn et al. (1985), Cr(VI) stability increases with increasing pH. Because of the high Cr content of all the Apollo samples (Table 6–4) in combination with the near-neutral pH of the lunar dust (Table 6–5), Cr toxicity will be a likely problem in lunar soils.

Nickel toxicity in lunar soils appears to be less likely than Cr toxicity, but the potential still exists in those lunar soils containing the higher levels of Ni. According to Mitchell (1964), Ni found in ultrabasic rocks is in a form that is readily released upon weathering. In Scotland, it has accumulated in amounts toxic to cereals and root crops in poorly drained areas near such rocks. Concentrations as high as 8000 mg kg^{-1} total Ni and 100 mg kg^{-1} acetic acid soluble Ni were found in these infertile patches. According to Slingsby and Brown (1977), increasing the soil pH and organic matter content may reduce Ni availability to plants. Because of this relationship, Ni toxicity in lunar soils may only be a problem if the pH of the soils decreases over time and high total Ni concentrations are present in a readily released form.

The slightly acid pH of the lunar dust sample studied by Keller and Huang (1971) indicates a favorable environment for plant growth. The presence of iron sulfides in the lunar soils, however, creates the potential for this initial pH to be lowered. Upon exposure to an earthlike atmosphere, oxidation of the sulfides will result in the formation of iron oxides and sulfuric acid followed by a drop in pH (Bohn et al., 1985). Because the lunar soil will have a relatively low-buffering capacity, materials containing large quantities of sulfides will be especially susceptible to this problem. The resulting lower pH will increase the chance for Ni toxicity as well as introduce the possibility for Al and Mn toxicities. A further reduction in soil pH may also result in Fe toxicity.

CONCLUSION

The composition of lunar soils and the dissolution properties of the lunar dust studied by Keller and Huang (1971) indicate that a lunar soil has the potential to be an excellent medium for the growth of higher plants. The lunar soil, when exposed to a moist, aerobic, Earth-like environment, can be the source for many of the plant essential nutrients. Additions of N and to a lesser degree, P and K, will be necessary for optimum growth of higher plants. Higher natural concentrations and potential dissolution of certain trace metals, particularly Cr and Ni, may prove toxic to plants and bacteria.

REFERENCES

Baker, D.E., and M.C. Amacher. 1982. Nickel, copper, zinc, and cadmium. *In* A.L. Page et al. (ed.) Methods of soil analysis, Part 2. Chemical and microbiological properties. Agronomy 9:323–336.

Barber, S.A. 1962. A diffusion and mass flow concept of soil nutrient availability. Soil Sci. 93:39.

Bartlett, R.J., and B.R. James. 1979. Behavior of chromium in soils: III. Oxidation. J. Environ. Qual. 8:31–35.

Bartlett, R.J., and J.M. Kimble. 1976. Behavior of chromium in soils: I. Trivalent forms. J. Environ. Qual. 5:379–383.

Bohn, H.L., B.L. McNeal, and G.A. O'Connor. 1985. Soil chemistry. John Wiley and Sons, New York.

Buol, S.W., F.D. Hole, and R.J. McCracken. 1980. Soil genesis and classification. Iowa State Univ. Press, Ames.

Cox, F.R., and E.J. Kamprath. 1972. Micronutrient soil tests. *In* J.J. Mortvedt et al. (ed.) Micronutrients in agriculture. Soil Sci. Soc. Am., Madison, WI.

Gibson, E.K., Jr. 1975. Distribution, movement, and evolution of the volatile elements in the lunar regolith. The Moon 13:321–326.

Keller, W.D., and W.H. Huang. 1971. Response of Apollo 12 lunar dust to reagents simulative of those in the weathering environment of Earth. Proc. Lunar Sci. Conf. 2nd, 1971 1:973–981.

Lindsay, W.L. 1979. Chemical equilibria in soils. John Wiley and Sons, New York.

Mitchell, R.L. 1964. Trace elements in soils. *In* F.E. Bear (ed.) Chemistry of the soil. Reinhold Publ. Co., New York.

Morris, R.V., R. Score, C. Dardano, and G. Heiken. 1983. Handbook of lunar soils: Parts I & II. NASA Planetary Materials Branch. Publ. 67. Houston, TX.

Page, A.L., R.H. Miller, and D.R. Keeney (ed.). 1982. Methods of soil analysis, Part 2. Agronomy 9:323–336.

Ross, D.S., R.E. Sjogren, and R.J. Bartlett. 1981. Behavior of chromium in soils: IV. Toxicity to microorganisms. J. Environ. Qual. 10:145–148.

Slingsby, D.R., and D.H. Brown. 1977. Nickel in British serpentine soils. J. Ecol. 65:597–618.

Thornton, I., and J.S. Webb. 1980. Regional distribution of trace element problems in Great Britain. *In* B.E. Davies (ed.) Applied soil trace elements. John Wiley and Sons, New York.

Tisdale, S.L., W.L. Nelson, and J.D. Beaton. 1985. Soil fertility and fertilizers. Macmillan Publ. Co., New York.

Turkevich, A.L. 1973. The average chemical composition of the lunar surface. Proc. Lunar Sci. Conf. 4th, 1973 2:1159–1168.

Williams, R.J., and J.J. Jadwick (ed.). 1980. Handbook of lunar materials. NASA Rep. RP-1057. NASA, Washington, DC.

7 Manufactured Soils for Plant Growth at a Lunar Base

Douglas W. Ming

NASA Johnson Space Center
Houston, Texas

The newly created Office of Exploration (OEXP) at the National Aeronautics and Space Administration (NASA) Headquarters is considering several missions that the agency may undertake as a part of the human exploration of our inner solar system. These scenarios include: (i) expeditions to establish the first human presence on another planet (e.g., Mars); (ii) lunar outposts to conduct extraterrestrial science; and (iii) evolutionary expansion to establish self-sufficient human presence beyond low Earth orbit (LEO). Evolutionary expansion is a step-by-step program away from LEO. The first step will probably be the establishment of a lunar outpost that will lead to a self-sufficient lunar base. A self-sufficient lunar base will require the utilization of in situ resources for construction materials, propellants, and life-support systems. Plant growth at a lunar base will be essential to sustain a self-sufficient human colony. Several systems exist in which to grow plants, e.g., hydroponics, aeroponics, and solid-support substrates. Most of the plant growth research in controlled ecological life support systems (CELSS) has been aimed toward hydroponic systems. Soils also may be viable plant growth systems; however, our knowledge of how lunar materials will react as soil is virtually unknown.

Lunar materials, as they now exist, may not provide an adequate growth medium for higher plants for several reasons. The poor physical structure of lunar regolith may prevent proper aeration and water movement through the lunar materials. Minerals and materials that comprise the lunar surface may have little nutrient supplying power or retention capacity for plant-essential elements. Also, the toxicity of lunar materials to plants is not well known. For example, high Cr and Ni contents in lunar materials may be toxic to plants. It may be necessary, therefore, to alter lunar materials or prepare "soils" that will be productive for plant growth.

ADVANTAGES AND DISADVANTAGES OF A SYNTHETIC SOIL

Synthetic soils are attractive because the physical, chemical, and mineralogical properties of lunar materials may be altered to best suit plant growth.

Lunar regolith is a fine-grained deposit, which is primarily a result of meteorite impacts on the surface. The mean grain size of lunar soils ranges between 0.04 and 8 mm; however, the majority of the particles fall between 0.045 and 0.1 mm (Williams & Jadwick, 1980). The regolith is poorly sorted with no soil structure and average bulk densities range from 1.50 to 1.74 Mg m^{-3}. These soil physical conditions may not be suited for plant growth due to the high-bulk density of the regolith. A synthetic or manufactured soil could be sized and prepared to best suit the physical and chemical properties necessary to maximize plant growth (see Chapter 15 in this book).

Lunar regolith consists of mineral (e.g., plagioclase feldspar, pyroxene, olivine, and ilmenite) and glass phases. The glass phases have formed primarily by meteorite impact melting and should be the most reactive regolith phases in water. Elements toxic to plants may exist in high enough concentrations in glasses to be detrimental to plant growth. For example, the Cr content in most lunar soils is higher than in terrestrial soils (see Chapter 6 in this book). Chromium may be released into solution from glass phases in lunar regolith-derived soils and pose a potential toxicity to plants. Lunar materials may be found on the surface that do not contain high levels of plant-toxic elements. However, the easiest way to avoid toxicity problems may be to synthesize a soil without these elements. One of the first resources to be extracted from lunar materials will be O$_2$, which will be used for rocket propellent and in life support systems. A number of by-products will be produced during O$_2$ production that may not contain plant-toxic elements; therefore, these by-products may be desirable materials to use as solid-support substrates for plant growth.

Minerals that exhibit cation exchange as a result of isomorphic substitution (e.g., phyllosilicates) do not exist on the Moon. The lunar regolith, therefore, lacks retention for plant-essential cations. It may be possible to alter regolith glasses to form minerals with cation exchange capacities (CECs). With this in mind, Ming and Lofgren (1989) have synthesized minerals with CECs (e.g., smectites, zeolites, and tobermorites) from lunar-analog glasses (Fig. 7–1). These reactive minerals may be added to other regolith materials to create a soil that will have a retention capacity for plant-essential elements.

There are drawbacks to manufacturing a soil at a lunar base. Special equipment to synthesize soils may need to be shipped to the Moon. Initially, the cost of shipping equipment and resources required to produce a soil may not be economically feasible. The synthesis of most minerals that exhibit CECs will require water as a solvent; however, there may be methods to synthesize minerals that exhibit CEC without water as a solvent (vide infra). Because the Moon is void of water, materials to make water (e.g., H) may have to be shipped to the Moon; or water will have to be made on the lunar surface from regolith O$_2$ and solar-implanted H. Oxygen production on the lunar surface should not be a problem; however, small concentrations of H in lunar materials (generally <50 mg kg^{-1}) may prevent the economical production of water using solar-implanted H.

No doubt, the samples returned by the Apollo missions have provided us with a wealth of information on the physical, chemical, and mineralogi-

Fig. 7–1. Scanning electron micrograph of Zeolite A (twinned cubes in upper center of micrograph) and Zeolite P_t (bipyramidal-like crystals in lower center of micrograph). These zeolites were synthesized under mild hydrothermal conditions from synthetic basaltic glass with chemical compositions analogous to basaltic glass of the lunar highlands. (Hydrothermal conditions: temperature = 150 °C, solution = 0.1 M $Na_2SiO_3 \cdot 5H_2O$, duration = 76 h, pressure = ~0.3 MPa, glass/solution mass ratio = 0.015). See Ming and Lofgren (1989) for experimental procedures.

cal properties of lunar materials. However, finding the starting materials that may be best suited for synthetic soils may have to wait until we return to the Moon and conduct a thorough survey of lunar resources. Until then, we will have to use existing chemical and mineralogical data on lunar materials to design synthetic soil systems.

SYNTHETIC SOILS

Four primary functions of a root media should be considered when developing a synthetic lunar soil—nutrient retention, aeration, moisture retention, and mechanical support. Most terrestrial potting media have organic matter that provides nutrient and water retention. The production of organic components for a root media at a lunar base will be difficult because of the small quantities of organic compounds present in the regolith. Therefore, several hypothetical, plant growth systems in inorganic, solid-support substrates are proposed in this chapter; however, a considerable amount of basic research is necessary before these systems may be used to their fullest potential at a lunar base or in terrestrial applications.

Zeoponics

Parham (1984) first used the term *zeoponics* to describe an artificial soil consisting of zeolites, peat, and vermiculite that Bulgarian researchers used. Zeolites are hydrated aluminosilicates of alkali and alkaline-earth cations (e.g., K^+, Na^+, Ca^{2+}, and Mg^{2+}) that possess infinite, three-dimensional crystal structures (i.e., tektosilicates). The primary building units of the zeolite crystal structure are $(Al,Si)O_4$ tetrahedra. When Al^{3+} and sometimes Fe^{3+} substitute for Si^{4+} in the central cation position of the tetrahedron, a net-negative charge is generated. This negative charge is counterbalanced primarily by monovalent and divalent cations (generally called *exchange cations*). Zeolites have the ability to exchange most of their constituent exchange cations as well as hydrate/dehydrate without major change of the structural framework. About 50 zeolites have been found in nature, and several hundred synthetic species have been made in the laboratory. Natural zeolites may have CECs of 200 to 300 $cmol_c$ kg^{-1}, whereas some synthetic zeolites may have CECs as high as 600 $cmol_c$ kg^{-1} (Table 7–1). Ming and Mumpton (1989), Gottardi and Galli (1985), and Breck (1974) have reviewed the chemical and mineralogical properties of zeolites.

Most zeolites have large channels and/or cages that allow easy access of exchange cations, including plant-essential cations, to sites of charge (Fig. 7–2). Zeolites have unique cation selectivities that depend upon a number of factors: (i) framework topology (channel configuration and dimensions), (ii) size and shape of the exchange ion(s), (iii) charge density in the channels and cages, (iv) valence and charge density of the exchange ion(s), and (v) electrolyte composition and concentration in the external solution (Barrer, 1978).

Table 7–1. Representative unit-cell formulae and selected physical and chemical properties of minerals that exhibit cation exchange.

Special purpose minerals	Representative unit-cell formulae[†]	Typical void volume	Theoretical CEC
		%	$cmol_c$ kg^{-1}
Zeolites			
Analcime	$Na_{16}\{Al_{16}Si_{32}O_{96}\}\cdot16H_2O$	18	460
Chabazite	$(Na_2,Ca)_6\{Al_{12}Si_{24}O_{72}\}\cdot40H_2O$	47	420
Clinoptilolite	$(Na_3K_3)\{Al_6Si_{30}O_{72}\}\cdot24H_2O$	34	220
Mordenite	$Na_8\{Al_8Si_{40}O_{96}\}\cdot24H_2O$	28	220
Phillipsite	$(Na,K)_5\{Al_5Si_{11}O_{32}\}\cdot20H_2O$	31	380
Linde Type A	$Na_{96}\{Al_{96}Si_{96}O_{384}\}\cdot216H_2O$	47	540
Linde Type X	$Na_{86}\{Al_{86}Si_{106}O_{384}\}\cdot264H_2O$	50	470
Phyllosilicates			
Vermiculite	$Mg_{0.4}(Mg,Fe^{2+})_3(Al_{0.8}Si_{3.2})O_{10}(OH)_2\cdot nH_2O$	--	160[§]
Smectite[‡]	$Ca_{0.25}(Al_{1.5}Mg_{0.5})Si_4O_{10}(OH)_2\cdot nH_2O$	--	110[§]

[†] Taken mainly from Breck (1974).   [§] Alexaides and Jackson (1965).
[‡] Montmorillonite.

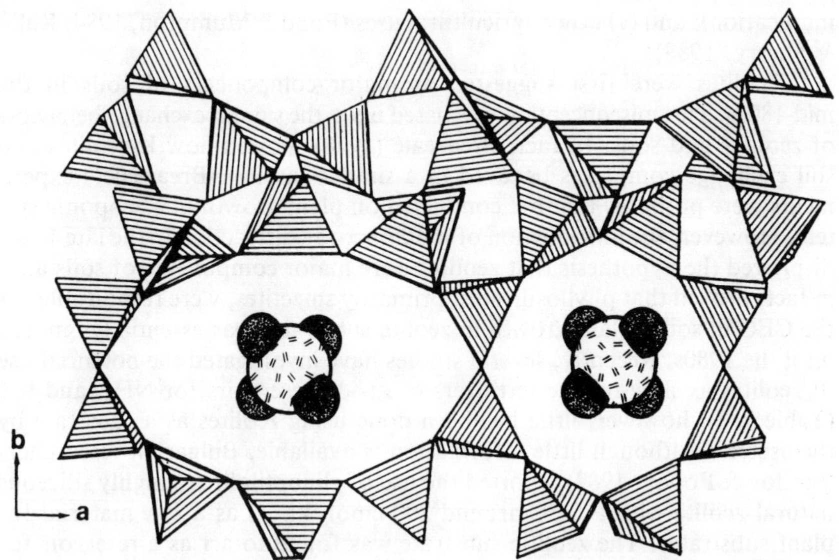

Fig. 7-2. Clinoptilolite structure showing 10- and 8-ring channels (modified from Koyama & Takeuchi, 1977). Ammonium ions readily access channels.

Based upon these unique properties of natural and synthetic zeolites, a number of agricultural scientists have examined potential use of zeolites, including (i) slow-release fertilizers (Table 7-2), (ii) traps for heavy metal ions (e.g., Cd, Pb, and Zn) in soils (Nishita & Haug, 1972; Nishita et al., 1968; Weber et al., 1984; Tuck & Ming, 1988, Grube & Herrmann, 1988), (iii) dietary supplements in animal nutrition (Mumpton & Fishman, 1977; Pond & Mumpton, 1984); (iv) nutrient retention and odor control in animal manures (Torri, 1978; Miner, 1984; P. van Straaten, 1985, personal com-

Table 7-2. Recent studies using zeolites as slow-release fertilizers or in-soil reservoirs for several plant essential cations.

Cation(s)/Zeolite-Mineral	Reference
NH$_4$/clinoptilolite	Pirela et al., 1984
	MacKown & Tucker, 1985
	Ferguson et al., 1986
	Ferguson & Pepper, 1987
	Bartz & Jones, 1983
	Lewis et al., 1984
	Weber et al., 1983
NH$_4$,K/clinoptilolite	Iskenderov & Mamedova, 1988
K/zeolites	Hershey et al., 1980
NH$_4$,K,Zn/clinoptilolite	Lewis, 1981
NH$_4$/clinoptilolite-apatite	Lai & Eberl, 1986; Barbarick et al., 1988
NH$_4$/zeolites-apatite	Chesworth et al., 1988
K,NH$_4$/clinoptilolite-apatite	Allen & Hossner, 1988
NH$_4$,K,Mn,Cu,Fe,Zn,Ca,Mg/clinoptilolite	Ming et al., 1987
Zeolitic substrates	Stoilov & Popov, 1982

munication); and (v) other agricultural uses (Pond & Mumpton, 1984; Kalló & Sherry, 1988).

Zeolites were first suggested as major components of soils in the mid-1800s. This misconception was based upon the similar exchange behaviors of zeolites and soils. In fact, Breazeale (1928) showed how K-zeolites and soil exchange complexes behaved in a similar fashion. Breazeale's experiments were probably the first conducted on plant growth in a zeoponic system. However, the introduction of modern x-ray diffraction in the late 1920s disproved the hypothesis that zeolites were major components of soils and, in fact, proved that phyllosilicates, primarily smectites, were responsible for the CEC in soils. Plant growth in zeolite substrates was essentially ignored until the 1980s. Recently, several studies have investigated the potential use of zeolites as slow-release fertilizers or in-soil reservoirs for NH_4^+ and K^+ (Table 7–2); however, little has been done using zeolites as a substrate by themselves. Although little information is available, Bulgarian researchers (Stoilov & Popov, 1982) reported the use of clinoptilolite (a highly siliceous natural zeolite with a CEC around 200 $cmol_c$ kg^{-1}) as a raw material for plant substrates. The zeolitic substrate was found to act as a reservoir for nutrient cations, to have desirable strength and other physical properties, to be sterile with respect to pathogenic microorganisms, and to be aesthetically pleasing. Depending on the plant variety, 20 to 150% increases in yields over control plots were observed for tomato (*Lycopersicon esculentum* L.), strawberry (*Fragaria chiloensis* L.), pepper (*Capsicum frutescens* L.), and rice (*Oryza sativa* L.). Also, the ripening of rice, cotton, and tomato was accelerated in the zeolite substrate.

Zeoponics is only in its developmental stages at the NASA Johnson Space Center. In this work, zeoponics has been defined as the cultivation of plants in zeolite substrates that (i) contain essential, plant-growth cations on their exchange sites, and (ii) have minor amounts of mineral phases (e.g., apatite)

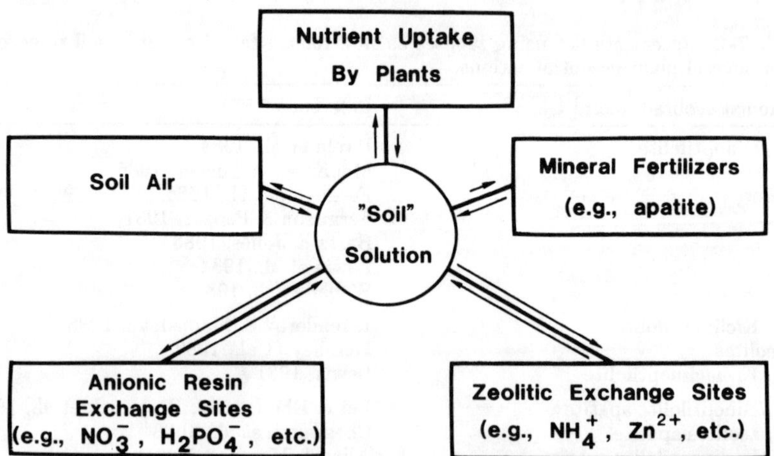

Fig. 7–3. Dynamic equilibria for a zeoponics system. The reactions in soil solution should (theoretically) be driven towards the root-soil interface by the uptake of nutrients by the plant.

and/or anion-exchange resins (e.g., activated aluminum resins) that supply essential plant growth anions (e.g., $H_2PO_4^-$). A zeoponics system is illustrated in Fig. 7–3. It is possible to saturate partially or wholly plant-essential cations (e.g., NH_4^+, K^+, Mn^{2+}, Cu^{2+}, Fe^{2+}, Zn^{2+}, Ca^{2+}, and Mg^{2+}) on clinoptilolite. Ming et al. (1987) have found the apparent selectivity of plant-essential cations and Na for native cations in clinoptilolites from Oregon and Wyoming to be:

$$K^+ > NH_4^+ \gg Na^+ > Mn^{2+}$$

$$= Cu^{2+} = Fe^{2+} > Zn^{2+} > Ca^{2+} > Mg^{2+}.$$

The dissolution of apatite is accelerated in a zeolite system (Lai & Eberl, 1986; Barbarick et al., 1988; Chesworth et al., 1988). Apparently, the zeolitic exchange sites act as sinks for Ca^{2+} released into solution by the dissolution of apatite. Once Ca^{2+} is removed from solution by the zeolite, the dissolution of apatite proceeds, thereby causing the release of phosphate into solution. In zeoponic systems, Ca^{2+} released from apatite should compete with plant-essential cations located at zeolitic sites. Once released into solutions, these plant-essential cations become available for plant uptake. Since most S compounds are more soluble than P compounds, it should not be difficult to find a S-containing material to supply necessary S for plant growth. Undoubtedly, the redox potential of the system will have an important role in selecting the S compound best suited for a zeoponic system.

Alternatively, anion-exchange materials may be used to supply plant essential anionic elements (e.g., $H_2PO_4^-$, SO_4^{2-}, NO_3^-). For example, Checkai et al. (1987) developed a mixed-resin hydroponic system that buffered the activities of cations and phosphate in solution without appreciably affecting pH or ionic strength. The P content was controlled by a cation-exchange resin (Dowex 50W-X4) containing adsorbed polynuclear hydroxyaluminum. Even though this resin was designed for hydroponic systems, similar resins may provide anion sinks in solid-support substrates.

Zeoponic systems have the potential to be regenerated with plant-essential nutrients and reused over and over. Nutrient solutions that contain experimentally determined concentrations of the plant growth elements will be passed through zeoponic materials until the desired type and amount of each nutrient has been adsorbed on zeolitic-exchange sites and anion-exchange resins.

Several problems may develop in zeoponic systems. First, the plant requirements for essential, plant growth elements may not be completely satisfied by a single ionic phase. For example, plants are known to take up N in the form of NH_4^+ as well as NO_3^-. However, it is not well known to what extent plants will take up and use NH_4^+ and/or NO_3^-. Because the zeolite substrate will supply NH_4-N, it may be necessary to amend the zeoponic system with nitrates (possibly by anion-exchange resins saturated with NO_3^-) to facilitate maximum yields. Another problem may be in establishing a buffered pH that will best suit the growth of a particular plant species. Because of the abundance of exchange reactions occurring in a zeoponic system, it may be difficult to maintain a constant pH and ionic strength of solution; however, the addition of a mineral phase (e.g., apatite) will help buffer the solution pH and ionic strength.

It will be essential to understand the exchange behavior for zeolites and/or anion-exchange resins used in zeoponic systems; therefore, it will be necessary to establish ion-exchange isotherms for these exchange resins at specific electrolyte concentrations. Zeoponics must have the ability to buffer ionic strength and pH. Once the exchange behavior of plant-essential elements for the synthetic or natural zeolites and anion-exchange materials is understood, plant growth experiments must be conducted to determine economic feasibility and how plant production in zeoponics compares to other plant growth systems (e.g., hydroponics).

Preparation of zeoponic systems at a lunar base will not be an easy task. Raw materials will have to be found on the lunar surface that will be best suited for the syntheses of zeolites and other exchange resins. However, Ming and Lofgren (1989) have shown that zeolites can be synthesized from lunar-analog glass subjected to mild hydrothermal conditions. Plant-essential elements will have to be extracted from the regolith or transported from Earth as may be the case for N. Because small traces of apatite and metal sulfides occur in the regolith (Williams & Jadwick, 1980), the extraction of PO_4 and S should not be a major problem.

The use of synthetic zeolites at lunar bases should not be limited to agricultural purposes. Based on their unique adsorption, hydration/dehydration, molecular-sieving, ion-exchange, and catalytic properties, synthetic zeolites may be used (i) as adsorption media for the separation of various gases, (ii) as catalysts, (iii) as molecular sieves, and (iv) as cation exchangers in sewage-effluent treatment, in radioactive-waste disposal, and in pollution control (Ming, 1989).

Other Cation-Exchange Materials

Smectite, vermiculite, and organic matter are a few of the materials used in terrestrial greenhouses to increase the CEC and nutrient retention in plant growth substrates. The addition of organic matter to a lunar soil will be nearly impossible due to the low abundance of organic molecules in the regolith. However, wastes from various processes (e.g., crop by-products, composted garbage) may be used as organic additives to lunar soils. A more likely candidate for increasing the CEC of these lunar soils will be the addition of inorganic phases that exhibit cation exchange.

Expanded vermiculites are in wide-spread use as terrestrial-potting media. Water between vermiculite particles (or quasi-crystals) causes permanent expansion between particles upon heating. The expanded volume of the vermiculite may be up to 16 times larger than the original material. Expanded vermiculites are desirable solid-support substrates for plant growth because of their nutrient and water retention, good root aeration, and low bulk densities. Lunar regolith amended with expanded vermiculite may act as excellent soils for plant growth at a lunar base; however, it may be difficult to synthesize vermiculites from lunar materials. In nature, vermiculites are thought to be alteration products of micas (Douglas, 1977). In the laboratory, vermiculization of chlorite can be readily achieved by thermal treatments of chloride (Ross & Kodama, 1974, 1976; Senkayi et al., 1981). The direct synthesis of vermiculite from solution, however, is rather difficult to achieve in the laboratory. It may be possible to synthesize high-charged vermiculite-type silicates in the absence of water by heating lunar-starting materials (e.g., pyroxenes) and NaF, MgF_2, or CaF_2; then allow the melt to slowly cool to promote crystallization. Gregorkiewitz and Rausell-Colom (1987) synthesized a high-charged, mica-type silicate from the reaction of augite in NaF-MgF_2 melts, which were allowed to slowly cool. The interlayer of the mica-type silicate, $Na_{4.0}(Mg_{6.0}Ti_{0.05})$ $[Fe_{0.1}Al_{3.4}Si_{4.5}O_{20.7}F_{3.3}]$, readily hydrated and Na^+ in the interlayer was easily replaced by K^+ in solution, indicating that the product exhibited cation-exchange properties.

Smectites, which are responsible for a large portion of the CEC in terrestrial soils (Borchardt, 1977), may be a more realistic material than vermiculites to amend lunar soils to increase the CEC. Smectites may be easier to synthesize from lunar regolith than chlorites and micas. Ming and Lofgren (1989) have synthesized smectites from hydrothermally altered glass, which has a chemical composition similar to lunar basaltic glasses. In terrestrial soils, smectitic cation-exchange sites create sinks to hold fertilizer cations such as K^+, NH_4^+, Ca^{2+}, Mg^{2+}, Zn^{2+}, and Fe^{2+}. The addition of smectite to lunar regolith will increase the nutrient and water-retention capacity of the manufactured soil.

As with zeolites, the use of synthetic smectites at lunar bases should not be limited to agricultural purposes. Smectites may be used as (i) adsorption media for waste renovation, (ii) cation exchangers, and (iii) adsorption media for organic molecules (Ming, 1989).

Nutriculture

Nutriculture is the cultivation of plants in inert substrates (e.g., water, sand, and air). Hydroponics (cultivation of plants in water) has been considered by plant physiologists as a leading candidate for plant growth systems in CELSS. Since hydroponic systems for CELSS have been discussed elsewhere (see Chapter 8 in this book; Averner et al., 1987; Salisbury et al., 1987; Bugbee & Salisbury, 1985; Salisbury & Bugbee, 1985), only solid-support substrates used in nutriculture will be discussed.

Nutriculture systems using a solid-support substrate have been suggested as viable growth systems for potato (*Solanum tuberosum* L.) in CELSS. Tibbitts and Wheeler (1987) found that tuberization was normal when recirculating nutrient solutions were passed through calcined clay particles or sphagnum moss; however, when solution was recirculated through containers filled with nutrient solution and the plant roots immersed, tuberization was delayed and the plants failed to tuberize normally. If potato and other edible tubers are to be grown in CELSS, it may be necessary to have a solid support substrate to promote tuberization.

Several inorganic materials and plastics are used terrestrially as inert, solid support substrates, including perlite (crushed siliceous volcanic rock that expands to a lightweight cellular material 10 to 20 times the original volume when heated to high temperatures), rockwool (mass of fine, intertwined fibers formed by passing molten coke, basalt, limestone, and possibly slag through a high-speed rotor), and polystyrenes (plastic made by polymerization of the hydrocarbon styrene). These materials, unless they are chemically altered, e.g., chloromethylation of polystyrene, should not contribute to or alter plant nutrients. Unfortunately, these materials do not exist on the Moon, or they may be difficult to synthesize on the lunar surface; the synthesis of polystrene would be difficult because of the lack of organic molecules in lunar materials. However, other materials produced from lunar materials may serve as excellent, inert, solid support substrates for nutriculture.

A resource-processing pilot plant will probably be the first industry to be built on the Moon that will use in situ resources (see Chapter 2 in this book). The primary products of a lunar-processing plant will include: (i) O_2 for rocket propellant and life support systems, (ii) construction materials (e.g., concrete, ceramics, and glasses), (iii) volatiles (e.g., H_2 and CO_2) for life support systems, and (iv) metals for construction. Undoubtedly, a number of useful by-products will be produced by a resource-processing pilot plant. Some of these products and by-products (e.g., ceramics and concrete) may provide excellent solid support substrates for plant growth. For nutriculture systems, special attention will have to be given to the particle size and the reactivity of the substrate. Ideally, the substrate in nutriculture systems should be inert and have a particle size that allows proper root aeration and drainage between the addition of the nutrient solution. With the addition of a nutrient solution (e.g., Hoaglands solution), plant growth systems using inert, solid support substrates are productive and fairly well understood; thereby, these systems should become attractive for use as solid support substrates for plant growth at a lunar base.

SUMMARY

It appears that a soil capable of growing plants can be produced from lunar materials. However, the synthesis of a lunar soil is only the beginning of a complex and advanced research effort. Many factors (e.g., source of water and plant nutrients, growth modules, effects of radiation, plant varieties, microbial populations, and reduced gravity) must be thoroughly examined before plants can be grown in lunar materials.

There may be a variety of materials that can be used as solid support substrates at lunar bases; some of the more likely candidates may include: (i) native lunar soils, (ii) sized lunar regolith amended with synthetic materials (e.g., zeolites and smectites) that provide nutrient and water retention, (iii) synthetic, inorganic, highly reactive substrates (e.g., zeoponics), and (iv) sized lunar materials or industrial by-products used as inert, solid-support substrates with nutriculture systems. The design of new, plant growth substrates may have tremendous terrestrial applications. For example, zeoponic systems may be used in commercial greenhouses. These systems are only in their design states and will require further research before they can be used to their fullest potential.

REFERENCES

Alexaides, C.A., and M.L. Jackson. 1965. Quantitative determination of vermiculite in soils. Soil Sci. Soc. Am. Proc. 29:522–527.

Allen, E.R., and L.R. Hossner. 1988. Supplying nitrogen, phosphorus, and potassium through ion exchange using a zeolite:apatite substrate. p. 193. *In* Agronomy abstracts. ASA, Madison, WI.

Averner, M.M., R.D. MacElroy, and D.T. Smernoff. 1987. Controlled ecological life support systems: Development of a plant growth module. p. 11–40. *In* R.D. MacElroy et al. (ed.) Controlled ecological life support system: Design, development, and use of a ground-based growth module. NASA-CP-2479. Ames Res. Center, Moffett Field, CA.

Barbarick, K.A., T.M. Lai, and D.D. Eberl. 1988. Response of sorghum-sudangrass in soils amended with phosphate rock and NH_4-exchanged zeolite (clinoptilolite). Colorado State Univ. Tech. Bull. TB88-1, Fort Collins.

Barrer, R.M. 1978. Cation-exchange equilibria in zeolites and feldspathoids. p. 385–401. *In* L.B. Sand and F.A. Mumpton (ed.) Natural zeolites: Occurrence, properties, use. Pergamon Press, New York.

Bartz, J.K., and R.L. Jones. 1983. Availability of nitrogen to sudangrass from ammonium-saturated clinoptilolite. Soil Sci. Soc. Am. J. 47:259–262.

Borchardt, G.A. 1977. Montmorillonite and other smectite minerals. p. 293–330. *In* J.B. Dixon and S.B. Weed (ed.) Minerals in soil environments. SSSA, Madison, WI.

Breazeale, J.F. 1928. Soil zeolites and plant growth. p. 499–520. Univ. of Arizona Agric. Exp. Stn. Tech. Bull. 21.

Breck, D.W. 1974. Zeolite molecular sieves. John Wiley and Sons, New York.

Bugbee, B.G., and F.B. Salisbury. 1985. Studies on maximum yield of wheat for the controlled environments of space. p. 447–486. *In* R.D. MacElroy et al. (ed.) Controlled ecological life support systems: CELSS '85 Workshop. NASA-TM-88215. Ames Res. Center, Moffett Field, CA.

Checkai, R.T., L.L. Hendrickson, R.B. Corey, and P.A. Helmke. 1987. A method for controlling the activities of free metal, hydrogen, and phosphate ions in hydroponic solutions using ion exchange and chelating resins. Plant Soil 99:321–334.

Chesworth, W., P. van Straaten, P. Smith, and S. Sadura. 1987. Solubility of apatite in clay and zeolite bearing systems: Applications to agriculture. Appl. Clay Sci. 2:291–297.

Douglas, L.A. 1977. Vermiculites. p. 259–292. *In* J.B. Dixon and S.B. Weed (ed.) Minerals in soil environments. SSSA, Madison, WI.

Ferguson, G.A., and I.L. Pepper. 1987. Ammonium retention in sand amended with clinoptilolite. Soil Sci. Soc. Am. J. 51:231–234.

Ferguson, G.A., I.L. Pepper, and W.R. Kneebone. 1986. Growth of creeping bentgrass on a new medium for turfgrass growth: Clinoptilolite zeolite-amended sand. Agron. J. 78:1095–1098.

Gottardi, G., and E. Galli. 1985. Natural zeolites. Springer-Verlag New York, New York.

Gregorkiewitz, M., and J.A. Rausell-Colom. 1987. Characterization and properties of a new synthetic silicate with highly charged mica-type layers. Am. Mineral. 72:515–527.

Grube, W.E., and J.G. Herrmann. 1988. Use of naturally occurring zeolite minerals for pollutant fixation. p. 307. *In* Agronomy abstracts. ASA, Madison, WI.

Hershey, D.R., J.L. Paul, and R.M. Carson. 1980. Evaluation of potassium-enriched clinoptilolite as a potassium source of potting media. Hortic. Sci. 15:87–89.

Iskenderov, I. Sh., and S.N. Mamedova. 1988. The utilization of natural zeolite in Azerbaijan SSR for increasing yield of wheat. p. 717–720. *In* D. Kalló and H.S. Sherry (ed.) Occurrence, properties, and utilization of natural zeolites. Akadémiai Kiadó, Budapest.

Kalló, D., and H.S. Sherry (ed.). 1988. Occurrence, properties, and utilization of natural zeolites. Akadémiai Kiadó, Budapest.

Koyama, K., and Y. Takeucki. 1977. Clinoptilolite: The distribution of potassium atoms and its role in thermal stability. Z. Kristallogr. 145:216–239.

Lai, T.M., and D.D. Eberl. 1986. Controlled and renewable release of phosphorus in soils from mixtures of phosphate rock and NH_4-exchanged clinoptilolite. Zeolites 6:129–132.

Lewis, M.D. 1981. Clinoptilolite as a N, K, and Zn source of plants. M.S. thesis, Colorado State Univ., Fort Collins.

Lewis, M.D., F.D. Moore, III, and K.L. Goldsberry. 1984. Ammonium-exchanged clinoptilolite and granulated clinoptilolite with urea as nitrogen fertilizers. p. 105–111. *In* W.G. Pond and F.A. Mumpton (ed.) Zeo-agriculture: Use of natural zeolites in agriculture and aquaculture. Westview Press, Boulder, CO.

MacKown, C.T., and T.C. Tucker. 1985. Ammonium nitrogen movement in a coarse-textured soil amended with zeolite. Soil Sci. Soc. Am. J. 49:235–238.

Miner, J.R. 1984. Use of natural zeolites in the treatment of animal wastes. p. 257–262. *In* W.G. Pond and F.A. Mumpton (ed.) Zeo-agriculture: Use of natural zeolites in agriculture and aquaculture. Westview Press, Boulder, CO.

Ming, D.W. 1989. Applications for special-purpose minerals at a lunar base. *In* W.W. Mendell (ed.) Second symposium on lunar bases and space activities of the 1st century. Lunar and Planetary Inst., Houston, TX. (In press.)

Ming, D.W., D.L. Henninger, and G.E. Lofgren. 1987. Ion-exchange selectivities of essential plant nutrient cations in clinoptilolite. p. 172. *In* Agronomy abstracts. ASA, Madison, WI.

Ming, D.W., and G.E. Lofgren. 1989. Crystal morphologies of minerals formed by hydrothermal alterations of synthetic lunar basaltic glass. *In* L.A. Douglas (ed.) Proceedings of the Soil Micromorphology Workshop. Elsevier Sci. Publ., Amsterdam. (In press).

Ming, D.W., and F.A. Mumpton. 1989. Zeolites in soils. p. 873–911. *In* J.B. Dixon and S.B. Weed (ed.) Minerals in soil environments. 2nd ed. SSSA, Madison, WI.

Mumpton, F.A., and P.H. Fishman. 1977. The application of natural zeolites in animal science and aquaculture. J. Anim. Sci. 45:1188–1203.

Nishita, H., and R.M. Haug. 1972. Influences of clinoptilolite on Sr90 and Cs137 uptakes by plants. Soil Sci. 114:149–157.

Nishita, H., R.M. Haug, and M. Hamilton. 1968. Influence of minerals on Sr90 and Cs137 uptake by bean plants. Soil Sci. 105:237–243.

Parham, W.E. 1984. Future perspectives for natural zeolites in agriculture and aquaculture. p. 283–286. *In* W.G. Pond and F.A. Mumpton (ed.) Zeo-agriculture: Use of natural zeolites in agriculture and aquaculture. Westview Press, Boulder, CO.

Pirela, H.J., D.G. Westfall, and K.A. Barbarick. 1984. Use of clinoptilolite in combination with nitrogen fertilization to increase plant growth. p. 113–122. *In* W.G. Pond and F.A. Mumpton (ed.) Zeo-agriculture: Use of natural zeolites in agriculture and aquaculture. Westview Press, Boulder. CO.

Pond, W.G., and F.A. Mumpton (ed.). 1984. Zeo-agriculture: Use of natural zeolites in agriculture and aquaculture. Westview Press, Boulder, CO.

Ross, G.J., and H. Kodama. 1974. Experimental transformation of a chlorite into a vermiculite. Clays Clay Miner. 22:205–211.

Ross, G.J., and H. Kodama. 1976. Experimental alteration of a chlorite into a regularly interstratified chlorite/vermiculite by chemical oxidation. Clays Clay Miner. 24:184–190.

Salisbury, F.B., and B.G. Bugbee. 1985. Wheat farming in a lunar base. p. 635–645. *In* W.W. Mendell (ed.) Lunar bases and space activities of the 21st century. Lunar and Planetary Inst., Houston, TX.

Salisbury, F.B., B.G. Bugbee, and D. Bubenheim. 1987. Wheat production in controlled environments. p. 121–130. *In* R.D. MacElroy and D.T. Smernoff (ed.) Controlled ecological life support system: Regenerative life support system in space. NASA-CP-2480. Ames Res. Center, Moffett Field, CA.

Senkayi, A.L., J.B. Dixon, and L.R. Hossner. 1981. Transformation of chlorite to smectite through regularly interstratified intermediates. Soil Sci. Soc. Am. J. 45:650–656.

Stoilov, G., and N. Popov. 1982. Agricultural uses of natural zeolites in Bulgaria. p. 38. *In* Zeo-agriculture '82. A conference on the use of natural zeolites in agriculture and aquaculture. Rochester, NY, 1–4 June. International Committee on Natural Zeolites, Brockport, NY.

Tibbitts, T.W., and R.M. Wheeler. 1987. Utilization of potatoes in bioregenerative life support systems. p. 113–120. *In* R.D. MacElroy and D.T. Smernoff (ed.) Controlled ecological life support system: Regenerative life support system in space. NASA-CP-2480. Ames Res. Center, Moffett Field, CA.

Torri, K. 1978. Utilization of natural zeolites in Japan. p. 441–450. *In* L.B. Sand and F.A. Mumpton (ed.) Natural zeolites: Occurrence, properties, use. Pergamon Press, Elmsford, NY.

Tuck, L., and D.W. Ming. 1988. Clay liner materials amended with zeolites to adsorb heavy metals. p. 309. *In* Agronomy abstracts. ASA, Madison, WI.

Weber, M.A., K.A. Barbarick, and D.G. Westfall. 1983. Ammonium adsorption by a zeolite in a static and a dynamic system. J. Environ. Qual. 12:549–552.

Weber, M.A., K.A. Barbarick, and D.G. Westfall. 1984. Application of clinoptilolite to soil amended with municipal sewage sludge. p. 263–271. *In* W.G. Pond and F.A. Mumpton (ed.) Zeo-agriculture: Use of natural zeolites in agriculture and aquaculture. Westview Press, Boulder, CO.

Williams, R.J., and J.J. Jadwick (ed.). 1980. Handbook of lunar materials. NASA Ref. Publ. 1057. NASA, Scientific and Tech. Info. Office, Washington, DC.

8 Controlled Environment Crop Production: Hydroponic vs. Lunar Regolith

Bruce G. Bugbee and Frank B. Salisbury
Utah State University
Logan, Utah

This chapter discusses two aspects of controlled-environment crop production in a lunar colony. First, we report findings about the effects of optimal aerial and root-zone environments on plant growth. Second, liquid hydroponic systems are compared with lunar regolith as substrates for plant growth.

GROWING WHEAT IN OPTIMAL ENVIRONMENTS

Since 1981, the National Aeronautics and Space Administration (NASA) controlled ecological life support system (CELSS) project at Utah State Univ. has been studying maximum yields of wheat (*Triticum aestivum* L.) in controlled environments (Bugbee & Salisbury, 1988). We control the foliar environment in three kinds of facilities: small growth chambers, modified greenhouses, and walk-in growth rooms. All facilities provide CO_2 enrichment, temperature control, irradiation primarily from high-pressure sodium lamps (Bubenheim et al., 1988), and control of humidity and of air velocity. Control of the root-zone environment is as important as the foliar environment. We use recirculating liquid hydroponic culture in all growth chambers and a soilless medium in the greenhouses. These facilities are similar to those of other NASA-CELSS projects.

Potential Yield per Mole of Photons

An ultimate goal of higher plant research in a CELSS is to remove environmental constraints such that the photosynthetic photon flux (PPF) is the only factor limiting growth and yield. This approach maximizes the energy efficiency of a CELSS. One important measure of yield is thus grams of edi-

[1] Research supported by NASA Cooperative Agreement no. NCC 2-139, Administered by the Ames Res. Ctr., Moffett Field, CA, and by the Utah Agric. Exp. Stn. This is Utah Agric. Exp. Stn. Paper no. 3736.

ble food per mole of photosynthetic photons. It is useful to calculate what might be achieved in a CELSS. Assuming a quantum requirement of 13, 95% absorption of the incident PPF, a respiratory C-use efficiency of 75%, and 40% C in the plant dry mass results in a maximum growth rate of 1.64 g mol^{-1} of photosynthetic photons (Bugbee & Salisbury, 1988, p. 869). If the harvest index is 50%, the potential yield of edible food is 0.82 g mol^{-1}. A typical PPF during the summer months is about 45 mol m^{-2} d^{-1}, so the potential growth rate (total biomass) of field-grown crops is about 74 g m^{-2} d^{-1}. Potential yield in g m^{-2} depends on the length of the life cycle. The longer the growth period, the higher the final yield.

Current Yield per Mole of Photons

Averaged over the life cycle, our highest yields of total biomass and of dry grain, per mole of photons, are 1.4 g mol^{-1} and 0.56 g mol^{-1}. These yields are about 20% higher than the highest short-term growth rates measured in the field and are about double the highest long-term growth rates in the field.

Current and Potential Yield per Unit Area per Unit Time

The size of the space farm is also important, so yield per unit area must be evaluated. Yield per unit volume may be a more meaningful measure of CELSS efficiency, but the required volume depends on many engineering factors that are difficult to estimate. A preliminary estimate is possible, however. We are selecting for short plant height so the combined height of the plants and the lighting system can be about a meter tall (0.4-m tops, 0.1-m roots, 0.5-m lighting system). This means that yield m^{-2} might also be expressed as yield m^{-3}, but until we know more about the final CELSS design (lighting system, etc.), we prefer to report yields per unit area. Other studies can then use the yield per area to project volume requirements.

Yield in a CELSS must be measured per unit time because planting one crop would almost immediately follow the harvest of the preceding crop (continuous cropping). The most appropriate measure of yield is thus g m^{-2} d^{-1}.

Our experience has identified some of the problems and potentials in CELSS research. Until mid-1986, our best yields were 24 g m^{-2} d^{-1} (Bugbee & Salisbury, 1985), considerably higher than high field yields of 7 and record yields of 14 g m^{-2} d^{-1}. Soviet reports, however, indicated yields of 77 g m^{-2} d^{-1} in a trial that used light that was two to three times the irradiance of sunlight (4000–6000 μmol m^{-2} s^{-1}; Polonskiy & Lisovskiy, 1980, 1986). While we were harvesting large quantities of biomass, the harvest index (proportion of total biomass as edible grain) was only about 25% compared with 40 to 45% in the field. Attempting to increase yield per day by shortening the life cycle with warm temperatures (27 °C) was inhibiting seed set. Lengthening the life cycle by lowering temperature to 20 °C greatly improved seed set, and the harvest index increased to 44%. Grain yields increased to 60 g m^{-2} d^{-1} at a PPF of 2000 μmol m^{-2} s^{-1}. In terms of energy input, this yield was more efficient than that achieved by the Soviets. Yield increased almost linearly with irradiance with no sign of saturation, so we do not yet know the potential yield m^{-2} d^{-1}. Efficiency of light conversion to chemical energy decreased linearly with irradiance from about 10%, close to the potentially achievable efficiency, at 500 μmol m^{-2} s^{-1}, to about 7.0% at 2000 μmol m^{-2} s^{-1}. The last tillers to form were least efficient; eliminating late-forming tillers could potentially increase the harvest index to more than 50% (Bugbee & Salisbury, 1988).

Size of the Lunar Farm

These results (as well as reports from the Soviet Union) meant that a CELSS farm of only 13 to 50 m^2 could supply enough calories for an individual; the exact size would vary with the productivity of the crops, the desired margin of safety, and the PPF level. In a lunar colony, a CELSS farm about the size of a U.S. football field (ca. 5000 m^2) could feed 100 people, and that would include a large safety factor (see discussion in Salisbury & Bugbee, 1988).

Genetic Effects

Wheat is grown in thousands of locations around the world, and specific cultivars have been developed for each location. Some cultivars are also much better adapted to controlled environments. The green revolution increased wheat yields by developing shorter, semidwarf cultivars that are 70- to 100-cm tall in the field and about 20% taller in optimal conditions in controlled environments. Semidwarf cultivars are too tall for the confined volumes of many controlled environments. We screened more than 1000 wheat cultivars from around the world to find high-yielding, full-dwarf wheat. We are continuing to evaluate and develop new germplasm, but our best-yielding cultivars are currently Veery 10 (40-cm tall) and Yecora Rojo (50-cm tall), both from the International Center for Maize and Wheat Improvement (Centro Internacional para Mejoramiento de Maize y Trigo; CIMMYT) in Mexico. Our breeding lines show promise of obtaining high yields in a cultivar that is only 20- to 30-cm tall.

The high-light, high-CO$_2$, high-N environment of a CELSS promotes photosynthesis and source capacity. Conventional wheat cultivars have been developed for the relatively stressful environment in the field and appear to have too much source capacity (leaves) and too little sink strength. The ideal wheat cultivar for CELSS would be extremely short (100 mm) and have fewer (and smaller) leaves. Achieving this goal may require genetic engineering. To this end, Carman (1988), in an adjunct project, has made impressive progress with in vitro propagation of wheat. This effort not only opens the door to genetic engineering, but allows propagation of hybrid wheat and means that crops could be planted with cultured explants, thus saving the seed for food.

Tillering

Wheat plants in the field typically form several stems (called *culms* or *tillers*), each of which produces a single head of wheat. Tiller formation, which increases with more favorable growing conditions, typically increases yields in the field. Tillering is a variable plant response that is important in the variable field environment but is neither necessary nor desirable in uniform controlled environments. Furthermore, late-forming culms often have a much lower harvest index, which reduces system efficiency. It would thus be advantageous if each seed produced only one culm, to reduce competition among plants at high densities. Wade Dewey (a co-investigator on this project) developed several genetically uniculm (nontillering) lines (using germplasm from Israel) that produce only one culm in the field, but all of these produce several culms in our optimized conditions—a phenomenon that warrants further study. Because genetic selection has not eliminated tillering and based on the results of other studies (Kasperbauer & Karlen, 1986), we are manipulating red/far-red radiation ratios to control tillering.

Plant Density

Competition for water and nutrients in natural ecosystems usually limits plant density. Plant density can increase when fields are cultivated, irrigated, and fertilized, but factors other than PPF still limit production. Flowing hydroponic culture should eliminate water and nutrients as limiting factors, while CO_2 enrichment of the foliar environment should eliminate this barrier to growth. We anticipated that increasing plant densities might improve productivity in controlled environments, but it was surprising to find that the optimum plant density was at least 10 times higher than the optimum plant density in the field. The optimum plant density for wheat in the field is 250 to 500 plants per m^2 (depending on availability of water and nutrients) while the optimum density in controlled environments is somewhere between 2500 and 10 000 plants per m^2! At these densities, up to 5% of the harvest must be used to replant the next crop, but it may be possible to use tissue cultured plants rather than conventional seed (Carman, 1988).

UTAH STATE UNIVERSITY HYDROPONIC SYSTEMS

Design: Ratio of Plant Mass to Solution Mass

The mass and volume of the root-zone environment is of particular interest in a CELSS. Increasing growth rates increase the demand on the root-zone environment so the ratio of plant mass to solution mass is a more appropriate measure than the size of the root-zone. Our basic system uses four, 0.1-m deep tubs in each chamber (Fig. 8–1). These tubs (20 L each) are connected to an 80-L reservoir located outside the chamber; the total volume of the system is 160 L. Each of the tubs has a surface area of 0.2 m^2 so there is 0.8 m^2 of plant growing area per system. When high PPF levels are used, up to 8 kg of dry plant mass have been grown in the system. The resulting ratio of plant mass to solution mass (at harvest) is 0.05 (8 kg of dry plants/160 kg of solution). We have not attempted to minimize solution volumes, but we anticipate that the nutrient solution reservoir could be removed, which would increase the plant mass/solution mass ratio to 0.1. Further decreases in the mass of solution would mean decreasing the depth of the root-zone, which would increase the density of the root system per unit volume of solution. At some point, the root length density (m of root length per m^{-3} of solution) would become too dense and prevent adequate flow of nutrient solution over all root surfaces. The minimum root-zone volume is thus determined by the maximum root density at which adequate solution flow can occur.

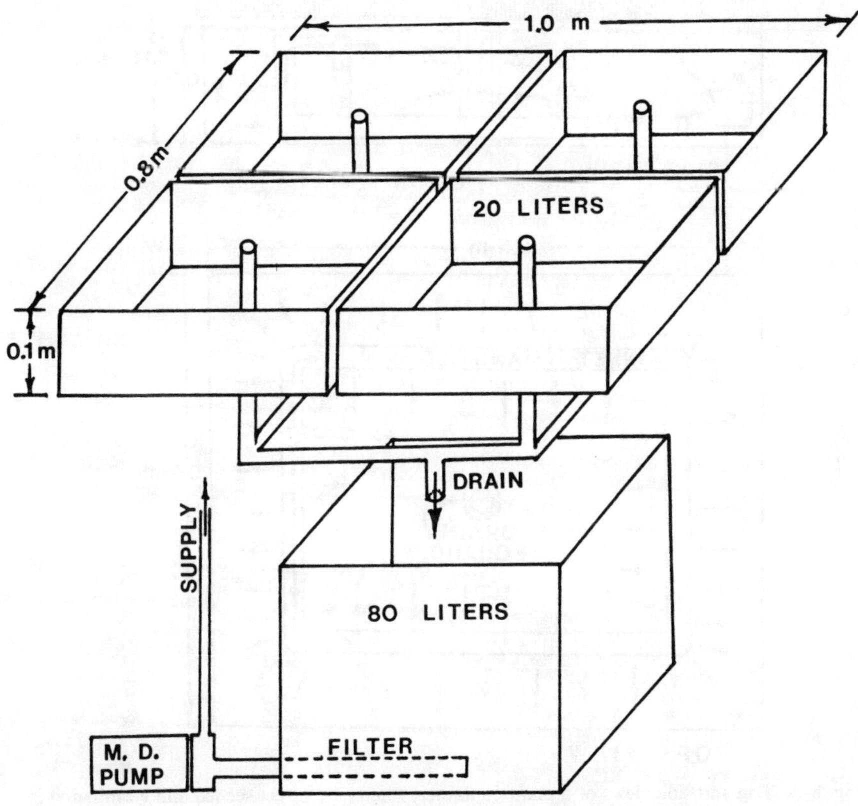

Fig. 8–1. The Utah State Univ. hydroponic system. The nutrient solution is supplied to four, 20-L plant containers with an epoxy-coated magnetic-drive pump. Solution cascades into a drain tube in the center of each tub and returns, by gravity, to an 80-L reservoir. The pump intake filter is made with fiberglass window screen glued over holes in PVC pipe. Additional filtering has not been necessary. There are no metal or rubber parts in the system. The total system volume is 160 L; the surface area is 0.8 m^2. See additional discussion in text.

Flow Rates

Rapid flow rates decrease the rhizosphere boundary layer and deliver O_2 and nutrients to the root surface. Oxygen is the element with the most demanding flow rate requirements. Oxygen is soluble in solution as a micronutrient (220 mmol O_2 m^{-3} in Logan, UT; 250 mmol O_2 m^{-3} at sea level), yet its uptake rate is usually more rapid than any other element in solution (up to 400 μmol O_2 h^{-1} g^{-1} dry wt.; Lambers, 1985, p. 459). Slow flow rates may provide high levels of dissolved O_2 in the bulk solution but can result in a large, stationary boundary layer around individual roots. Oxygen must cross this boundary layer by diffusion, which is an extremely slow process in liquids. Drew (1979) points out that the steady-state rate of diffusion of O_2 in water is 3 million times slower than through an equivalent distance in air. This difference is principally due to the smaller diffusion

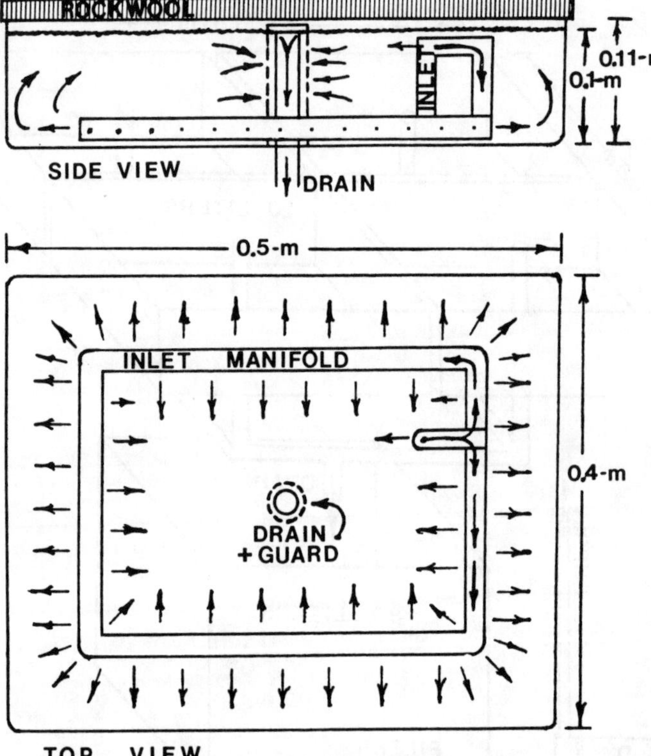

Fig. 8–2. Top and side views of a plant container. Plants are direct seeded and germinated in rockwool, which is supported by an inert grid above the flowing nutrient solution. An inlet manifold helps to uniformly distribute the incoming nutrient to all parts of the container. Roots are prevented from growing into the center drain by a guard, which is made with fiberglass window screen. Backward siphoning of the solution, which would occur during a power loss, is prevented by including a small hole in the inlet near the top of the container. When the power is interrupted, the solution siphons backward until it reaches this hole. Air then enters the inlet line and the siphon action stops. The plant roots are thus never without water.

coefficient but also because of the low solubility of O_2 in water. Because of differences in the size of the boundary layer, the minimum concentration of O_2 required in bulk solution depends heavily on solution flow rate. Griffin (1968) analyzed these concerns in detail. In a review of the factors limiting nutrient uptake, Robinson (1986) pointed out that boundary layers are never zero and demonstrated, theoretically, the large difference between boundary layers of 10 and 20 μm. Asher and Edwards (1983) discussed the importance of flow rate and reviewed the flow rates that were used in experiments with flowing systems. Many systems had inadequate flow rates.

Our experience indicates that achieving *uniform* solution flow is far more difficult than getting *adequate* total flow. Liquids flow through areas of least resistance (this is called *channeling*). Unfortunately, the root-zone areas with the highest root density need the most rapid flow and yet have the highest resistance to flow. Typically, the root system is most dense at the bottom of the container. We use a distribution manifold to maximize the flow rate through this densely rooted, bottom area (Fig. 8-2).

The nutrient solution is pumped into this distribution manifold in each compartment at a rate of 0.08 L s^{-1} [0.4 L m^{-2} (surface area) s^{-1}]. The manifold is designed so the maximum distance from the inlet to the outlet is < 0.2 m (Fig. 8-2). These rapid flow rates minimize gradients of O_2 (and other nutrients) in the rhizosphere during rapid growth. Dissolved O_2 is introduced into the solution by allowing it to cascade by gravity flow back into a storage tank. We measure above 85% of saturation at all locations in the bulk solution and rapid flow rates mean that O_2 should also be high at the root surface.

Although originally designed to enhance O_2 supply, a major problem with the nutrient film technique (NFT) of hydroponic culture is flow rates that are too slow to deliver adequate O_2 to plant roots (Graves, 1983; Jones, 1983). This problem is compounded in NFT by channeling of the solution through areas with lower root density (Resh, 1983). Maher (1977) reported that increasing the flow rate in NFT from 1 to 4 L per minute improved plant appearance and dissolved O_2 level in solution. Modern NFT designs have attempted to resolve the O_2 problem either by increasing flow rates or by reducing the distance from the inlet to the outlet (Cooper, 1984; Jensen & Collins, 1985). The minimum adequate flow rate is not a constant. It varies with system design and crop growth rate. A hydroponic system that provides adequate O_2, water, and nutrients for relatively slow crop growth rates may not be adequate to support the larger plant and root biomass and more rapid growth rates that might occur in a CELSS.

Minimum Root-Zone Volume

It is possible to calculate a minimum volume for the root zone of rapidly growing CELSS crops. This calculation assumes that in an optimum root-zone environment, maximum nutrient and water uptake per unit surface area of roots will occur, and water and nutrients will be supplied from the smallest possible root system. The size of the root system depends on the demand of the top for water and nutrients, which is determined by the crop growth rate, which depends on the PPF level. In our highest PPF, highest-yielding trial, the root system fresh mass was 3 kg m^{-2} on Day 24 and stayed at about this level until harvest (calculated from data in Bugbee & Salisbury, 1988). The mass of the foliar plant parts continued to increase after Day 24, but growth occurred at a constant rate per day. The root system that was formed by Day 24 was sufficient to supply water and nutrients for this growth rate during the final 55 d of growth. This is possible in hydroponic culture because all parts of the root system retain their water and nutrient absorption capabilities (Glass, 1989, p. 61). Water and nutrients are resupplied to the existing root system, and new root growth to get additional resources is not necessary (as it is in the field).

For the purposes of this analysis, we will assume that the root system that was formed in our most productive trial was the smallest possible size to support the crop growth rate. The root length density associated with the 3 kg m^{-2} of roots in this trial can be calculated by assuming a typical value for the average diameter of wheat roots (0.2 mm; Barber & Silberbush, 1984, p. 71; de Willigen & van Noordwijk, 1987). It is also necessary to assume that roots are cylinders and have a specific gravity of 1. These assumptions may not be absolutely correct, but they have been widely used by other workers and are necessary to proceed. Substituting the root radius (0.1 mm) into the equation for cylinder volume indicates that the average root length is about 32 km kg^{-1} of root fresh mass. This means that there was 96 km of roots m^{-2} surface area. The nutrient solution in our systems is 0.1-m deep so the root length density was 1 Mm m^{-3} (1000 km m^{-3}; 100 cm cm^{-3}). This density is 20 to 100 times higher than root-length densities in cultivated fields (1–5 cm cm^{-3}; de Willigen & van Noordwijk, 1987). Although this denity is unusually high, it occupies only about 3% (root volume = 3 L per 100 L of solution) of the total root-zone volume in our system (100-mm deep). This small percentage (3%) is not apparent upon viewing the root mass because the roots are so fine that much air remains trapped between them, making their volume appear larger than it is. The root system is so dense it cannot be easily compressed to <50 mm.

If an adequate flow of nutrient solution could be forced through a dense array of roots, it would be theoretically possible to have a root zone only 10- to 20-mm deep. In this severely confined, root-zone volume a large pressure drop would occur between the solution inlet and the outlet, which would require high-inlet pressures and careful sealing of the stem bases of the plants. Such a pressurized root zone might result in unusually high plant water potentials.

Considerable research has been done in recent years on the effect of restricted root-zone volumes (Krizek et al., 1985; Krizek & Dubik, 1987; Carmi, 1986). Plants grown in small containers usually develop a highly branched root system and are more dwarfed than control plants. This effect may be mediated by altered hormone production from the root system (Ruff et al., 1987). Although small root-zone volumes are theoretically possible using hydroponic culture, there are many complex interactions, and considerable research will be required to fully establish the effects on plant growth.

Maintenance: Recirculating the Nutrient Solution

During the life cycle, each crop (0.8-m^2 surface area) typically absorbs about 300 g of mineral elements from the solution and transpires 0.6 m^3 (600 L) of water. A dilute refill solution (Table 8–1) replaces nutrients and water lost via transpiration, but the entire solution is replaced only between studies (Bugbee, 1987a, b).

The composition of the nutrient solution has been successfully maintained during the life cycle of the plants, which is 12 wk in some studies. The ratio of ions in the original and make-up solutions is based on ideal concentrations in plant tissues. The concentration of ions is determined by the transpiration/photosynthesis ratio (the ratio of water that passes through the plants to demand for nutrients, based on photosynthetic incorporation of C), and it varies with the CO_2 concentration and the leaf-to-air vapor-pressure gradient (Table 8–1). We have measured transpiration/photosynthesis ratios as high as 400:1 (kg of water per kg dry plant mass) in hydroponic culture with ambient CO_2. High transpiration/photosynthesis ratios cause the electrical conductivity of the solution to increase, but the ratios of the ions absorbed and incorporated into biomass stays about the same. The con-

Table 8–1. Composition of nutrient solution. The ionic composition of the nutrient solution is based on the principle of mass balance. The ratio of nutrients in plant tissue that results in good growth is adequately established and salts were added to the nutrient solution in this ratio.

Element	Desired concentration in plant tissue		Concentration in nutrient solution†
	%	mol kg^{-1}	mmol kg^{-1} water
N	4.0	2.9	14.4
P	0.6	0.2	1.0
K	3.1	0.8	4.0
Ca	1.0	0.25	1.25
Mg	0.3	0.13	0.65
S	0.3	0.11	0.55
	mg kg^{-1}	mmol kg^{-1}	μmol kg^{-1} water
Fe	112	2.0	10
B	44	4.0	20
Mn	55	1.0	5
Zn	32	0.5	2.5
Cu	10	0.15	0.75
Mo	0.5	0.005	0.025
Cl	--	--	11
Si	420	15.0	75
Na	--	--	150

† This column is calculated from the mol kg^{-1} concentration in plant tissue by assuming that 200 kg of water are transpired in the process of creating 1 kg of plant dry mass (transpiration to photosynthesis ratio with CO_2 enrichment).

centration of the refill solution is altered as needed to maintain the electrical conductivity of the recirculating solution at 0.12 S m^{-1} (1.2 mmhos cm^{-1}).

The circulating solution is not sterile, but counts have shown bacterial numbers in solution to be extremely low. One indication of bacterial populations in solution is that 12-wk-old solutions are visually clear and look identical to freshly made nutrient solution. Studies by Rovira and Davey (1974) and others have indicated that nearly all the bacteria in solution are present on the root surfaces. The only significant energy source for bacteria in hydroponic solutions are the exudates of organic C from plant roots. Bacteria on root surfaces have immediate access to these exudates.

Measurements of total organic C in our solutions are typically <10 mg kg^{-1}. This extremely low level of organic C in solution is probably caused by an equilibrium between microbes on the root surface and efflux from plant roots (Rovira & Davey, 1974). Estimates of the quantity of organic compounds exuded by roots vary widely, but there is considerable evidence indicating that C efflux increases when plants are under stress (Barber & Gunn, 1974; Smucker, 1984; Haller & Stolp, 1985). Bowen and Rovira (1976) found that roots in solution culture appear to produce much smaller quantities of exudate than in soil. Trolldenier and Hecht-Buchholz (1984) found that reduced root growth due to lack of aeration in hydroponic culture was accompanied by a remarkably higher root-microbe population, which they attributed to increased C exudation from roots.

Routine, automated monitoring for signs of incipient stress would be extremely valuable in a CELSS. Monitoring organic C in nutrient solutions might provide an early indication of plant stress. We hope to test this hypothesis in future studies.

pH Control

The solution pH is maintained between 5.5 and 5.8 with an in-line electrode (Orion, model 91-06) and pH controller (Omega, model PHCN-36). When the pH increases to 5.8, the controller opens a solenoid that allows nitric acid (HNO$_3$) to flow into the bulk solution. After 20 d, the increase in plant growth causes the pH to change more rapidly, and ammonium nitrate is added to the pH control solution in a 1:1 molar ratio with HNO$_3$; after 30 d the ratio is increased to 2:1. Ammonium ions (NH$_4^+$) are taken up more rapidly than nitrate ions (NO$_3^-$), which causes the pH to decrease about 0.2 pH units during a 1-h period after which the NH$_4^+$ is at low levels in solution. Nitrogen uptake is then predominately from NO$_3^-$, and the pH slowly increases until the set-point pH is reached and the cycle begins again. This pH control method provides about 25% of the total N as NH$_4^+$ after the first 30 d of growth.

Some laboratories use a dual solenoid system that adds acid on pH rise and base on pH decrease. This type of control is called *active control*, as opposed to passive control with only one solenoid. We have not found that active control is necessary if the acid is added in close proximity to the pH electrode. This results in frequent, small additions of the pH control solution and a highly stable pH in the bulk solution.

Plant Support

Many different techniques are used to support plants in flowing hydroponic systems. Initially, we transplanted into foam plugs, but transplanting became impractical at higher plant densities. Therefore, a system for direct seeding was developed. Seeds are planted directly into rockwool (Grodan, 18-mm thick) with a vacuum seeder. The rockwool is kept moist for 20 d after germination to allow seminal roots to develop. The rockwool is suspended 10-mm above the surface of the flowing solution, and the air space between the rockwool and the solution is sealed to maintain 100% humidity. This prevents desiccation of emerging root tips. This system allows virtually unlimited plant densities.

Soilless Medium

In the greenhouse, plants are grown in a bench containing a soilless medium (a 1:1:1 mixture by volume of peat:perlite:vermiculite) that is automatically watered with a complete nutrient solution. We use a three-component soilless media mix to obtain the beneficial contribution of each component and to minimize potential toxicities that can result from any of the components used alone. The medium is leached with nutrient solution before planting.

ROOT-ZONE ENVIRONMENTS ON THE MOON: HYDROPONIC VS. LUNAR REGOLITH

Although hydroponic culture is widely used to grow high-input specialty crops on the Earth, the vast majority of the world food supply is grown in field soil. Large amounts of food will be required to support a lunar colony, and it has been suggested that the lunar regolith could be modified to provide a better root-zone environment than flowing hydroponic systems. Economic factors determine the amount of environmental control used to grow crops on the Earth (see review by Jensen & Collins, 1985), but there are other critical considerations on the Moon. Three questions need to be asked:

1. Which environment can maximize energy efficiency by providing the most optimum conditions for growth?
2. Which environment is the most stable and reliable?
3. Which environment is the most cost effective?

OPTIMUM ROOT-ZONE ENVIRONMENT

Many unfounded claims have been made about hydroponic culture. Hydroponics moved out of the laboratory and into commercial production in the 1930s when W.F. Gericke at the Univ. of California grew a wide variety of crops in large tanks. Plant growth was impressive, and tomato [*Lycopersicon esculentum* (L.)] plants were harvested with ladders (Resh, 1983). The press called hydroponics the discovery of the century, and entrepreneurs began marketing special nutrient formulas and equipment to a public that has always wanted to believe in magic potions to promote plant growth. Claims about the quantitative values of hydroponic culture continue to be made. Resh (1983) points out that most of the spectacular yield increases are the result of the more favorable foliar environment usually associated with hydroponic root-zone environments. When root-zone environments are compared in identical foliar environments, Resh says that hydroponic tomato yields are usually 20 to 25% higher than controls grown in soil. A production increase of this magnitude (25%) probably results from a less than optimum root-zone environment in solid media, but it is far more difficult to create an optimum root-zone environment in solid media than in hydroponic culture, so comparisons favor hydroponic culture. Can the root-zone environment in solid media be as favorable as a properly managed hydroponic environment? The following sections discuss the physiological processes that are affected by the root-zone environment.

Carbon Partitioning

It is well established that plants respond to root-zone stress by partitioning more C to the root system (Gales, 1979; Hoogenboom et al., 1988; Wild et al., 1987). The root dry weight fraction of hydroponically grown plants is typically lower (3–6% of the dry wt. at harvest) than in the field (10–30% at harvest). There are several reasons why the root systems of hydroponic plants can be smaller than soil-grown plants.

Nutrient Availability

Nutrients in solid media reach plant roots by mass flow and diffusion. Transpiration removes water from the root zone and creates mass flow from the soil solution to the root surface. The concentration of elements in the soil solution (in the field) is often too low to be supplied at adequate levels to the plant by mass flow, so the concentration of the element in the rhizosphere is reduced below the concentration of the element in the bulk solution. This creates a concentration gradient and diffusion occurs. For some elements, the combination of mass flow and diffusion is inadequate to provide nutrients to the plant root. The deficiency is made up by the growth of new roots into new soil regions (see discussion in Marschner, 1986, p. 414–427). This provides adequate nutrients, but more C must be partitioned to the root system, which reduces yield. Mass flow to the root surface supplies even less nutrients when the CO_2 concentration in the foliar environment is increased. High-CO_2 levels increase growth and nutrient demand and, at the same time, close stomates and decrease transpiration.

In a properly designed hydroponic system, flow rates are fast enough to deliver all nutrients to the root surface by mass flow. An electric water pump moves the nutrients to the roots instead of relying on biological energy (carbohydrates) to move the roots to the nutrients. Thus, the size of the root system in flowing hydroponic culture can be reduced.

It is not impossible to provide a balanced nutrient supply to the root surface in solid media. A highly porous medium, watered frequently (or continuously) with nutrient solution, can achieve a similar effect, but the slower movement of solution causes larger root boundary layers and makes ion control more difficult. It also means that nutrient concentrations in the bulk solution must be higher to obtain the same rate of uptake (Wild et al., 1987).

Nutrient Absorption Zones

It is often assumed that the root tip or region of root hair development is the principal region of nutrient absorption. While this is true in field soil, experiments with numerous species in hydroponic culture have shown that nutrient absorption can occur for considerable distances (more than 50 cm) behind the tip (Glass, 1989, p. 61–63). Although older roots have the ability to absorb nutrients, depletion of nutrients in the rhizosphere greatly reduces uptake. Mass flow in hydroponic solutions resupplies nutrients to all parts of the root system, which reduces the need for additional root growth.

Root-Zone Oxygen

It might appear that O_2 availability would always be reduced in liquid culture relative to soil culture. About 25% of the volume of well-aerated soils is air-filled pores. As mentioned earlier, the movement of O_2 (by diffusion) is about 3 million times faster in air than in water (Drew, 1979), so O_2 can be at high levels for about 25% of the root-surface area. However, O_2 must move by diffusion to the other 75% of the root surface that is not adjacent to an air-filled pore.

This presents a conflicting requirement for soil-root systems: root-soil contact has to be large to allow efficient uptake of nutrients and water, but sufficient root-air contact is required for aeration. Soils can quickly become anaerobic from overwatering because the soil solution is stationary and gasses must move by diffusion. Three recent papers discuss the difficulty of simultaneously providing adequate O_2 and water in solid media (Milks et al., 1989a, b, c). The roots of several crop plants form aerenchyma (internal air-channels) to increase the O_2 supply to roots. There is considerable variation among species in this ability. A few percent air-filled, internal porosity in roots can decrease the external O_2 requirement by 50%, but aerenchyma greatly reduce the ability of roots to penetrate soils. The trade-offs among aeration, nutrient and water uptake, and root morphology have been discussed in detail by de Willigen and van Noordwijk (1987, p. 244–248).

Oxygen is at low levels in hydroponic solutions (relative to soil pores filled with air), but rapid flow rates minimize boundary layers and uniformly deliver O_2 by mass flow to all parts of the root system. We have studied the effects of aerating nutrient solutions with 100% O_2 gas. This treatment resulted in a four-fold increase in dissolved O_2 in solution (because the partial pressure of O_2 in the aerating gas was 4.8 times higher than air), but there was no increase in plant growth rates or C partitioning. These data suggest that normal levels of dissolved O_2 in solution are adequate to support maximum growth rates if the flow rates are fast enough.

Effects of Soil Structure

Investigators have long known that soil physical properties can restrict root growth (Taylor, 1974). The rate of root elongation can be dramatically reduced by surprisingly small external pressures. Scott-Russell and Goss (1974) and Goss (1977) used a solid substrate (glass ballotini), an unlimited supply of nutrients and water, and found that a pressure of only 0.05 MPa inhibited root extension in barley (*Hordeum vulgare* L.) by 80%. Marschner (1986, p. 440–444) reviewed mechanical impedance effects and concluded that the stress factor of mechanical impedance seems to be transformed on the root-cap cells into a hormonal signal that leads to a decrease in cell expansion of the main axes and depression of lateral root initiation.

The sensitivity of plants to soil structure was exhibited when a group of investigators at different laboratories studied the uniformity and reproducibility of lettuce (*Lactuca sativa* L.) growth in controlled environments. This group found that small increases in the compaction of soilless media significantly decreased plant growth (Hammer et al., 1978).

Roots grow as cell division and elongation push the root tip forward. Because turgor pressure is the driving force, the roots of soil-grown plants are likely to have a higher osmotic concentration than roots grown in nutrient solution. This may contribute to the relative increase in dry root mass of soil-grown plants.

Plant Water Potential

If the transpiration rate and the hydraulic conductivity of the xylem are equal, plant water potential is determined by the root-zone water potential. The root-zone water potential in hydroponic culture is determined by the osmotic potential of the solution, which is -0.03 MPa for half-strength nutrient solution. The root-zone water potential at field capacity is about the same as hydroponic solution (-0.03 MPa) because the contribution of matric forces is negligible. Theoretically, plants in well-watered media should have the same plant water potential as plants in liquid culture. However, the leaves of hydroponically grown plants are frequently larger than leaves of the same plants growing in solid media. Even when abundant water is available, resistance in the xylem causes leaf water potentials to be -0.8 MPa (or lower) during midday periods of rapid transpiration (Kramer, 1983). This apoplastic water potential is low enough to reduce turgor pressure in the cells, which is the driving force for cell enlargement. Cell enlargement is one of the most sensitive parameters to plant water potential, so reduced water potential thus reduces leaf size (e.g., Kramer, 1983, p. 355–359). We are not aware of published studies comparing leaf size in hydroponic culture and in solid media, but the finite hydraulic conductivity of solid media means that it may be almost impossible to provide the same high plant water potential as in hydroponic culture.

Hydroponically grown plants typically transpire more water per unit dry weight than equivalent plants in solid media. Erickson and Kirkham (1979) studied the effects of soil matric potential on wheat water relationships and growth and found that roots in well-watered soil lost 2.5 times less water than plants grown hydroponically. During periods of rapid transpiration, water potential at the root surface quickly becomes lower than in the bulk soil. Thus, Erickson and Kirkham (1979) suggest that the high value for maximum growth is zero matric potential. Newman (1974) points out that, although water potential gradients are small when the soil is at field capacity, studies have shown that slight decreases in the soil matric potential below field capacity (occasional decreases to -0.1 MPa) can significantly decrease plant growth (cited in Newman, 1974). The difficulty is in maintaining zero matric potential with a rapid removal of water from the root zone. Eavis and Taylor (1979) found that the transpiration rate of soybean [*Glycine max* (L.) Merr.] (per unit leaf area) increased linearly as soil water content increased up to field capacity. These observations indicate the difficulty of avoiding mild water stress in plants grown in solid media.

A root-zone environment that is adequate at low PPF levels may not be adequate at high PPF levels. Transpiration rates are increased at high irradiances. Photoperiod is also important. Plants recover from low water potentials at night, but there is no chance for recovery in the 24-h photoperiods that are common in CELSS research (Bugbee & Salisbury, 1989b).

Root-Produced Hormones

Roots provide shoots with hormones and other organic compounds essential for growth. In recent years, several studies have suggested that hormone mediated root-shoot interaction may be particularly important (Lang & Thorpe, 1986). Blackman and Davies (1985) found that the roots of corn (*Zea mays* L.) plants communicate incipient soil drying to the shoots before plant water potential changes. Their findings suggest that there may be a reduction in cytokinin translocation from the root before plant water potential decreases. Future studies of hormone-mediated root-shoot interactions may help to elucidate physiological benefits of a consistently high root-zone water potential.

Rhizosphere pH Control

The ability to control pH in the rhizosphere environment can be used to great advantage in a CELSS. It may be desirable to recycle N in the reduced (NH_4^+) form, yet many studies have demonstrated severe toxic effects of high levels of NH_4^+ in the root-zone (e.g., Magalhaes & Wilcox, 1984). Ammonium uptake creates an H^+ ion efflux, which can cause the rhizosphere pH to be as much as 2 pH units lower than the bulk soil (Marschner & Rom-held, 1983; Nye, 1984). It now appears that NH_4^+ toxicity is associated with this reduction in rhizosphere pH. When the pH of flowing hydroponic solution was carefully controlled, there was no reduction in growth, even when 100% of the N was supplied in the NH_4^+ form (Rufty et al., 1983; Peet et al., 1985). The necessary level of rhizosphere pH control would be extremely difficult to achieve in solid media.

The root-zone environment is dynamic. Although it is possible to provide a favorable root-zone environment in solid media, it is difficult to maintain it. Flowing hydroponic culture sets a standard for nutrient and water availability that is difficult to match. In a review of watering techniques for controlled environment research, Rawlins (1979) states that "It is doubtful that any technique can control the water, nutrient, and aeration status of plants as well as nutrient solution culture."

STABILITY AND RELIABILITY

It has been suggested that lunar regolith can provide a more stable and highly buffered root-zone environment. This hypothesis warrants careful scrutiny. The rhizosphere of roots in field soil reaches a chemical and microbial equilibrium that allows plant growth without human intervention. Humans intervene, however, with fertilizers, chelating compounds, and pesticides to make the rhizosphere more favorable for plant growth. Similarly, plant growth occurs in hydroponic culture without any monitoring or control, but we intervene to improve plant growth. The lunar food production system must be predictable, and plants are more predictable when they are unstressed. Models of crop growth are often developed by first predicting performance in favorable environmental conditions and are then expanded to predict stress responses. By virtue of being more optimal, hydroponic systems may be more predictable. The reliability of lunar regolith is completely untested and may not be as great as some have imagined.

Hydroponic systems have a reputation for being inherently unstable, but this is only true when they are designed incorrectly. Unfortunately, there are many examples of poorly designed hydroponic systems. For example, the nutrient film technique (NFT) requires continuous electric power to supply water to plant roots, which means that brief power outages can be fatal. A properly designed hydroponic system keeps the roots submerged in solution at all times and requires continuous electric power only to circulate the solution. Although interruptions of the power supply in a CELSS will affect the foliar environment more than the root-zone environment, hydroponic systems should always be designed to minimize reliance on a continuous supply of electric power.

Buffering Capacity

The buffering capacity (stability) of all controlled root-zone environments is determined by their design. Systems can be made as stable as desired. Hydroponic root-zone environments can be stabilized in four ways:

1. By increasing the volume of nutrient solution.
2. By increasing the liquid contact with nutrients adsorbed onto a solid phase. (Ion exchange resins are used in liquid hydroponic culture to buffer ion concentrations.)
3. By providing sensitive monitoring and control systems.
4. By adding chelating compounds in excess of micronutrient cations.

Some combination of these four options would probably be used to stabilize the hydroponic root-zone environment in a lunar CELSS. The first two methods are the primary reasons for stability in solid media.

Hydroponic culture facilitates trade-offs among options. For example, increasing the sensitivity of the monitoring and control system makes it possible to decrease the mass and volume of the nutrient solution. The mass of electronic instrumentation is insignificant relative to the mass of solution. Automated monitoring and control technology to stabilize hydroponic solutions has already been developed and is being adapted for commercial use (Glass et al., 1987; Glass, 1989; Wild et al., 1987).

Buffering nutrient concentrations with chelating agents is a recent development in hydroponic culture. The principle is similar to adsorption of nutrients onto a solid phase in that cations are bound in chelates when concentrations increase and are released as concentrations decrease. Chelates are often used to bring precipitated elements into solution, but ions bound in chelates have a lower activity than unbound ions in solution. Chaney (1988, 1989) has determined binding constants for several chelate-metal complexes and has modified a computer program (GEOCHEM-PC) to more accurately calculate ion activities in hydroponic solutions with different chelating agents. Tong et al. (1986) have shown that chelate-buffered systems can protect against inadvertent contamination from toxic metals. Chelate-buffered hydroponic solutions will probably be more widely used as researchers and others become more familiar with the system.

Stability in hydroponic and solid media root-zone environments can be achieved by using large volumes and solid phase contact. Hydroponic systems are not inherently less stable than solid phase systems. The strength of flowing hydroponic culture is that it provides precise control of the environment at, or close to, the root surface. Thus, the rhizosphere environment can be better defined and reproduced. Reproducibility improves predictability, which is crucial to a CELSS.

ECONOMICS

Transporting material to the Moon is expensive. The estimated cost of launching 1 kg of mass into space ranges from $5000 to $25 000 (U.S. currency). This means that 1 g of water on the Moon may be more valuable than 1 g of gold on the Earth. Energy may also be a precious commodity. High levels of damaging radiation mean that the lunar farm will be located underground rather than in a greenhouse on the lunar surface as many peo-

Table 8–2. Typical composition of oven-dry biomass from wheat (stems, leaves, roots, and seeds).

Oxygen	48.0%
Carbon	36.0
Hydrogen	6.0
Nitrogen	3.0
Potassium	2.7
Calcium	2.0
Magnesium	0.8
Sulfur	0.8
Phosphorus	0.6
Silica	0.05
Iron	0.01
Micronutrients	0.02

ple have visualized. An underground CELSS would use either fiber optics (during the lunar day) or electric lamps to provide photosynthetic radiation for plant growth (Mori et al., 1986; Oleson et al., 1987; Olson et al., 1988). Optimum plant growth may require 100 times more photosynthetic radiation than that needed for humans (PPF of 2000 vs. 20 μmol m^{-2} s^{-1}). The efficiency with which photosynthetic energy is converted into food may have important economic effects, depending on the value of energy. High photosynthetic efficiency cannot be achieved without an optimum root-zone environment.

The mass of inorganic salts required to grow a crop may be an important consideration in deciding whether plants should be grown in lunar regolith (as a soilless medium watered with nutrient solution or as a developing soil to which organic matter is added). About 90% of plant dry mass is C, H, and O. In a well-fertilized crop plant, about 10% of the dry mass is inorganic nutrients (Table 8–2). If a human on the Moon requires 10 MJ d^{-1} (2400 kcal d^{-1}), this can be supplied by about 570 g of dry plant biomass such as wheat. Oven-dry wheat grain has about 17.6 MJ kg^{-1} (4.2 kcal g^{-1}) of digestible biomass (see discussion in Bugbee & Salisbury, 1988). This equals about 57 g of inorganic elements per day, but some of these elements must be supplied as anions that contain large amounts of O$_2$ (notably NO$_3^-$, SO$_4^{2-}$, and PO$_4^{3-}$), and hydrated salts might be used. About 200 g d^{-1} of inorganic salts (including the O and water of hydration) might be required to grow plants for one person. In a functioning lunar CELSS, these salts would mostly be recycled as part of the waste management system, but some initial quantity would have to be taken to the Moon to initiate the system, and some would be lost during recycling. On this basis, a lunar station with 10 inhabitants would require about 200 kg of inorganic salts to grow plants for 100 d. Although this is a significant amount, it is a small fraction of the mass of water required for plant growth and other needs. As discussed earlier, hydroponic systems facilitate reducing the volume of water in the system.

A CELSS farm for 10 inhabitants might have an area of 300 m^2; with the hydroponic system we now use (which could be improved) and with efficient recycling of water, about 60 000 kg (liters) of water would be required, 300 times the mass of the required salts. Of course there would be much more equipment, clothing, and structures, so the salts are truly trivial. Nevertheless, certain elements might be extracted from the lunar regolith, but the value of these extracted nutrients would have to exceed the cost of the extraction equipment.

OTHER CONSIDERATIONS

Several other technical questions have been raised about the utility of hydroponic culture in a lunar colony. It has been suggested that not all crops can be grown in hydroponic culture, yet there are no examples of crops that cannot be grown hydroponically. The productivity of some important crop plants (like peanut, *Arachis hypogaea* L.) has not yet been widely studied in hydroponic culture, but the productivity of tuber crops (requiring special hydroponic systems) has been increased partly because of hydroponic culture. The CELSS research indicates that the yield and photosynthetic efficiency of such tuber crops as white potato and sweet potato is higher in controlled environments than in the field. This productivity is primarily the result of changes in the foliar environment, but it is clear that most technical problems of hydroponic culture have been solved. Innovative cultural techniques will be required to grow groundnut (i.e., peanut) crops hydroponically. As these techniques are developed, it is likely that the productivity of groundnuts will increase.

Some people criticize hydroponic systems because unique nutrient solutions are required for each crop. Even though unique nutrient solutions can be beneficial, they are not required for individual crops. Hydroponic culture often improves plant growth partly because individually tailored nutrient solutions can easily be provided. This is one of the advantages of hydroponic culture. Another advantage is that the root-zone environment can be precisely adjusted to match different stages of the crop life cycle (Bugbee & Salisbury, 1989a). For example, boron (B) is translocated with the transpiration stream and accumulates in leaves as they age. This causes marginal necrosis and kills leaf tips in wheat leaves at high irradiance. Reducing B concentrations can minimize the problem, but B must be maintained at a higher level (10 μM) during the first half of the life cycle to promote normal pollination. After pollination is complete, B can be reduced to 0.5 μM to prevent accumulation in the leaf tips. This manipulation would be difficult in a regolith-based medium.

High-purity nutrient salts are not required in hydroponic culture. Reagent grade salts and water are used in all highly controlled hydroponic research, but these are not required. Tap water and fertilizer-grade salts are used in commercial hydroponic culture (Jones, 1983; Resh, 1983), and many commercial hydroponic systems recirculate and replenish the solution, without replacement, for periods longer than a year (N. Davis, 1988, personal communication).

The accumulation of organic compounds in nutrient solutions has not been widely studied, but the available data indicate that organic compounds are not necessarily toxic. Organic compounds (other than iron chelates) are sometimes deliberately added to nutrient solutions in special studies. Marschner (1986, p. 437–440) has reviewed some of the effects of low-molecular-weight organic solutes and suggests that growth reductions are often associated with anaerobic conditions around the roots. If flow rates and O_2 levels are adequate, adding organic compounds to solutions may not affect plant growth.

Many people have suggested that aeroponics (spraying roots that hang in high humidity air with a nutrient-solution mist) is ideal for a CELSS because it would reduce the volume of solution, which would reduce system mass. As mentioned earlier, the volume of solution is an important part of the buffering capacity of a system. Uniform contact with all parts of the root system is also important. Flowing hydroponic systems can operate with a small volume of nutrient solution, but this requires more sensitive monitoring and control apparatus. The potential stability of lunar regolith is partly a function of the large mass of solution in the system, but this is an expensive way to enhance system stability. Aeroponics is an excellent research system when frequent observation of the roots is important. However, the size of the root system in aeroponics is typically larger than equivalent hydroponically grown plants. The increased partitioning of C to the root system is usually attributed to the reduced water and nutrient availability associated with aeroponic culture.

Two areas will need extensive study before the chemical and physical properties of lunar regolith are understood. The nutrient-release characteristics of regolith have not been studied. Some of these studies can be conducted on the Earth, but lunar soil is a precious commodity. Simulated lunar soil is available but it differs from the real thing and may thus provide misleading information about something as complex as lunar regolith. Studies of the chemistry of lunar regolith will not be complete until they include higher plants. Plant roots excrete a wide variety of siderophores and other compounds to alter the rhizosphere environment, and the effect of these compounds on terrestrial soils is not well understood (see review by Curl & Truelove, 1986). Their effects on lunar regolith are unknown. Root exudates might cause the release of toxic substances as well as important nutrients.

The second area that must be studied is the movement of fluids through porous media (hydraulic conductivity). This has been widely studied in a 1-g environment, but it would not be the same at one-sixth gravity. Modeling of this critically important phenomenon would have to be verified on the Moon. The effect of gravity quickly diminishes in importance as the matric potential becomes more negative, but maximum plant water potential occurs only when the matric potential is near zero (Erickson & Kirkham, 1979). Research can probably solve problems associated with the unique chemical and physical characteristics of lunar regolith, but much of the research will need to be done on the Moon. Additional research on hydroponic culture will also need to be conducted on the Moon, but the pressurized flow of liquids through pipes and the performance of monitoring sensors will not be substantially different in one-sixth gravity.

CONCLUSION

There appear to be few significant problems associated with hydroponic culture. It has been shown to provide a highly optimum environment for plant roots. Compared with solid media, hydroponics provides more uniform aeration, high reliability, and can reduce the water volume that would cycle in the life support system. Hydroponics is a proven, time-honored technique. There is no immediate incentive to discontinue studies to refine it.

Plant growth in lunar-derived soils is a most interesting topic, but it will be extremely challenging to modify these soils to provide the same root-zone control and optimization that is readily obtained with hydroponic culture. We do not yet know, however, what level of control and optimization will be necessary to sustain high productivity and energy efficiency in a life support system. Furthermore, it is difficult to predict what types of efficiency will be important on the Moon. If future developments indicate that a lower-input/lower-output food production system is desirable, lunar regolith may be an excellent alternative. Lunar soil may also provide valuable options for microbial waste recycling. Predicting the future has always been difficult. Lord Kelvin, in 1895, said "Heavier than air flying machines are impossible."

REFERENCES

Asher, C.J., and D.G. Edwards. 1983. Modern solution culture techniques. Encycl. Plant Physiol. 15A:94–119.

Barber, D.A., and K.B. Gunn. 1974. The effect of mechanical forces on the exudation of organic substrates by the roots of cereal plants grown under sterile conditions. New Phytol. 73:39–45.

Barber, D.A., and M. Silberbush. 1984. Plant root morphology and nutrient uptake. p. 65–88. *In* Roots, nutrient and water influx, and plant growth. ASA Spec. Publ. 49. ASA, CSSA, and SSSA, Madison, WI.

Blackman, P.G., and W.J. Davies. 1985. Root to shoot communication in maize plants of the effects of soil drying. J. Exp. Bot. 36:39–48.

Bowen, G.D. and A.D. Rovira. 1976. Microbial colonization of plant roots. Ann. Rev. Plant Phytopathol. 14:121–144.

Bubenheim, D.L., B. Bugbee, and F.B. Salisbury. 1988. Radiation in controlled environments: Influence of lamp type and filter material. J. Am. Soc. Hortic. Sci. 113:468–474.

Bugbee, B.G. 1987b. Design of recirculating hydroponic systems. HortScience 22:1053.

Bugbee, B.G., and F.B. Salisbury. 1985. Wheat production in the controlled environments of space. Utah Sci. 46:145–151.

Bugbee, B.G., and F.B. Salisbury. 1988. Exploring the limits of crop productivity. I. Photosynthetic efficiency of wheat in high irradiance environments. Plant Physiol. 88:869–878.

Bugbee, B.G., and F.B. Salisbury. 1989a. The role of phasic environmental control in lunar food production efficiency. *In* W.W. Mendell (ed.) Lunar bases and space activities in the 21st century. Lunar and Planetary Inst., Houston. (In press.)

Bugbee, B.G., and F.B. Salisbury. 1989b. Current and potential productivity of wheat for a controlled environment life support system. Adv. Space Res. (In press.)

Bugbee, D.B. 1987a. Maintenance of recirculating hydroponic systems. HortScience 22:1000.

Carman, J.G. 1988. Improved somatic embryogenesis in wheat by partial simulation of the in-ovulo oxygen, growth-regulator and desiccation environments. Planta 175:417–424.

Carmi, A. 1986. Effects of root-zone volume and plant density on the vegetative and reproductive development of cotton. Field Crop Res. 13:25–32.

Chaney, R.L. 1988. Metal speciation and interactions among elements affect trace element transfer in agricultural and environmental food-chains. p. 219–260. *In* J.R. Kramer and H.E. Allen (ed.) Metal speciation: Theory, analysis and application. Lewis Publ., Chelsea, MI.

Chaney, R.L. 1989. Screening strategies for improved nutrient uptake and utilization. Hort-Science (In press.)

Cooper, A.J. 1985. New ABC's of NFT. p. 180–185. *In* Hydroponics worldwide: State of the art in soilless crop production. Int. Center for Spec. Studies, Honolulu, HI.

Curl, E.A., and B. Truelove. 1986. Root exudates. p. 54–92. *In* The rhizosphere. Springer-Verlag, New York.

de Willigen, P., and M. van Noordwijk. 1987. Roots, plant production and nutrient use efficiency. Inst. Bodemvruchtbaarheid, Netherlands.

Drew, M.C. 1979. Plant responses to anaerobic conditions in soil and solution culture. Curr. Adv. Plant Sci. 36:1–14.

Eavis, B.W., and H.M. Taylor. 1979. Transpiration of soybeans as related to leaf area, root length, and soil water content. Agron. J. 71:441–445.

Erickson, P., and M. Kirkham. 1979. Growth and water relations of wheat plants with roots split between soil and nutrient solution. Agron. J. 71:361–364.

Gales, K. 1979. Effect of water supply on partitioning of dry matter between roots and shoots in *Lolium perenne*. J. Appl. Ecol. 16:863–877.

Glass, A.D.M. 1989. Plant nutrition. Jones and Bartlett, Boston.

Glass, A.D.M., M. Saccomani, G. Crookall, and M.Y. Siddiqi. 1987. A microcomputer-controlled system for the automatic measurement and maintenance of ion activities in nutrient solutions during their absorption by intact plants in hydroponic facilities. Plant Cell Environ. 10:375–381.

Goss, M.J. 1977. Effects of mechanical impedance on root growth in barley (*Hordeum vulgare* L.). J. Exp. Bot. 28:96–111.

Graves, C.J. 1983. The nutrient film technique. Hortic. Rev. 5:1–44.

Griffin, D.M. 1968. A theoretical study relating the concentration and diffusion of oxygen to the biology of organisms in soil. New Phytol. 70:85–96.

Haller, T., and H. Stolp. 1985. Quantitative estimation of root exudation of the maize plant. Plant Soil 86:207–216.

Hammer, P., T. Tibbitts, R. Langhans, and J.C. McFarlane. 1978. Base-line growth studies of 'Grand Rapids' lettuce in controlled environments. J. Am. Soc. Hortic. Soc. 103:649–655.

Hoogenboom, G., M.G. Huck, and C.M. Peterson. 1988. Predicting root growth and water uptake under different soil water regimes. Agric. Systems 26:263–290.

Jensen, M.H., and W.L. Collins. 1985. Hydroponic vegetable production. Hortic. Rev. 7:483–558.

Jones, J.B., Jr. 1983. A guide for the hydroponic and soilless culture grower. Timber Press, Portland, OR.

Kasperbauer, M.J., and D.L. Karlen. 1986. Light mediated bioregulation of tillering and photosynthate partitioning in wheat. Physiol. Plant. 66:159–163.

Kramer, P.J. 1983. Water relations of plants. Academic Press, New York.

Krizek, D.T., A. Carmi, R.M. Mirecke, F.W. Snyder, and J.A. Bunce. 1985. Comparative effects of soil moisture stress and restricted root-zone volume on morphogenetic and physiological response of soybean. J. Exp. Bot. 36:25–38.

Krizek, D.T., and S.T. Dubik. 1987. Influence of water stress and restricted root volume on growth and development of urban trees. J. Aboric. 13:47–55.

Lambers, H. 1985. Higher plant cell respiration. *In* R. Douce and D. Day (ed.) The encyclopedia of plant physiology. Springer-Verlag, New York.

Lang, A., and M.R. Thorpe. 1986. Water potential, translocation, and assimilate partitioning. J. Exp. Bot. 37:495–503.

Magalhaes, J.R., and G.E. Wilcox. 1984. Ammonium toxicity development in tomato plants relative to nitrogen form and light intensity. J. Plant Nutr. 7:1497–1509.

Maher, M.J. 1977. The use of hydroponics for the production of greenhouse tomatoes in Ireland. p. 161–169. *In* Proc. 4th Int. Congr. soilless culture. I.W.O.S.C., Las Palmas.

Marschner, H. 1986. Mineral nutrition of higher plants. Academic Press, New York.

Marschner, H., and V. Romheld. 1983. In-vivo measurement of root-induced pH changes at the soil-root interface: Effect of plant species and nitrogen source. Z. Pflanzenphysiol. 111:241–251.

Milks, R., W. Fonteno, and R. Larson. 1989a. Hydrology of horticultural substrates: I. Mathematical models for moisture characteristics of horticultural container media. J. Am. Soc. Hortic. Sci. 114:48–52.

Milks, R., W. Fonteno, and R. Larson. 1989b. Hydrology of horticultural substrates: II. Predicting physical properties of media in containers. J. Am. Soc. Hortic. Sci. 114:53–56.

Milks, R., W. Fonteno, and R. Larson. 1989c. Hydrology of horticultural substrates: III. Predicting air and water content of limited volume plug cells. J. Am. Soc. Hortic. Sci. 114:57–61.

Mori, K., N. Tanatsugu, and M. Yamashita. 1986. Visible solar-ray supply system for space station. Acta Astronaut. 13:71–79.

Newman, E.I. 1974. Root soil water relations. VII. Matric potential gradients in the rhizosphere. *In* E.W. Carson (ed.) The plant root and its environment. Univ. Press of Virginia, Charlottesville.

Nye, P.H. 1984. pH changes and phosphate solubilization near roots—an example of coupled diffusion processes. p. 89–100. *In* Roots, nutrient and water influx, and plant growth. ASA Spec. Publ. 49. ASA, CSSA, and SSSA, Madison, WI.

Oleson, M.W., T.J. Slavin, and R.L. Olson. 1987. Lighting considerations in a controlled environmental life support system. Soc. Automotive Eng. Tech. Paper Ser. 871435.

Olson, R.L., M.W. Oleson, and T.J. Salvin. 1988. CELSS for advanced space mission. HortScience 23:275–286.

Peet, M.M., C.D. Raper, Jr., L. Tolley, and W. Robarge. 1985. Tomato responses to ammonium and nitrate nutrition under controlled root-zone pH. J. Plant Nutr. 8:787–798.

Polonskiy, V.I., and G.M. Lisovskiy. 1980. Net production of wheat crop under high PhAR irradiance with artificial light. Photosynthetica 14:177–181.

Polonskiy, V.I., and G.M. Lisovskiy. 1986. Techniques for creating rapid growth strains of wheat for human life support system. Kosmicjeskaya Biologiya i Aviakosmicheskaya Meditsina 20:96 (abstract) *In* L.R. Hooke (ed.) USSR Space Life Sciences Digest, Issue 6. NASA Contractor Rep. 3922(06):54.

Rawlins, S.L. 1979. Watering. p. 271–289. *In* T.W. Tibbitts and T.T. Kozlowski (ed.) Controlled environment guidelines for plant research. Academic Press, New York.

Resh, H.M. 1983. Hydroponic food production. Woodbridge Press, Santa Barbara, CA.

Robinson, D. 1986. Limits to nutrient inflow rates in roots and root systems. Physiol. Plant. 68:551–559.

Rovira, A.D., and C.B. Davey. 1974. Biology of the rhizosphere. p. 153–204. *In* E.W. Carson (ed.) The plant root and its environment. Univ. Press of Virginia, Charlottesville.

Ruff, M.S., D.T. Krizek, R.M. Mirecki, and D.W. Inouye. 1987. Restricted root-zone volume: Influence on growth and development of tomato. J. Am. Soc. Hortic. Sci. 112:763–769.

Rufty, T.W., C.D. Raper, and W.A. Jackson. 1983. Growth and nitrogen assimilation of soybeans in response to ammonium and nitrate nutrition. Bot. Gaz. 144:466–470.

Salisbury, F.B., and B. Bugbee. 1988. Plant productivity in controlled environments. HortScience 23:293–299.

Scott-Russell, R., and M.J. Goss. 1974. Physical aspects of soil fertility—The response of roots to mechanical impedance. Neth. J. Agric. Sci. 22:305–318.

Smucker, A.J.M. 1984. Carbon utilization and losses by plant root systems. p. 27–46. *In* Roots, nutrient and water influx, and plant growth. ASA Spec. Publ. 49. ASA, CSSA, and SSSA, Madison, WI.

Taylor, H.M. 1974. Root behavior as affected by soil structure and strength. *In* E.W. Carson (ed.) The plant root and its environment. Univ. Press of Virginia, Charlottesville.

Tong, Y.A., F. Fan, R. Korcak, R. Chaney, and M. Faust. 1986. Effect of micronutrients, phosphorous and chelator to iron ratio on growth, chlorosis, and nutrition of apple seedlings. J. Plant Nutr. 9:23–41.

Trolldenier, G., and C. Hecht-Buchholz. 1984. Effect of aeration status of nutrient solution on microorganisms, mucilage and ultrastructure of wheat roots. Plant Soil 80:381–390.

Wild, A., L.H.P. Jones, and J.H. Macduff. 1987. Uptake of mineral nutrients and crop growth: The use of flowing nutrient solutions. Adv. Agron. 41:171–219.

9 Microorganisms and the Growth of Higher Plants in Lunar-Derived Soils

G. Stotzky

New York University
New York, New York

Life on a lunar base will be affected—in fact, controlled—by microbes, as is life on Earth (e.g., pathogenesis, contamination, spoilage, decay, and biogeochemical cycles, including degradation of wastes). Consequently, the factors that influence the activity, ecology, and population dynamics of microbes on Earth should be similar on the Moon (Table 9-1) (Stotzky, 1974). This, of course, assumes that a lunar base will be an enclosed environment, wherein temperature, pressure, electromagnetic and particulate radiation, atmospheric composition, and other critical environmental factors will be controlled.

On Earth, microbes are the major biological agents responsible for the conversion of primary minerals into soil. The basic similarities in the composition of the terrestrial bedrock and the lunar parent bedrock and regolith (Williams & Jadwick, 1980; Henninger et al., 1987) suggest that the latter could be converted into an acceptable soil to support plant growth in an enclosed environment on the Moon if water, energy sources (e.g., from plant and animal wastes), and other factors necessary for the growth and activity

Table 9-1. Factors affecting the activity, ecology, and population dynamics of microorganisms in natural habitats.

Carbon and energy sources
Mineral nutrients
Growth factors
Ionic composition
Available water
Temperature
Pressure
Atmospheric composition
Electromagnetic radiation
pH
Oxidation-reduction potential
Surfaces
Spatial relations
Genetics of the microorganisms
Interactions between microorganisms

Table 9–2. Physicochemical factors of an environment that affect the toxicity of heavy metals and other pollutants to biological systems.

pH (acidity/alkalinity)
Buffering capacity
E_h (oxidation-reduction potential)
Aeration status (aerobic, microaerobic, anaerobic)
Inorganic anionic composition
Inorganic cationic composition
Water content
Clay mineralogy
Hydrous metal oxides
Organic matter
Cation exchange capacity
Anion exchange capacity
Temperature
Nutrients
Hydrostatic pressure
Osmotic pressure
Solar radiation

of microbes are provided (many of these factors will probably have to be brought from Earth, at least initially).

Despite their general similarities, however, the regolith is higher in some heavy metals, notably Ni and Cr than is soil on Earth (Williams & Jadwick, 1980), and these metals can be toxic to microbes, as well as to plants and animals. The toxicity of most heavy metals is primarily dependent on their chemical speciation form (Babich & Stotzky, 1980, 1983a), and appropriate manipulation of the environment (i.e., changes in such physicochemical factors as pH, E_h, aeration, addition of inorganic or organic ligands) can alter the speciation to yield forms of the metals that are lower in toxicity or are nontoxic (Table 9–2; Babich & Stotzky, 1983b, 1985; Babich et al., 1985). In addition, microbes could be used in the beneficiation of the regolith for the recovery of these heavy and other metals for industrial purposes, with the less toxic residues then being used as a growth medium for higher plants (Stotzky & Babich, 1986b).

IMPORTANCE OF CLAY MINERALS

One of the components of natural terrestrial soils that is important to the growth of microbes and plants are clay minerals, which appear to be absent in the lunar regolith. These clay minerals, primarily crystalline hydrous aluminosilicates, are essentially solid-state crystals that contain both negative (cation-exchange) and positive (anion-exchange) charges. These crystals are important in retaining cationic and anionic plant nutrients in soil (Huang & Schnitzer, 1986). These charged particulates also retain water against gravity. It is in this water that microbes reside and metabolize (Stotzky, 1986). In addition to retaining nutrients on their exchange complexes, clay minerals remove protons, heavy metals, and other toxicants from their associated water, thereby enabling the continuous growth of microbes (Babich & Stotzky,

1980, 1983a; Stotzky, 1986). Clays also bind organic molecules and viruses, in addition to undergoing surface interactions (e.g., adhesion) with microbes, which affect microbial events in terrestrial soil (Stotzky, 1986). Clay minerals also are important in maintaining the structure and, hence, the aeration of terrestrial soils. Consequently, the retention of water and nutrients and effective aeration will be limited in the regolith in the absence of these surface-active particulates, which, in turn, will reduce the efficient growth of microbes and plants.

It has been suggested (e.g., Henninger et al., 1987; Ming & Lofgren, 1989) that zeolites, crystalline hydrous aluminosilicates that share many of the properties of clay minerals, and even clay minerals could be synthesized on the lunar surface by hydrothermal alteration of regolith components. Such synthetic zeolites could have two to three times the cation-exchange capacity (CEC) of terrestrial clay minerals, which would enhance the retention of cationic nutrients. Insufficient data appear to be available on the ability of such synthetic zeolites to retain anionic nutrients (e.g., PO_4^{2-} and NO_3^-) and water and on their stability and production costs. Regardless whether zeolites are synthesized on the Moon or easily mined terrestrial clay minerals with high ion exchange and water-retention capacities (e.g., montmorillonite) are brought initially to a lunar base, the growth of both plants and microbes will be enhanced, and the costs of recycling nutrients, water, and air will be reduced, if the regolith is amended with such particulates.

WHICH MICROBES?

In Table 9–2, the last five physicochemical factors probably do not affect the chemical speciation form of heavy metals but influence their toxicity by affecting the physiological state of the exposed microbes. This raises the question of the types of microbes that should be introduced into an enclosed lunar base. Although the initial temptation might be to select microbes that are capable of conducting specific beneficial functions (e.g., N_2 fixation, rapid degradation of cellulosic and other relatively recalcitrant wastes, generation of H_2 and CH_4), this approach could have dire consequences in an isolated and insulated environment, such as a lunar base. On Earth, soils, as well as most other natural microbial habitats, contain a mixture of microbial populations (i.e., the species diversity is high), and their composition is steered primarily by the physicochemical characteristics of the specific habitat. This mixture of microbial populations confers a degree of homeostasis and stability on the particular habitat and enables the microbial community to perform numerous diverse functions, as well as to reduce the establishment and enrichment of pathogens. An enclosed lunar base will be continuously challenged with potential pathogens of both plants and animals, including human beings, which will be derived from seeds and other planting stock, base personnel and visiting crews, and matériels. Inasmuch as it will be impossible and, perhaps more important, counterproductive to maintain a sterile lunar base, the best defense against the enrichment of pathogens is to maintain as great a species diversity in the ambient microbial communities (e.g., in the regolith, on plants, and in wastes) as possible. Such species diversity can probably be attained by inoculating the various microbial habitats on a lunar base with appropriate mixtures of either soil, water, and wastes brought from Earth. Considerable experimentation on Earth will, of course, be necessary to concoct mixtures of these inocula that will be most successful in surviving, colonizing, and growing in the environmental conditions that will prevail on a lunar base. No amount of experimentation on Earth, however, will cover all contingencies that might occur on the Moon. Hence, it will be necessary to maintain reservoirs of terrestrial soils and waters on the Moon for periodic reinoculation, especially in the event of serious decreases in species diversity and drift toward monocultures. Moreover, comparable reservoirs should be maintained on Earth as a contingency in the event that it is not possible to maintain these reservoirs on the Moon.

GENETICALLY ENGINEERED MICROORGANISMS

There will probably also be the temptation to introduce into a lunar base genetically engineered microorganisms (GEMs) that have been constructed to perform specific functions (e.g., mineralization of halogenated hydrocarbons, enhanced degradation of other wastes). Although GEMs may have potentially important roles in rendering the lunar surface more hospitable for earthlings, too little information on the potential hazards of such "superbugs" is available, even on Earth, to consider seriously their introduction into the sterile lunar environment now. There have been few releases of GEMs into the terrestrial environment (e.g., ice-minus strains of *Pseudomonas syringae*; a root-colonizing strain of *P. aureofaciens* that contains the *lac ZY* genes from *Escherichia coli*). Studies in the laboratory, however, have shown that GEMs not only survive in both nonsterile and sterile terrestrial soils and waters, but proliferate to high numbers (e.g., ca. 10^9 cells/g of soil) in sterile systems (Devanas et al., 1986; Devanas & Stotzky, 1986; Stotzky & Babich, 1986a). Furthermore, transfer of genetic information among bacteria by conjugation (intraspecific chromosomal transfer [Weinberg & Stotzky, 1972; Krasovsky & Stotzky, 1987] and intra- and interspecific plasmid transfer [Devanas & Stotzky, 1988]), transduction (Zeph et al., 1988), and transformation (Graham & Istock, 1981) have been demonstrated in soil (Stotzky & Babich, 1986a; Stotzky, 1989). Consequently, there is the potential that a GEM or its novel gene may proliferate in a natural habitat, especially if it is enriched by, for example, the presence of a substrate that only the product of the novel gene is capable of transforming. Such proliferation could significantly alter the structure of microbial communities in the isolated and insulated environment of a lunar base.

Nevertheless, the potential benefits of GEMs in converting the regolith into soil capable of supporting plant growth should be considered. Some genetic constructs that may facilitate this conversion are suggested in Table 9-3.

Table 9-3. Some microbes and enzymes of possible importance for introduction into lunar soil and environment for enhanced primary productivity and general health.

Hydrogenomonas (*Alcaligenes*): $2H_2 + O_2 \longrightarrow 2H_2O$
Aquamonas: $2H_2$ (solar winds) $+ O_2$ (metal oxides) $\longrightarrow 2H_2O$
Methanogens: $4H_2 + CO_2 \longrightarrow CH_4 + 2H_2O$
Ice-plus microbes (e.g., *Pseudomonas syringae*) (condensation nuclei for rain)
Cellulose and lignin degraders (high production of appropriate hydrolases)
Thermophilic actinomycetes for aerobic composting of sewage sludge and other wastes
    Complexation of heavy metals
    Recycling of heat produced
Nitrifiers probably not necessary (cation- vs. anion-exchange capacity)
Mixed populations rather than single species
    Greater stability (activity, ecology, and population dynamics)
    Less danger of establishing microbial monocultures (e.g., pathogens)
    Unavoidable introduction of microbes from crew, seeds, and matériels
Genetically engineered microbes (consider potential hazards)
    "Superbugs"
    Possible novel genes that code for enzymes to perform specific functions
        Combustase: release CO_2 from regolith
        Vitroclase: break bonds of glass-binding cement to enhance structure
        Agglutinase: break down agglutinate particles to release trapped volatiles of
            biological importance (e.g., N_2, H_2, and CO_2)
        Weatherase: convert regolith into soil
        Clay synthetase: form clay minerals from primary minerals in regolith (increase
            cation- and anion-exchange capacity, buffering ability, water-holding
            capacity, structure and porosity); of value in hydroponic, aeroponic, and
            soil culture
        Heavy metal immobilase: eliminate heavy metal toxicity
        Sortase: separate regolith into supportive and nonsupportive plant growth media

---

RESEARCH NEEDS

It is apparent from even this brief discussion that much research is needed now on the potential roles of microbes on a lunar base, especially in the conversion of the regolith into functioning soil. Such research should include inoculation of different types of regoliths with various soil-derived and other microbial communities and measuring, over time, such parameters as microbial activity (e.g., respiration, transformations of N and other elements); species diversity (including micro- and mesofauna), solubilization of elements, pH, cation- and anion-exchange capacities, water tension profiles, particle-size distribution, and plant growth. In these studies, the regoliths should be amended with different clay minerals or zeolites, C and energy sources, and essential mineral nutrients and maintained at various water and O_2 tensions, with and without plants. The experiments should be so designed that the survival of selected pathogens and viruses, the stability and transfer of genetic information and the dispersal of GEMs, potential radiation-induced mutagenesis, and toxicity of heavy metals can be evaluated. The information that is needed is probably infinite. Although the initial design and testing of research protocols could be conducted with simulants of regoliths, the definitive studies should be conducted with actual samples of the regoliths brought to Earth by the Apollo mission of 1972.

The current reluctance to use the limited amount of regolith available (ca. 333 kg) for such experimentation is understandable. However, the risks are too great to base the designs and predictions of a successful lunar base, that will have to grow its own food and dispose efficaciously of its wastes, solely on studies with simulants. Furthermore, there will, undoubtedly, be additional lunar missions before the establishment of such a base, and the stock of regoliths will be replenished.

SUMMARY

Microbes will be of prime importance in any colonization and exploitation of the lunar surface. Consequently, despite the relatively long lead-time before such colonization will be realized, it is imperative that studies on ways to control and manipulate microbial activities on the Moon be initiated now. Furthermore, the knowledge and experience gained in the framework of the Moon will probably be applicable to the eventual colonization of Mars and, perhaps, other planetary bodies and will increase knowledge about microbial activities and ecology on Earth.

ACKNOWLEDGMENT

The preparation of this chapter and some of the studies discussed were supported, in part, by cooperative agreements CR812484, CR813431, and CR813650 from the U.S. Environmental Protection Agency-Corvallis Environmental Res. Lab. The opinions expressed herein are not necessarily those of the Agency.

REFERENCES

Babich, H., and G. Stotzky. 1980. Environmental factors that influence the toxicity of heavy metals and gaseous pollutants to microorganisms. Crit. Rev. Microbiol. 9:99–145.

Babich, H., and G. Stotzky. 1983a. Influence of chemical speciation on the toxicity of heavy metals to the microbiota. p. 1–46. *In* J.O. Nriagu (ed.) Aquatic toxicology. John Wiley and Sons, New York.

Babich, H., and G. Stotzky. 1983b. Developing standards for environmental toxicants: The need to consider abiotic environmental factors and microbe-mediated ecologic processes. Environ. Health Perspect. 49:247–260.

Babich, H., and G. Stotzky. 1985. Heavy metal toxicity to microbe-mediated ecologic processes: A review and potential application to regulatory policies. Environ. Res. 36:111–137.

Babich, H., M.A. Devanas, and G. Stotzky. 1985. The mediation of mutagenicity and clastogenicity of heavy metals by physicochemical factors. Environ. Res. 37:253–286.

Devanas, M.A., and G. Stotzky. 1986. Fate in soil of a recombinant plasmid carrying a *Drosophila* gene. Curr. Microbiol. 13:279–283.

Devanas, M.A., and G. Stotzky. 1988. Survival of genetically engineered microbes in the environment: Effect of host/vector relationship. p. 287–296. *In* G.E. Pierce (ed.) Developments in industrial microbiology. Soc. Ind. Microbiol., Baltimore.

Devanas, M.A., D. Rafaeli-Eshkol, and G. Stotzky. 1986. Survival of plasmid-containing strains of *Escherichia coli* in soil: Effect of plasmid size and nutrients on survival of hosts and maintenance of plasmids. Curr. Microbiol. 13:269–277.

Graham, J., and C. Istock. 1981. Parasexuality and microevolution in experimental populations of *Bacillus subtilis*. Evolution 36:954–963.

Henninger, D.L., C.W. Lagle, and D.W. Ming. 1987. A lunar derived "soil" for the growth of higher plants. p. 1–20. *In* G.M. Andrus (ed.) Symposium '86: The first lunar development symposium. S86-35. Lunar Develop. Counc., Pitman, NJ.

Huang, P.M., and M. Schnitzer (ed.). 1986. Interactions of soil minerals with natural organics and microbes. SSSA Spec. Publ. 17. SSSA, Madison, WI.

Krasovsky, V.N., and G. Stotzky. 1987. Conjugation and genetic recombination in *Escherichia coli* in sterile and nonsterile soil. Soil Biol. Biochem. 19:631–638.

Ming, D.W., and G.E. Lofgren. 1989. Crystal morphologies of minerals formed by hydrothermal alterations of synthetic lunar basaltic glass. *In* L.A. Douglas (ed.) Proc. Soil Micromorphology Workshop. Elsevier, Amsterdam, Holland (In press.)

Stotzky, G. 1974. Activity, ecology, and population dynamics of microorganisms in soil. p. 57–135. *In* A.I. Laskin and H. Lechevalier (ed.) Microbial ecology. Chemical Rubber Co., Cleveland.

Stotzky, G. 1986. Influence of soil mineral colloids on metabolic processes, growth, adhesion, and ecology of microbes and viruses. p. 305–428. *In* P.M. Huang and M. Schnitzer (ed.) Interactions of soil minerals with natural organics and microbes. SSSA Spec. Publ. 17. SSSA, Madison, WI.

Stotzky, G. 1989. Gene transfer among bacteria in soil. p. 165–222. *In* S.B. Levy and R.V. Miller (ed.) Gene transfer in the environment. McGraw-Hill Book Co., New York.

Stotzky, G., and H. Babich. 1986a. Survival of, and genetic transfer by, genetically engineered bacteria in natural environments. p. 93–138. *In* A.I. Laskin (ed.) Advances in applied microbiology. Academic Press, New York.

Stotzky, G., and H. Babich. 1986b. Physicochemical environmental factors affect the response of microorganisms to heavy metals: Implications for the application of microbiology to mineral exploration. p. 238–264. *In* D. Carlisle et al. (ed.) Mineral exploration: Biological systems and organic matter. Prentice Hall, Englewood Cliffs, NJ.

Weinberg, S.R., and G. Stotzky. 1972. Conjugation and genetic recombination of *Escherichia coli* in soil. Soil Biol. Biochem. 4:171–180.

Williams, R.J., and J.J. Jadwick (ed.). 1980. Handbook of lunar materials. NASA Rep. RP 1057, NASA Sci. and Tech. Inf. Office, Washington, DC.

Zeph, L.R., M.A. Onaga, and G. Stotzky. 1988. Transduction of *Escherichia coli* by bacteriophage P1 in soil. Appl. Environ. Microbiol. 54:1731–1737.

10 Role of Microbes to Condition Lunar Regolith for Plant Cultivation

Henry L. Ehrlich

Rensselaer Polytechnic Institute
Troy, New York

Apart from CO_2 and water, plants need mineral elements in their nutrition. On Earth, they derive these elements from the soil in which they grow. Besides N, P, and S needed in somewhat larger amounts, plants also require other minerals in trace amounts. Table 10–1 lists elements that have been reported in the literature as either essential or growth stimulating (see also Shkolnik, 1984). These elements are not only needed in plant nutrition but also for the nutrition of the microbes required to support plant growth in

Table 10–1. Mineral elements in plant nutrition.

Element	Function	Reference
Aluminum	Essential or stimulatory to growth	Shkolnik (1984)
Boron	Growth promoting	Shkolnik (1984)
Calcium	Essential	Bowen (1979)
Chlorine	Growth stimulating	Bowen (1979)
Chromium	Accumulated by some plants	Bowen (1979)
Cobalt	Essential to growth	Shkolnik (1984)
Copper	Essential to growth	Shkolnik (1984)
Fluorine	Accumulated by some plants	Bowen (1979)
Iodine	Growth stimulating	Bowen (1979)
Iron	Essential to growth	Bowen (1979)
Lithium	Accumulated by some plants; stimulatory?	Shkolnik (1984)
Magnesium	Esential to growth	Bowen (1979)
Manganese	Essential to growth	Shkolnik (1984)
Molybdenum	Essential to growth; N_2 fixation	Shkolnik (1984)
Nickel	Essential to growth; needed by some bacteria	Shkolnik (1984)
Phosphorus	Essential to growth	Bowen (1979)
Rubidium	Accumulated by some plants	Shkolnik (1984)
Selenium	Essential to some plants	Bowen (1979)
Silicon	Accumulated; growth promoting in some	Shkolnik (1984)
Silver	Stimulates flowering in some plants	Shkolnik (1984)
Sodium	Essential to growth	Shkolnik (1984)
Strontium	Can partially replace Ca in some plants	Shkolnik (1984)
Sulfur	Essential to growth	Bowen (1979)
Titanium	Role in photosynthesis	Shkolnik (1984)
Vanadium	Role in N_2 fixation	Shkolnik (1984)
Zinc	Essential to growth	Shkolnik (1984)

a soil. Many of the elements listed in Table 10-1, while growth promoting at low concentrations, become toxic to plants and microbes when their concentrations exceed a certain limit.

MOBILIZATION OF TRACE ELEMENTS BY MICROBES

In terrestrial soil, microbes play a significant role in mobilizing essential or growth-promoting elements contained in soil minerals. As Table 10-2 shows, acidophilic mesophiles like *Thiobacillus ferrooxidans* and *Leptospirillum ferrooxidans*, and acidophilic thermophiles like *Sulfolobus* spp. and *Sulfobacillus thermosulfidooxidans* can oxidize metal sulfides to their corresponding metal sulfates, which are mostly water soluble in contrast to the corresponding metal sulfides (lead sulfate being an exception).

Since terrestrial plants meet their nutritional S requirements with sulfate (SO_4), microbes are even important in converting free forms of reduced S [e.g., sulfide, elemental S, thiosulfate, and tetrathionate] to SO_4 by oxidation (Table 10-2). These include the above acidophiles except *Leptospirillum*, as well as *Sulfolobus* spp. and neutrophilic thiobacilli (*Thiobacillus thioparus, T. neapolitanus,* and *T. denitrificans*).

A source of P for terrestrial plants may be insoluble apatite. Acid- and ligand-forming microbes can dissolve apatite (Table 10-2), thereby releasing the phosphate (PO_4) to plants. The acids may be mineral acids such as sulfuric (H_2SO_4) (from S oxidation), nitric (HNO_3) (from nitrification), or organic acids such as citric or oxalic formed by fungi, or lactic, acetic, formic, or succinic acids formed by certain bacteria. Some of these acids, like citric, oxalic, or 2-ketogluconic, may also act as ligands of Ca that is pulled out of the crystal lattice of apatite causing its disintegration.

The phosphate in iron phosphates such as strengite or vivianite, which may occur in terrestrial soils, can be solubilized by sulfate-reducing bacteria that produce H_2S. In turn, the H_2S precipitates the Fe as insoluble sulfide, leaving the PO_4 in solution (Table 10-2). This happens anaerobically because bacterial SO_4 reduction is strictly an anaerobic process.

In terrestrial soil, Na, K, Ca, and Mg can be mobilized by acid- or ligand-forming bacteria or fungi acting on aluminosilicates or silicates. Silicon may be solubilized by silicate bacteria from aluminosilicates, silicates, or even quartz. The silicate bacteria form an extracellular polysaccharide that can react chemically with the silicon in the respective compounds, thereby solubilizing it (Avakyan et al., 1985, 1986; Karavaiko et al., 1980).

Iron in ferric oxide minerals, which are insoluble at neutral pH in terrestrial soils, is solubilized under reducing conditions by enzymatic reduction to Fe(II) (Table 10-2). Under oxidizing conditions, it is mobilized by microbial production of specific ligands called *siderophores* that have an extremely high affinity for Fe(III), but a low affinity for Fe(II).

Manganese, which is important in plant photosynthesis, may occur as insoluble Mn oxide under oxidizing conditions in terrestrial soil. It can be made available to plants by Mn-reducing bacteria, which convert it to Mn(II)

Table 10-2. Examples of microorganisms that can mobilize trace elements from minerals that are nutritionally important to plants.

Sulfate-sulfur:

From metal sulfides:
    Acidophiles: *Thiobacillus ferrooxidans, Leptospirillum, Sulfolobus* spp., and
        Sulfobacillus thermosulfidooxidans
From hydrogen sulfide, S, thiosulfate, or tetrathionate:
    Acidophiles: *Thiobacillus thiooxidans, T. ferrooxidans,* and *Sulfolobus* spp.
    Neutrophiles: *Thiobacillus thioparus, T. neapolitanus,* and *T. denitrificans*

Phosphate:

From apatite and whitlockite (?):
    Acid- and ligand-forming bacteria and fungi
From strengite or vivianite:
    Sulfate-reducing bacteria

Na, K, Ca, Mg, and Si:

From aluminosilicates, silicates, and quartz:
    Ligand-forming bacteria and fungi; silicate bacteria

Iron:

From iron oxide minerals:
    Ferric iron-reducing bacteria; siderophore-forming bacteria
From iron sulfide minerals:
    Thiobacillus ferrooxidans and *Sulfolobus* spp.

Manganese:

From manganese oxides:
    Manganese oxide reducing bacteria
From pyroxenes and basalt:
    Acid- and ligand-forming bacteria; silicate bacteria

Co, Cu, Ni, and Zn:

From metal sulfides:
    Thiobacillus ferrooxidans, Sulfolobus spp.

Molybdenum:

From molybdenite:
    Sulfolobus spp.

Selenium:

From metal selenides:
    Thiobacillus ferrooxidans plus Se^0-oxidizing bacteria (e.g., *Bacillus megaterium*
    producing selenite and selenate from Se^0)

Chromium:

From chromite:
    Thiobacillus thiooxidans oxidizing S to sulfuric acid (Experimentally the amount
    of chromium leached by biogenic sulfuric acid from chromite at 25°C was
    <10%, probably sufficient to satisfy nutritional demands by plants [Ehrlich,
    1983]).

Aluminum:

From aluminosilicates:
    Acid-forming bacteria or fungi

Table 10-3. Lunar minerals or rock that could supply essential elements to plants.†

Sulfur
  Mackinawite, troilite [FeS]; pentlandite [(Fe,Ni)$_9$Se]; chalcopyrite [CuFeS$_2$];
  sphalerite [(Zn,Fe)S]
Phosphate
  Apatite [(Ca$_5$(PO$_4$)$_3$(F,Cl)$_3$]; whitlockite [Ca$_9$(Mg,Fe)(PO$_4$)$_7$(F,Cl)]
Na, K, Ca, Mg, and Si
  Pyroxene [(Ca,Mg,Fe)SiO$_3$]; plagioclase feldspars [(Ca,Na)Al$_2$Si$_2$O$_8$]; silica [SiO$_2$],
  basalt
Fe, Cu, and Ni
  Chalcopyrite; pentlandite; sphalerite; troilite; mackinawite; olivine [(Mg,Fe)$_2$SiO$_4$;
  ilmenite [FeTiO$_3$]; hematite [Fe$_2$O$_3$]; magnetite [Fe$_3$O$_4$]; goethite [FeO(OH)]
Mn
  Pyroxene, niningerite [(Mg,Fe,Mn)S, basalt
Cr
  Spinels [(Fe,Mg,Al,Cr,Ti)O$_4$]
Ti
  Spinels; armalcolite [(Fe$_2$TiO$_6$)]; rutile [TiO$_2$]; ilmenite
Zn
  Sphalerite
Co, Se, V
  Basalt

† From data in Henninger et al. (1987) and Williams and Jadwick (1980).

Table 10-4. Elements that are nutritionally important to plants, but are not apparently
available from lunar minerals or rocks.

Molybdenum
  Important for microbial N$_2$-fixing and NO$_3$-reducing activity
Boron
  Important for growth promotion in plants

Table 10-5. Relative reaction rates.

Geomicrobial process	Relative reaction rate
Rockweathering	Slow
Silicate attack	Slow
Aluminosilicate attack	Slow
Metal sulfide oxidation	Fast
Iron oxidation	Fast
Manganese (IV) oxide reduction	Fast
Sulfate reduction	Fast
Biosorption	Fast†
Bioprecipitation	Fast†
Phosphate solubilization	Intermediate

† Dependence on rate of formation of reactants.

Table 10-6. Microbial immobilization of solubilized elements of nutritional importance to plants.

By bacterial oxidation or reduction:
  $Fe(II) \longrightarrow Fe(III)$
  $Mn(II) \longrightarrow Mn(III), Mn(IV)$
  $H_2S \longrightarrow S^0$
  $Se(VI), Se(IV) \longrightarrow Se^0$

By precipitation with microbially produced counterions:
  $Fe(III), Fe(II), Al(III)$ + phosphate $\longrightarrow$ Fe or Al phosphate
  $Fe(II)$ + sulfide $\longrightarrow$ iron sulfide
  $Fe(II), Mn(II)$ + carbonate $\longrightarrow$ Fe or Mn carbonate

By biosorption:
  For example, binding by glycocalyx, peptidoglycan, teichoic, or teichuronic acids, cell membrane, or intracellular components (metallothioneins, etc.) of prokaryotes; chitosan of fungi (see, e.g., Beveridge, 1986; Volesky, 1986)

(Table 10-2). Manganous Mn may also be made available to plants by acid- and ligand-forming and silicate bacteria acting on pyroxenes (Table 10-2).

The microbial interactions with soil minerals, listed in Table 10-2, are a few of several reactions that may mobilize elements nutritionally important to plants (see Ehrlich, 1981 for some other reactions).

CONDITIONING OF LUNAR REGOLITH BY MICROBES

Table 10-3 lists minerals that occur in lunar regolith from which, in the presence of sufficient moisture and other nutritional requirements, nutritionally important elements may be mobilized by microbes through reactions such as those listed in Table 10-2. A significant absence of some elements in lunar minerals has been noted: namely Mo and B (Table 10-4), which are important in plant nutrition and microbial activity (e.g., N_2 fixation and NO_3 reduction). Unless lunar minerals containing these elements are found on further Moon exploration, these elements will have to be supplied from Earth.

As previously mentioned, a nutritionally beneficial element becomes toxic to plants and soil microbes when the concentration in soil solution exceeds an upper limit. Excessive concentrations of an element may be the result of a fast solubilizing reaction (Table 10-5). As a means of counteraction, nature seems to have provided for microbial reactions that can lower the concentration of the dissolved mineral species, thus assuring a form of homeostasis. Microbial reactions that can contribute to the lowering of the concentration of mineral elements include oxidations and reductions, precipitation by microbial products of metabolism, or biosorption (Table 10-6).

SUMMARY

The foregoing consideration should make it clear that under appropriate conditions, microbes should be able to promote weathering of minerals in regolith on the Moon, thereby making available to plants and themselves nutritionally important mineral elements.

REFERENCES

Avakyan, Z.A., N.P. Belkanova, G.I. Karavaiko, and V.P. Piskunov. 1985. Silicon compounds in solution during bacterial quartz degradation. Mikrobiologiya 54:301–307 (English transl., p. 250–256).

Avakyan, Z.A., T.A. Pivovarova, and G.I. Karavaiko. 1986. Properties of a new species, *Bacillus mucilaginosus*. Mikrobiologiya 55:477–482 (English transl., p. 369–374).

Beveridge, T.J. 1986. The immobilization of soluble metals by bacterial walls. p. 127–139. *In* H.L. Ehrlich and D.S. Holmes (ed.) Workshop on biotechnology for the mining, metal-refining and fossil fuel processing industries. Biotech. Bioeng. Symp. 16. John Wiley and Sons, New York.

Bowen, H.J.M. 1979. Environmental chemistry of the elements. Academic Press, London.

Ehrlich, H.L. 1981. Geomicrobiology. Marcel Dekker, New York.

Ehrlich, H.L. 1983. Leaching of chromite ore and sulfide matte with dilute sulfuric acid generated by *Thiobacillus ferrooxidans* from sulfur. p. 19–42. *In* G. Rossi and A.E. Torma (ed.) Recent progress in biohydrometallurgy. Associazone Mineraria Sarda, Iglesias, Italy.

Henninger, D.L., C.W. Lagle, and D.W. Ming. 1987. A lunar derived "soil" for the growth of higher plants. p. 1–20. *In* G.M. Andrus (ed.) Symposium '86: The first lunar development symposium. S86-35. Lunar Develop. Counc., Pitman, NJ.

Karavaiko, G.I., V.S. Krutsko, E.O. Mal'nikova, Z.A. Avakyan, and Yu. I. Ostroushko. 1980. Role of microorganisms in spodumene degradation. Mikrobiologiya 49:547–551 (English transl., p. 402–406).

Shkolnik, M.Ya. 1984. Trace elements in plants. Development in crop science (6). Elsevier, Amsterdam.

Volesky, B. 1986. Biosorbent materials. p. 121–126. *In* H.L. Ehrlich and D.S. Holmes (ed.) Workshop on biotechnology for the mining, metal-refining and fossil fuel processing industries. Biotech. Bioeng. Symp. 16. John Wiley and Sons, New York.

Williams, R.J., and J.J. Jadwick. 1980. Handbook of lunar minerals. NASA Reference Publ. 1057. NASA Sci. and Tech. Info. Office, Washington, DC.

11 Controlled Ecological Life Support System

Maurice M. Averner

NASA Headquarters
Washington, DC

The National Aeronautics and Space Administration (NASA) controlled ecological life support system (CELSS) program was initiated in 1978 by the Life Sciences Division, Office of Space Science and Applications (OSSA), with the premise that NASA's goals would eventually include extended-duration missions with sizeable crews requiring capabilities beyond the ability of conventional life support technology. Currently, as mission duration and crew size increase, the mass and volume required for consumable life support supplies also increases linearly. Under these circumstances, the logistics arrangements and associated costs for life support resupply will adversely affect the ability of NASA to conduct long-duration missions. A solution to the problem is to develop technology for the recycling of life support supplies from wastes. The CELSS concept is based upon the integration of biological and physicochemical processes to construct a system that will produce food, potable water, and a breathable atmosphere from metabolic and other wastes, in a stable and reliable manner (Fig. 11-1). A central feature of a CELSS is the use of green plant photosynthesis to produce food, with the resulting production of O_2 and potable water, and the removal of CO_2.

BIOREGENERATION CONCEPT FOR CONTROLLED ECOLOGICAL LIFE SUPPORT SYSTEM

Carbohydrate and fat make up nearly 90% of the energy sources of a normal human diet. Therefore, the processes used to produce these nutrients in space tend to have determining effects on the food sources and architecture of a CELSS. Edible fats can be synthesized through chemical processes, but techniques for synthesizing carbohydrates are only beginning to be researched. Crop plants produce large amounts of carbohydrates during photosynthesis and large percentages of edible plant biomass (up to 50% dry mass), although volume and energy requirements are high. Another prospective food source, algae, can produce similar quantities of carbohydrate and edible biomass, but has smaller volume and energy requirements (Averner

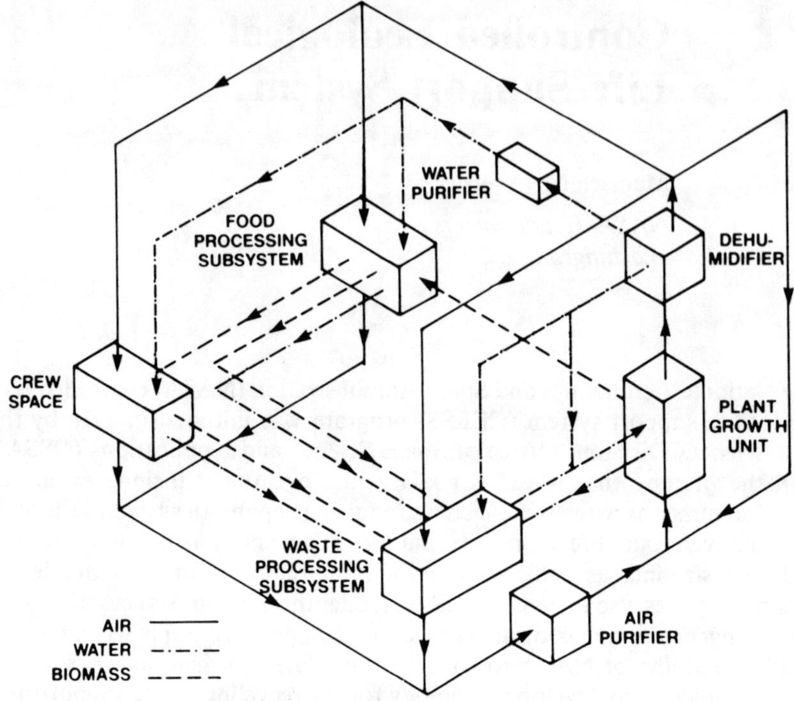

Fig. 11-1. The CELSS concept, a schematic CELSS system diagram.

et al., 1984). Inedible materials and wastes will be oxidized using photosyn-
thetically produced O_2 to produce those materials required for the continued
plant growth.

Central to the CELSS concept is the Plant Growth Unit (Fig. 11-2). Plant
biomass produced in the Plant Growth Unit is passed to the Food Processing System, where edible materials are harvested and converted to prepared
foods. Edible food products are consumed by the crew and metabolized into
CO_2, water, urine, fecal material, and minor amounts of other wastes. Solid
and liquid wastes produced by the crew and the Food Processing System are
transferred to the Waste Processing System and converted into chemical forms
that are usable as plant nutrients. Carbon dioxide and water vapor are recovered from waste solids by oxidation, using processes that leave soluble
mineral residue that can be returned to the nutrient solutions for the plants.

Water cycles between the Plant Growth Unit, Food Processing System,
crew, and Waste Processing System via a dehumidifier and water purifier.
A portion of the nutrient solution from the Plant Growth Unit is periodically circulated through the Waste Processing System, where the nutrient content is replenished and organic material is removed to minimize microbial
contamination of the nutrient solution. Air leaving the Plant Growth Unit
passes through a dehumidifier before being circulated through the system.
Crop plants produce large amounts of water vapor. The condensed water

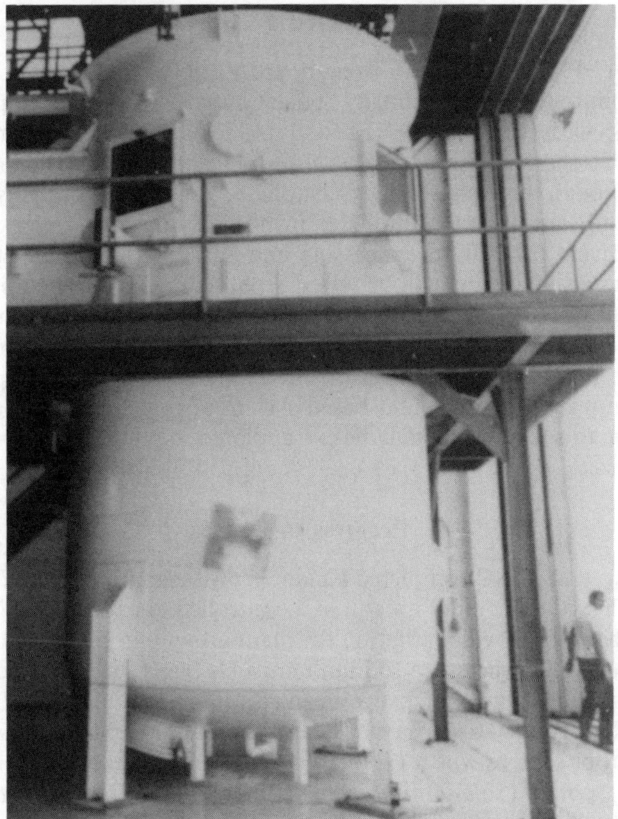

Fig. 11-2. The CELSS Plant Growth Chamber construction.

resulting from dehumidification is (i) returned directly to the plant nutrient solution to compensate for evaporation losses, (ii) purified to potable water suitable for crew consumption and household needs, or (iii) used in the Food-Waste Processing Systems. Evaporated water from the three subsystems returns to the Plant Growth Unit via air circulation, while liquid water from crew and food processing wastes is recycled to the plant nutrient solution after purification through the Waste Processing System.

Production of large amounts of water vapor by plants requires that air flow rates through the unit be high enough to maintain acceptable humidity levels. Upon leaving the Plant Growth Unit, the dehumidified air is distributed and used in the other subsystems of the life support system. Carbon leaving the Plant Growth Unit in the form of biomass is oxidized to CO_2 by either the crew or the Waste Processing System, with small amounts of C used in food processing operations. The CO_2, along with water vapor produced in the subsystems, is carried by the exhaust air stream through an air purifier that removes atmospheric contaminants. Purified air enriched with CO_2 is then circulated into the Plant Growth Unit, thereby closing the system's C and O_2 cycles.

BREADBOARD FACILITY

In early 1985, the CELSS Breadboard Facility (see Chapter 12 in this book) was initiated at the Kennedy Space Center (KSC) to develop a prototype CELSS with crop plants as the food source (CELSS Breadboard Project Plan, Life Sciences Division, OSSA, NASA Headquarters). The project will progress through three phases. The first phase, already underway, is the completion of a Plant Growth Unit. The purpose of the second phase is to design and develop the Food Processing and Waste Processing Subsystems. Each of these subsystems will be tested separately. During the third phase, the three subsystems will be integrated, tested, and evaluated as one system. Crew presence will be simulated. The Breadboard Facility will implement the basic techniques and processes required for a CELSS based on photosynthetic plant growth in a ground-based system of practical size. Results obtained from this simulation may be extrapolated to predict the performance of a full-sized, operational CELSS.

Progress to Date

During 1988, KSC completed a plant growth chamber that will provide 54 m^3 for hydroponic plant growth in a controlled environment (Fig. 11–2). The chamber will provide radiation for plant growth up to about 1000 μmol m^{-2} s^{-1} using high-pressure sodium lamps (J. Sager, 1988, personal communication). Environmental controls can adjust temperature from 16 to 30 °C and humidity from 60 to 70%. The chamber can hold up to 64 plant growth trays for crops such as wheat (*Triticum aestivum* L.), soybean [*Glycine max* (L.) Merr.], potato (*Solanum tuberosum* L.), and rice (*Oryza sativa* L.). The design and construction of these trays, racks to contain them, air flow systems, and lighting systems are in progress. Optimal parameters for temperature, radiation, lighting, and air control are being defined in the CELSS research program and adapted to use in the plant growth chamber. Once development has been completed, the chamber will be integrated with the Food and Waste Processing Systems. Life Sciences facilities at the KSC that support the Breadboard Project include small, commercial plant growth chambers, in which optimized plant growth methods are adapted for use in the Breadboard chamber; laboratories for analyzing nutritional and structural characteristics of plants harvested from the plant growth chamber; and engineering shops that construct the necessary hardware. Recent studies, concentrating on wheat, have developed the means to germinate seeds in the plant tray unit so that transplanting is unnecessary between germination and later phases of growth (R. Prince, 1988, unpublished data).

PLANT GROWTH TECHNIQUES

Studies to optimize plant growth techniques focus on maximizing productivity. Plant productivity studies define system parameters that will maximize crop yield, while minimizing the time from seedlings to harvest under controlled conditions.

Progress to Date

Experiments are being conducted and simulation models are being developed at academic laboratories and at the NASA Ames Research Center (ARC) to examine environmental parameters such as temperature, light intensity, photoperiodicity, radiation, CO_2 levels, and O_2 production, while comparing plant biomass production, time to harvest, and percentage of edible plant biomass under varying environmental conditions (MacElroy et al., 1986; see Chapter 13 in this book). Plants under study include wheat, soybean, lettuce (*Lactuca sativa* L.), and potato. Wheat studies to date have resulted in a harvest index (percentage of edible plant biomass to total plant biomass) of approximately 44%, with a time-to-harvest of approximately 79 d (Bugbee & Salisbury, 1988). Soybean studies have resulted in a harvest index of approximately 52%, with a time-to-harvest of approximately 97 d (Cure et al., 1985). Lettuce and potato studies have each resulted in harvest indices of 80%, with time-to-harvest of 110 to 147 d for potato and 19 d for lettuce (Knight & Mitchell, 1983, 1987; Tibbitts & Wheeler, 1987). Research areas at KSC include: (i) the appropriate nutrient solutions for each candidate crop plant; (ii) systems for delivering nutrients and water to the plants in a weightless environment; and (iii) methods for limiting and controlling bacterial and fungal growth in root systems (Fig. 11-3 and 11-4). Scientists are evaluating and modifying the Capillary Effect Root Environ-

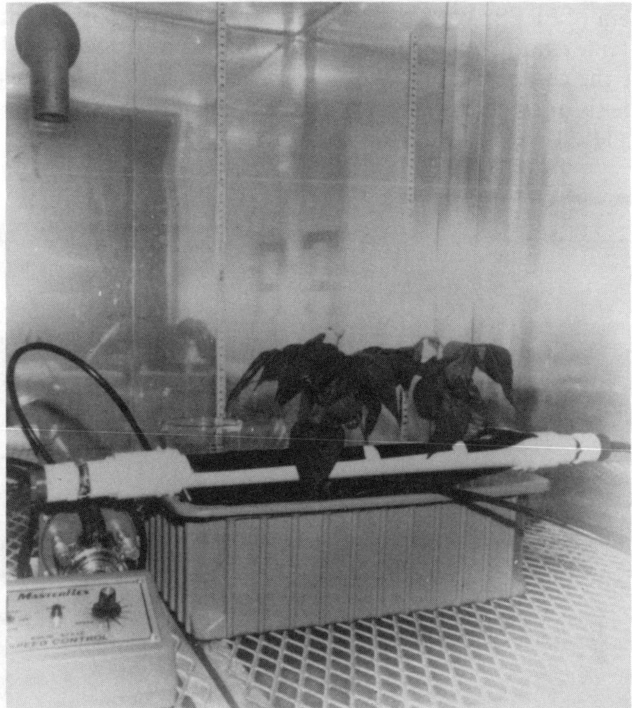

Fig. 11-3. Tubular membrane delivery system with bean plants.

Fig. 11-4. Tubular membrane delivery system with wheat.

ment System (CERES; W. Knott; NASA KSC), and a nutient film hydroponics system. Another controlled system, Minitron II, has been developed
to regulate the composition of the flowing atmosphere and to measure the
photosynthesis of small crop canopies growing hydroponically (Knight et al.,
1988; Fig. 11-5).

Fig. 11-5. Minitron II system.

FOOD SOURCES

Studies of food sources will lead to the selection of optimal plants for growth in a CELSS and selection of appropriate food production and processing techniques. The purpose of food selection studies is to determine which higher plant and microbial food sources are the best candidates for production in a CELSS.

Progress to Date

Plant characteristics under consideration are harvestability, palatability, processing requirements, nutritional content, growth habitat, and power requirements. In consideration of these characteristics, mathematical models are being developed to determine optimal plant combinations. Preliminary calculations have revealed that 99% of the crew's nutritional requirements can be met with relatively few plant types (e.g., wheat, soybean, and potato). Microbial food sources, particularly yeast and algae, are candidates for nutritional supplements.

Foods generated in a CELSS will require processing to make them suitable for human consumption. Studies are now underway to define a system that will harvest, prepare, and store plant biomass, and convert it into edible forms. Related research at KSC focuses on general food processing methods, and examines microbial, chemical, and enzymatic systems as candidate techniques for food conversion. Techniques have been developed to recover a pure protein concentrate from algae grown under controlled conditions (Kamarei et al., 1986).

WASTE MANAGEMENT AND CONTAMINANT CONTROL

Solid, liquid, and gaseous wastes are produced by the Plant Growth Unit, Food Processing System, and the crew. These must be regenerated so as to maintain the productivity of both plants and humans. Spacecraft volatile contaminants must also be controlled to maintain the health of both crew and plant crops.

Progress to Date

Simulation models are being developed to define system parameters pertaining to air and water reclamation, moisture condensation and vapor-liquid equilibrium, contaminant control, and O_2 removal (J. Bredt, 1988, unpublished data). Inedible plant biomass, urine, and fecal material represent the major portion of solid waste elements. The major liquid waste elements are water and urine, while the major gaseous waste elements are O_2, N, and CO_2. Additionally, mathematical models are being developed to simulate material flows with the purpose of defining CELSS monitoring and system parameters (Rummel & Volk, 1987; Volk & Rummel, 1987). Continued efforts at academic laboratories and NASA centers will lead to the complete specification of system requirements and controls. Related research seeks to determine the effects of plant-microbe interactions on plant health and growth, food production, and regeneration of atmospheric gases in a CELSS.

FUTURE THRUSTS

The National Commission on Space has reported that a highly reliable CELSS is mandatory for extended space travel. The basic and applied research described above will continue throughout the duration of the CELSS Program, while additional engineering and development is carried out to ensure the technological readiness of CELSS by the beginning of the 21st century.

Upon completion of the Breadboard Project, a ground-based, unmanned CELSS prototype will be used for testing CELSS performance. Initial space experiments will be performed in conjunction with NASA's Gravitational Biology Program and will address gravitational issues of plant growth in space, such as the effects of reduced gravity during the complete growth cycle of plants. Next, Space Station experiments will examine specific CELSS issues, such as the measurement of crop plant growth and productivity, along with the refinement of the measurement and control of CELSS environmental and system parameters. Subsequently, a manned CELSS prototype will be tested with a human crew to determine human interactions with CELSS, and vice versa. The purpose of these tests and experiments is to achieve a high-performance, reliable CELSS for future long-duration missions of human exploration. The development of an operational CELSS will provide economic, psychological, and mission operations benefits. For long-duration missions, such as permanent Lunar or Mars bases, where logistics supply is costly or impractical, the development of a fully integrated bioregenerative life support system will be enabling. As the duration of future manned space missions increases, a crossover point is reached where it will be more economical to provide life support supplies by the recycling of metabolic and hygiene wastes than to incur the repeating costs of resupply. In situ regeneration of life support consumables will protect the mission from unpredictable and potentially disastrous interruptions in the logistics train. Such a system will profoundly affect the success of future long-duration human missions involving the Moon and Mars. Indeed, such missions may not be possible without the security and autonomy implied by a biologically based regenerative life support system.

REFERENCES

Averner, M., M. Karel, and R. Radmer. 1984. Problems associated with the utilization of algae in bioregenerative life support systems. NASA Contractor Rep. 166615. Ames Res. Center, Moffett Field, CA.

Bugbee, B.G., and F.B. Salisbury. 1988. Exploring the limits of crop productivity: Photosynthetic efficiency of wheat in high irradiance environments. Plant Phys. 88: (in press).

Cure, J.D., C.D. Raper, Jr., and R.P. Patterson. 1985. Dinitrogen fixation in soybeans in response to leaf water stress and seed growth rate. Crop Sci. 25:52-58.

Kamarei, A.R., N. Nakhost, and M. Karel. 1986. Potential for utilization of algal biomass for components of the diet in CELSS. p. 13-22. *In* R.D. MacElroy et al. (ed.) Controlled ecological life support system: CELSS '85 Workshop. NASA Tech. Mem. 88215. NASA Ames Res. Center, Moffett Field, CA.

Knight, S.L., C.P. Akers, S.W. Akers, and C.A. Mitchell. 1988. Minitron II system for precise control of the plant growth environment. Photosynthetica 22:90-98.

Knight, S.L., and C.A. Mitchell. 1983. Enhancement of lettuce yield by manipulation of light and nitrogen nutrition. J. Am. Soc. Hortic. Sci. 108:750-754.

Knight, S.L., and C.A. Mitchell. 1987. Stimulating productivity of hydroponic lettuce in controlled environments with triacontanol. Hortic. Sci. 22:1307-1309.

MacElroy, R.D., N.V. Martello, and D.T. Smernoff (ed.). 1986. Controlled ecological life support system: CELSS '85 Workshop. NASA Tech. Mem. 88215. NASA Ames Res. Center, Moffett Field, CA.

Rummel, J., and T.Volk. 1987. A modular BLSS simulation model. p. 57-66. *In* R.D. MacElroy and D.T. Smernoff (ed.) Controlled ecological life support system. NASA Conf. Publ. 2480. NASA Ames Res. Center, Moffett Field, CA.

Tibbitts, T.W., and R.M. Wheeler. 1987. Utilization of potatoes in bioregenerative life support systems. Adv. Space Res. 7:115-122.

Volk, T., and J. Rummel. 1987. Mass balances for a biological life support system simulation model. p. 139-146. *In* R.D. MacElroy and D.T. Smernoff (ed.) Controlled ecological life support system. NASA Conf. Publ. 2480. NASA Ames Res. Center, Moffett Field, CA.

12 CELSS Breadboard Project at the Kennedy Space Center

R. P. Prince and W. M. Knott, III

Life Science Research Office, NASA
J. F. Kennedy Space Center, FL

The role plants play in the maintenance of our global environment has been recognized for many years. Undoubtedly, the seasonal changes that occur in many parts of our world contribute to the overall biological and atmospheric stability of planet Earth. Unfortunately, habitats considered for lunar and Mars bases and long-term space missions may not have available all of the natural ecological mechanisms to establish atmospheric equilibrium. Consequently, selected biological, physical, and chemical processes will be assembled into a workable unit that will be controlled by a computer system. An assortment of monitoring and feedback systems will have to be substituted to some extent for a part of the ecological control. Also, because of the size constraints on early missions, compartments for the storage of fresh and waste solids, liquids, and gases will be small forcing the turn-around time for many processes to be short.

A habitat for space missions lasting up to 6 yr for a crew of six will require a mass closure approaching 100%. Such a closure will be necessary for a lunar base and for the human exploration of Mars.

Without the digital computer, space travel by humans would not have been possible. The revitalization of air and water by physicochemical and biological processes for the survival of the crew also will be managed by computers. The production of food as a part of these cycling processes seems vital to the success of long-term presence in space. The requirements of a control system become those of selection, sizing, placement, and management of the separate pieces to make a controlled ecological life support system (CELSS) program function.

Fortunately, the photosynthetic process tends to balance CO_2 consumption with O_2 production that can be used in human metabolism. Also, an O_2 balance can be created by selecting crops that furnish enough food for the crew. It is this selection of food crops and the cultural practice chosen for production that continues to be of interest to researchers. Early work on algal systems by Ward et al. (1964) suggested photosynthetic systems of reasonably high photosynthetic efficiencies could be achieved. Additional work by Ward and Miller (1984) and Radmer et al. (1985) focused on continuous algal production systems.

Increased interest in higher plant use and efficiencies came about with the study by Gilt'son et al. (1975) in which biological regeneration systems were studied in relation to human food needs. In later studies by Gilt'son (1977) about 14 m^2 of cropping area would be required to support one crew member. MacElroy and Bredt (1984) have outlined many requirements for a CELSS including system control reservoirs and buffers, and bioregeneration of air and water. Oleson and Olson (1986) studied the components of a CELSS using a computer model and found that a volume of aprpoximately 56.9 m^3 per crew member will be required for growing crops, providing support equipment, and for acess. Nutritional requirements of many potential food crops for CELSS were studied by Hoff et al. (1982). They found that a reasonably nutritious diet could be obtained with 10 crops. They are soybean [*Glycine max* (L.)], lettuce (*Lactuca sativa* L.), peanut (*Arachis hypogaea* L.), wheat (*Triticum aestivum* L.), rice (*Oryza sativa* L.), potato (*Solanum tuberosum* L.), carrot (*Daucus carota* L.), chard (*Beta vulgaris* L.), cabbage (*Brassica aleracea* L.), and tomato (*Lycopersicon esculentum* Mill.). These and other crops were studied by Tibbitts and Alford (1982) to determine the crop yield as well as space and cultural practice requirements for CELSS.

THE KENNEDY SPACE CENTER BREADBOARD PROJECT

The NASA CELSS program encompasses in-house and extramural research aimed at providing information that will help lead to the long-term presence of humans in space habitats (e.g., lunar and Mars bases, space station, long-duration missions). As envisioned by MacElroy and Bredt (1984), a CELSS contains all of those processes and pieces of equipment to recycle O_2, water, C, and N for crew survival. A CELSS will undoubtedly contain a collection of vessels or chambers generally called *components*. These components are shown schematically in Fig. 12–1 and are known as biomass production, crew habitation, waste management, biomass processing, food preparation, product storage, air and water regeneration, and monitoring and control systems. These components are intended to function together in a variety of gravities, ranging from gravity on Earth (1 g) to the microgravity ($\sim 0\,g$) environment of space. All components are interfaced with the gas and liquid regeneration component, and the monitoring and control component. These subsystems serve to monitor and adjust the atmosphere and liquid for compatability with the new component or compartment. It assumes that some adjustment (removal of CO_2 from the airstream and organics from water, filtering pollen from the ventilating air) in air and water quality will be necessary. Solids (minerals, fiber, and organics) will be scheduled in and out of the different components as necessary to maintain a balance on reconstitution and distribution to points of use.

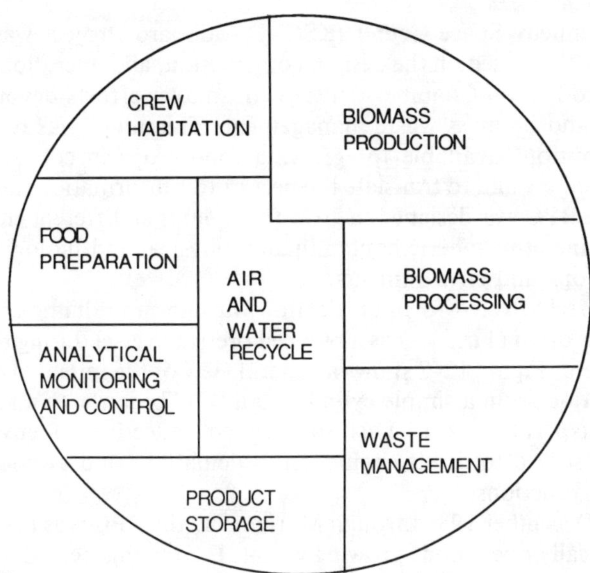

Fig. 12–1. The CELSS concept where air and water undergo regeneration from component to component.

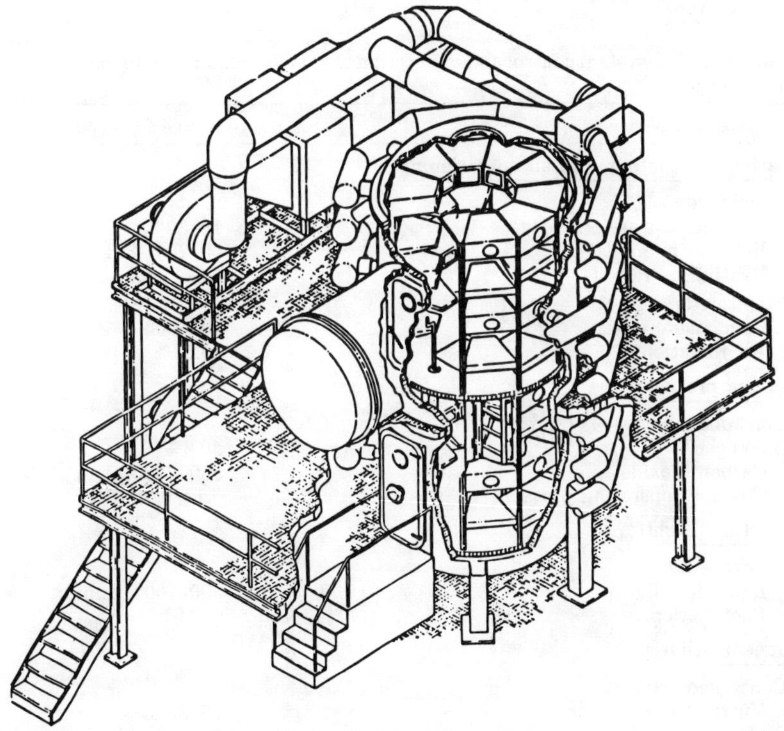

Fig. 12–2. Schematic of the Biomass Production Chamber (BPC) showing the interior configuration.

The Kennedy Space Center (KSC) Breadboard Project was approved in late 1985. It focused on the design, construction, and operation of a sealed Biomass Production Chamber (BPC) with smaller efforts devoted to food processing and biomass waste management. The intent was to investigate the many options available for growing food crops in commercial plant growth chambers and to translate the best of this information into the BPC. Further, the BPC was designed to grow food plants at different environments and determine atmospheric contaminants (physical and biological) for the different crops and environments.

The vessel selected for modification was 3.5 m in diameter and 7.5 m in height. Its original use was as a reduced pressure vessel during the Mercury spacecraft era. Figure 12–2 shows a general view of the intended design. The transformation from a simple cylinder to a BPC began with a set of design requirements, Table 12–1. This abbreviated collection of environmental parameters served to direct engineering calculations and provide the basis for control functions.

From December 1986 through March 1987 the BPC was operated in an atmospherically open mode growing wheat. During this period, the heating, ventilating, and air-conditioning subsystems were evaluated. Wheat growth was discontinued so that the other subsystems could be installed and made operational.

Table 12–1. Subsystem control and monitoring parameter requirements for the Biomass Production Chamber.

Subsystem	Range for type
Heating, ventilating, and air conditioning	
Controlled:	
Air temperature	18–30 °C
Relative humidity	60–70% RH
Ventilation rate	0.5–1 m s^{-1}
Monitored:	
Condensate water	400–500 L^{-1}
Air filtration	99.97% at 0.3 μ
Gas and pressure	
Controlled:	
Oxygen	20.8%
Carbon dioxide	350–2500 μL L^{-1}
Chamber operating pressure	0.10–0.25 kPa
Radiation (Light)	
Controlled:	
Radiation (light)	300–1000 μmol m^{-2} s^{-1}
Photoperiod	0–24 h
Nutrient delivery	
Controlled:	
Nutrient temperature	15–30 °C
pH	5.5–6.0 pH
Conductivity	100–250 uS m^{-1}
Flow rate	1 L min^{-1} tray^{-1}

Heating, Ventilating, and Air Conditioning

The major source of heat came from 96 400 W high pressure sodium (HPS) light bulbs located over the plant shelves. Two air-handling units located outside of the chambers removed this heat. Each has the capacity of 100% air circulation of 1200 m^3 h^{-1}; which resulted in a complete air change in the BPC every 17 s.

Air entered the chamber through the 32 diffusers (eight on each of four shelves) located underneath the lamp banks as shown schematically in Fig. 12-3. Sixteen lamp banks on each level provided the plenum for air return to the air-handling unit. A coarse filter was used at the entrance to the lamp bank to remove large particles of plant material. Each of the 400 W HPS bulbs was mounted under an adjustable stainless-steel parabolic reflector.

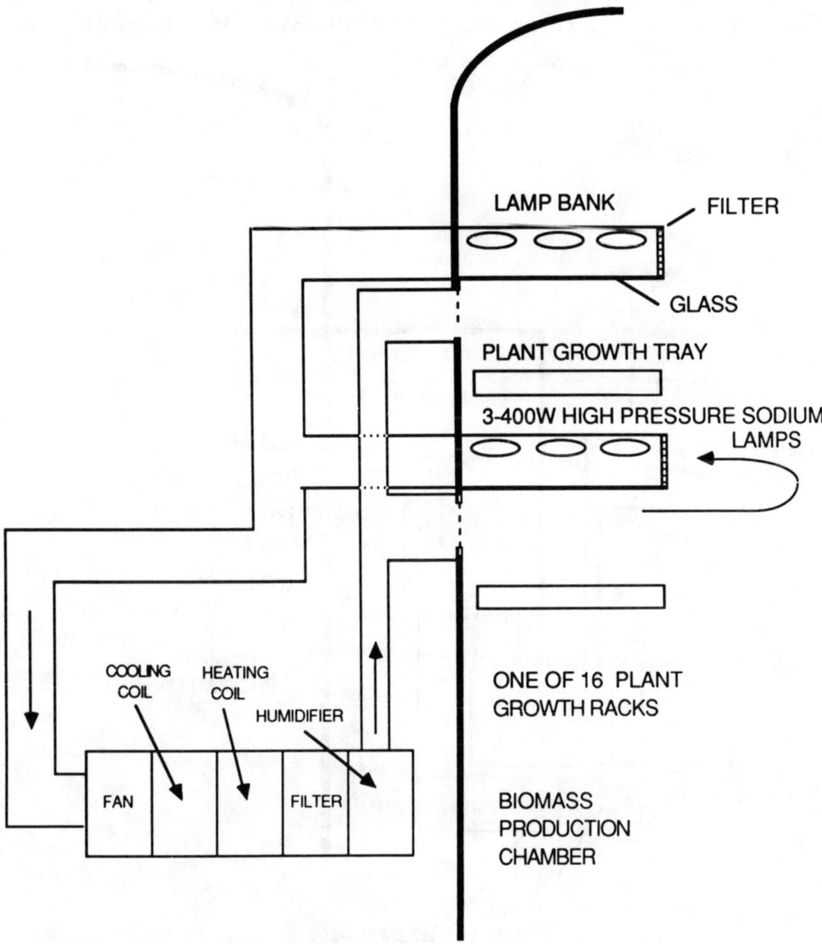

Fig. 12-3. Schematic of the heating, ventilating, and air-conditioning subsystem showing one growth rack with two shelves.

The three 400 W HPS bulbs located in each lamp bank were adequate to provide approximately 1000 μmol m^{-2} s^{-1} of photosynthetic photon flux (PPF) 0.5 m below the pyrex glass. The 96 ballasts for the HPS bulbs were located outside of the BPC. Two dimmer controls were provided for each of the four shelves for a total of eight dimmers. They were computer controlled and capable of reducing the irradiance to about 500 μmol m^{-2} s^{-1}.

Gas and Pressure

The air-handling subsystem provided the transport and mixing features for maintaining evenly distributed O$_2$ at 20.8% and CO$_2$ from atmospheric (350 μL L^{-1}) to 2500 μL L^{-1}. Pressure within the BPC was maintained at 12 mm of water by adjusting the resistance within the closed-loop air-handling circuit, adjusting breathing air, and maintaining CO$_2$. Oxygen will be maintained by the addition (if necessary) of compressed, bottled breathing air.

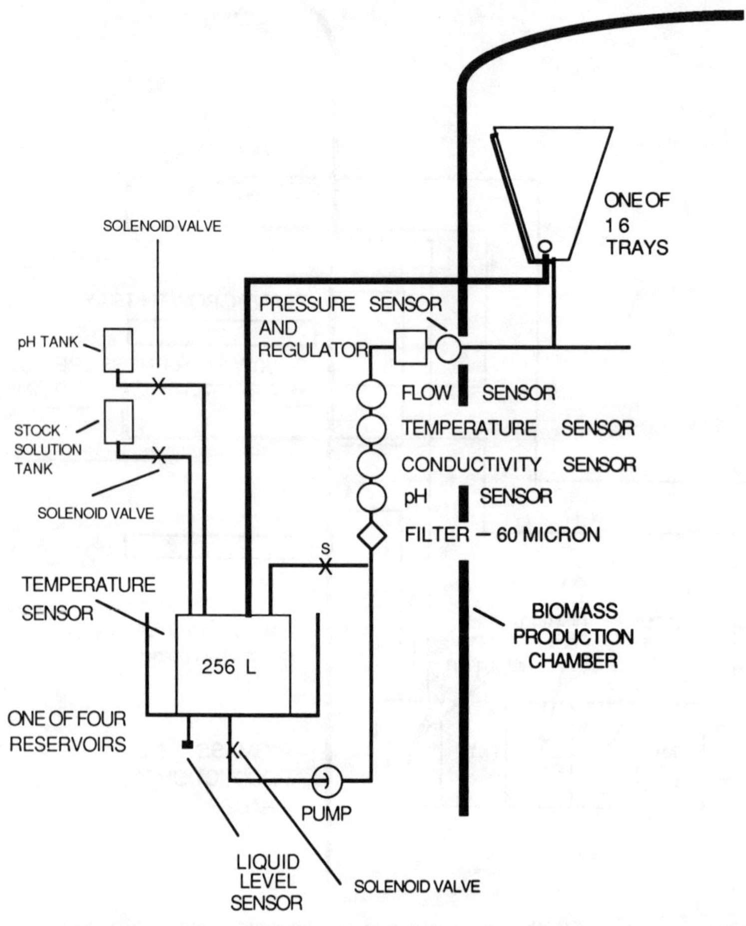

Fig. 12–4. Nutrient delivery systems for one shelf showing trays, reservoir, and plumbing layout.

Carbon dioxide will be maintained through the addition of bottled CO_2. The BPC atmosphere will be exhausted in the event of excess pressure and to maintain O_2 at 20.8%.

Nutrient Delivery

A continuously flowing, thin-film nutrient delivery system was installed. While not acceptable for a microgravity environment, the nutrient delivery system offered opportunities for plant movement during growth of some crops to obtain better space and irradiance utilization (Prince et al., 1978). It provided an opportunity to study high plant to reservoir ratios and to investigate nutrient reservoir maintenance procedures. Furthermore, the transition to a system suitable for microgravity might be minimal in terms of equipment and control features.

The BPC nutrient delivery subsystem consists of 64 polyvinyl chloride (PVC) trays complete with plumbing to distribute and adjust the flow of nutrient solution throughout the root mass, reservoirs, pipes, valves, and pumps as shown in Fig. 12–4. Each tray has a top that has been specially designed to support the crop being grown. For wheat, this top consisted of a capillary plant support (CPS) as described by Prince and Koontz (1984) and shown schematically in Fig. 12–5. The CPS contains sufficient lineal space for 400 wheat seeds in approximately 0.25 m^2. Once the tray bottom and tray top are in place, the germination hood is attached. This hood is constructed of clear plastic sides and a frosted top. Attached to the underside of the top is plastic screen net to hold distilled water for later evaporation. Heat from the lights evaporate the water on the screen to produce a warm humid atmosphere.

Once germination progresses to a point where the roots extend into the solution and shoots touch the screen (24–48 h), the hood is removed and a cage is installed (Fig. 12–5). The cage provides mechanical support to the crop throughout the life cycle. This wheat cage is light weight and the plastic frame supports a plastic net (100-mm grid) located approximately 150 mm above the tray top.

Monitoring and Control

The monitoring, control, and data-acquisition subsystem was designed to manage all subsystems and provide alarm signals to the operator (Prince et al., 1987). Six design requirements established for the monitoring, control, and data-acquisition subsystem are:

1. Separate sensors are used for control and for scientific data monitoring.
2. Additional subsystems can be added with minimum pertubation.
3. The control system is flexible enough to manage varied input/output requirements, both analog and digital.

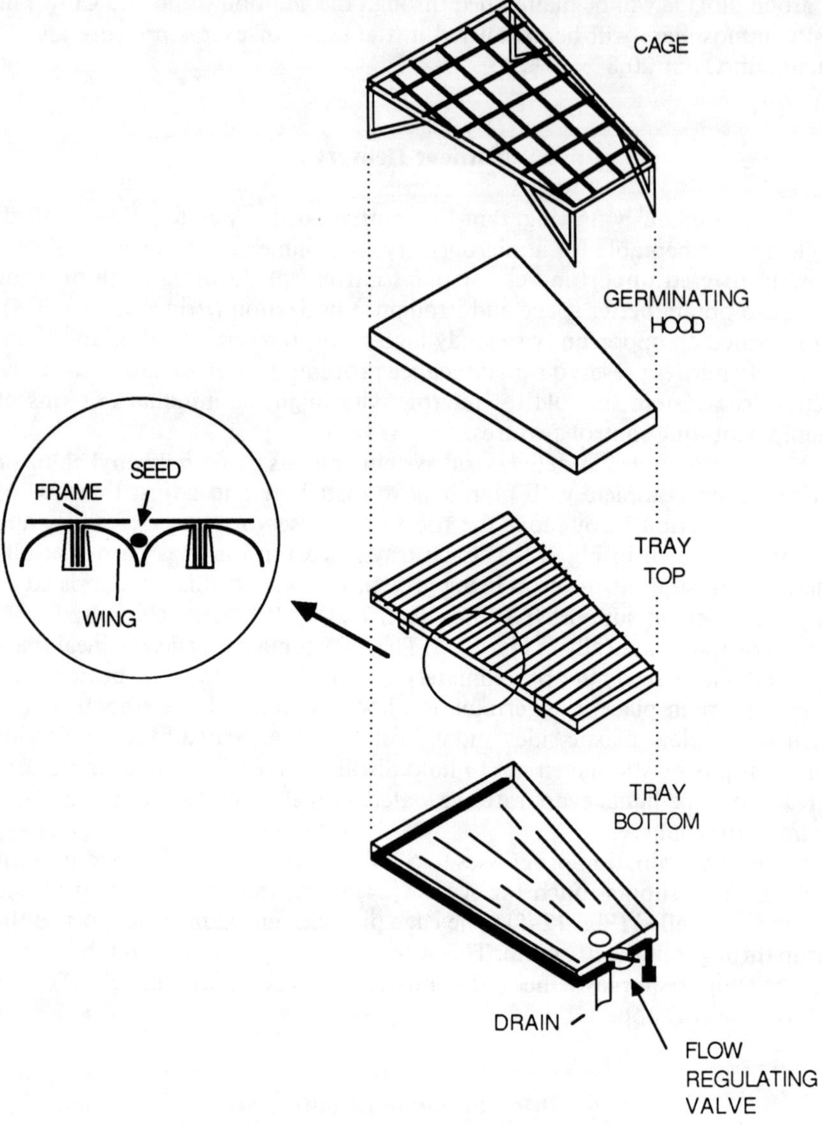

Fig. 12–5. Schematic of the thin film tray system for wheat growth.

4. The monitoring and control subsystem is capable of inter-machine communication.
5. Single-point failure places are minimized.
6. Redundancy measures are in place where there are potentially catastrophic single-point failures.

SUMMARY

The KSC Breadboard Project was initiated to study many aspects of a CELSS for long-term space missions. Food production, air and water regeneration, and engineering control were among the topics for intense study. Systems for growing food crops were investigated and led to the development of a Biomass Production Chamber as a central focus for this project.

This special chamber contained 20 m^2 of crop growing area under 96 400 W HPS lamps. Sixteen 0.25 m^2 plant growth trays were used on each of four growing shelves for a total of 64 trays. One 256-L nutrient solution reservoir with the appropriate continuous-flow, thin-film plumbing was provided for each shelf. A heating, ventilating, and air-conditioning system maintained atmospheric conditions and served to distribute O_2 at 20.8%, CO_2 between 350 and 2500 μL L^{-1}, and maintain pressure at 12 mm of water. A control and monitoring subsystem using a programmable logic controller manages the BPC subsystems.

REFERENCES

Gitel'son, I.I. 1977. Problems of creating biotechnical systems of human life support. NASA Tech. Translation TTF-17533.

Gitel'son, I.I., B.G. Kovrov, G.M. Lisovskiy, Yu. N. Okladnikov, M.S. Rerber, F. Ya. Sidko, and I.A. Terskov. 1975. Problems of space biology. Experimental ecological systems including man. Vol. 28. NASA Tech. Translation TTF-16993. Ames Res. Ctr., Moffett Field, CA.

Hoff, J.E., J.M. Howe, and C.A. Mitchell. 1982. Nutritional and cultural aspects of plant species selection for controlled ecological life support system. NASA Contractor Rep. 166324. Ames Res. Ctr., Moffett Field, CA.

MacElroy, R.D., and J. Bredt. 1984. Current concepts and future direction in CELSS. *In* Life support systems in space travel. NASA Conf. Publ. 2378. Ames Res. Ctr., Moffett Field, CA.

Oleson, M., and R.L. Olson. 1986. Controlled ecological life support systems (CELSS) conceptual design option study. NASA Contractor Rep. 177421. Ames Res. Ctr., Moffett Field, CA.

Prince, R.P., J.W. Bartok, Jr., W. Giger, Jr., T.L. Logee. 1978. Plant spacing for controlled environment plant growth. Trans. ASAE 21:332–336.

Prince, R., W. Knott, and P. Buchanan. 1987. Integration, design, and construction of a CELSS breadboard facility for bioregenerative life support system research. Proc. Int. Symp. on Biol. Sci. in Space. Najoya, Japan.

Prince, R.P., and H.V. Koontz. 1984. Lettuce production for a systems approach. Int. Soc. on Soilless Culture Proc. Wageningen, Netherlands.

Radmer, R., P. Behrens, E. Fernandez, and K. Arnett. 1985. An analysis of the productivity of a CELSS continuous algal culture system. Proc. CELSS '85 Workshop. NASA TM 88215. Ames Res. Ctr., Moffett Field, CA.

Tibbitts, T.W., and D.K. Alford. 1982. Controlled ecological life support system use of higher plants. NASA Conf. Publ. 2231. Ames Res. Ctr., Moffett Field, CA.

Ward, C.H., and R.L. Miller. 1984. Algal bioregenerative systems for space life support. The Physiologist 27:21–24.

Ward, C.H., J.E. Moyer, and G.R. Vela. 1964. Studies on bacteria associated with *Chlorella pyrenoidosa* TX71105 in mass culture. Develop. Ind. Microbiol. 6:213–222.

13 The CELSS Research Program: A Brief Review of Recent Activities

R. D. MacElroy, J. Tremor, and D. L. Bubenheim

NASA Ames Research Center
Moffett Field, California

J. Gale

Hebrew University of Jerusalem
Jerusalem, Israel

Between 400 and 500 million yr ago, vascular plants began to emerge from the Earth's oceans to colonize the land. This evolutionary change had a marked effect on the subsequent history of the Earth's biosphere because land plants injected large amounts of O_2 into the atmosphere. A relatively rapid evolution of land animals followed, including the appearance of mammals and eventually humans. Today the Earth's biosphere, which includes the surfaces of the land as well as the oceans, is complex and full of activity. Its existence depends on the continuous use and re-use of many chemical elements. The cycling of the major elements is made possible by the contributions of the biological, geological, hydrological, and atmospheric activities of the biosphere.

The biological component of the biosphere consists of organisms of two major kinds: producers, such as plants, photosynthetic bacteria, and algae, all of which use sunlight to make organic materials from inorganic compounds such as CO_2; and consumers, such as animals and heterotrophic bacteria, which consume much of the organic material made by the producers. The consumers convert organic materials back to minerals, for example, CO_2, through their metabolism. The Earth's physical processes, such as vulcanism, weathering, plate subduction, atmospheric photochemistry, all contribute in major ways and over various time scales to the Earth's element recycling processes.

Because humans evolved in the biosphere, they are completely dependent on it for their food, water, and O_2. When they travel outside of the biosphere, into space, people still require all of the same materials. These must be brought from Earth into space for human consumption, or they must be generated in place. Space travel in the future will involve extensive stays outside of the Earth's biosphere. The methods of life support that have been used to date (all life support materials are carried at the initial launch with the crew) will not be appropriate for space station, lunar bases, trips to Mars, and the establishment of bases there. This is because the costs of resupplying life support consumables will be high for orbital missions, and will be prohibitive for continuously occupied bases (MacElroy & Bredt, 1984). Instead, various methods for regenerating life support materials will be needed. In the future, increasingly autonomous life support systems will be needed. We understand enough about the processes of producing food, O_2, and water and removing wastes to know that we can create artificial biospheres for use in space. However, these biospheres will not resemble the Earth's except by analogy, although they will perform many of the same functions.

THE CONTROLLED ECOLOGICAL LIFE SUPPORT SYSTEM PROGRAM

The controlled ecological life support system (CELSS) program was initiated by the National Aeronautics and Space Administration (NASA) in the late 1970s to explore the use of bioregenerative methods of life support. It was assumed from the beginning that nonbiological methods might be required to perform some life support functions, but the program's intent was to focus on the processes involved in converting inorganic minerals and gases into life support materials using sunlight as the primary energy source.

The concepts that emerged from considerations of bioregeneration centered on maintaining biological processes in isolated environments in which cycling of materials would occur to the maximum extent possible. Because organisms can interact with one another at many levels, and because they also are sensitive, as are any chemical processes, to changes in the environment, the isolated system could behave as an ecological community and adjust its productivity in response to ecological and environmental parameters. To maintain required productivity rates, and the long-term stability of the system, it is considered essential that stringent control of the system must be practiced. The intent has been to consider the system, as a whole, as a farm that must produce materials (food, oxygen, and water) for the crew, and to minimize any potentially deleterious ecological or environmental influences. While the system will be an artificial ecological community that includes people, possibly other animals, plants, fungi, bacteria and viruses, one intent of the CELSS program is to ensure that the entire assemblage functions primarily to support humans within the constraints of the space environment.

The research, planning and technological development required by the CELSS program is conducted at NASA field centers (Ames, Kennedy, and Johnson), at various universities and by commercial organizations. Some of the work conducted by universities and Ames Research Center is briefly summarized below. Published reports on this work are available in Advances in Space Research (Oser, 1984; MacElroy & Smernoff, 1987; MacElroy et al., 1988), in NASA Conference Proceedings (MacElroy et al., 1986) and in NASA Contractor Reports. A list of CELSS Research Reports can be obtained from the authors on request.

RESEARCH ACTIVITIES AT UNIVERSITIES

To date, a significant fraction of the CELSS program's effort has been devoted to determining whether a bioregenerative life support system is appropriate for the space environment. Part of this effort has been directed toward exploring methods of reducing the size of the system, reducing its power requirements, understanding issues that are associated with its long-term stability, and identifying new technologies that might be useful in improving its efficiency.

The portion of the CELSS research program that is conducted at universities is addressing issues related to the constraints placed on the use of bioregenerative life support systems in space. One constraint that is considered to have a high research priority is reduction in the volume or planting area needed to grow higher plants. Such research has high priority because reducing the planting area results in decreases in both the mass of the plant growth system, and in its power demands for lighting.

The first level approach to this issue is to use well-known and characterized plant cultivars, with known food values, and to manipulate the environmental parameters in ways that allow maximal food production. Such parameters include light quantity, quality and cycling, temperature, nutrient delivery to the roots, gas concentrations, planting density, and other factors.

Bugbee and Salisbury (Utah State Univ.) have explored the growth of wheat in hydroponic solutions (Bugbee & Salisbury, 1985, 1988; Chapter 8 in this book; Bugbee, 1988). They have found: (i) that even at light levels three times that naturally received by field crops, the plants are not light saturated: i.e., that even higher light levels can be used by wheat plants; (ii) the grain yields produced under such conditions are four times higher than the world's record for field cultivation, on the basis of area and volume; and (iii) the above data clearly suggest that efficiency in the use of light should be considered.

The following is a brief summary of data on the efficiency of light use obtained by Bugbee (1988). At low light levels (400 μmol m^{-2} s^{-1}) the efficiency of photon use is about 10%. As the light irradiance increases, the efficiency of photon use drops almost linearly to about 6% (at 2000 μmol m^{-2} s^{-1}). However, the total biomass production rate increases linearly as the light irradiance increases, from about 60 (at 400 μmol m^{-2} s^{-1}) to about 130 g m^{-2} d^{-1} (at 2000 μmol m^{-2} s^{-1}). Edible biomass (wheat berries) similarly increases from 20 (low light) to 50 g m^{-2} d^{-1} at high light levels.

Work currently under way at Utah State is focusing on the use of temperature to maintain high light use efficiency under environmental conditions providing high biomass production rates (Bugbee & Salisbury, 1985, 1988). Surprisingly, photoperiod has been shown not to have an effect on efficiency. Subsequent work is expected to involve additional studies of crop growth efficiency as it is affected by interactions among various environmental parameters, such as radiation quality, irradiance, CO_2 concentrations and nutrient composition.

Another study of wheat, conducted by R. Huffaker (Univ. of California, Davis), has addressed the issue of plant nutrient composition, and in particular the rate at which N is taken up by plants. This is an issue of interest to the CELSS program because N within the system will be continuously converted, both biologically and chemically, from fixed forms (e.g., NH_4^+, NO_3^-) to N_2. Fixed-N availability is a major factor that influences plant growth.

Huffaker has examined the response of wheat plants to NH_3, NO_3^-, urea and NO_2 (Goyal & Huffaker, 1986a, b). The concentration of these chemical species, all of which are likely to be present in a plant growth solution, affects the rate at which each of the other compounds is absorbed. In a series of experiments, Goyal and Huffaker (1986b) explored the mutual effects on uptake of each N form, and have developed a multi-dimensional landscape description of the concentrations of each that are optimal for plant feeding. The results of these studies are that optimal concentrations of fixed N in the nutrient solution for wheat have been identified, and they are significantly lower than what has been used in standard plant growth experiments. The influence of each of the chemical species on the respective transporter enzymes in the roots has been elucidated, enabling the design of better nutrient solutions. This kind of information also aids in designing the machinery that will be essential to control mineral concentrations for optimal plant growth.

It is now possible to use the information developed by Huffaker to aid the work of Bugbee and Salisbury, and vice versa. The next year will involve increased interaction among these investigator groups. For example, it will be useful if a nutrient control system is developed that will provide plants with the forms of fixed N that they require, and at the concentrations that are appropriate for the various stages in plant growth.

Design of such a nutrient delivery system is one focus of the work conducted by D. Raper (North Carolina State Univ.). Raper has studied the uptake and allocation of fixed N by soybean (*Glycine max* L.) (Tolley-Henry & Raper, 1986a, b). He has been investigating the interaction of pH, CO_2 concentration and other factors on the uptake of fixed N, in particular NH_3, NH_4^+ and NO_3^-. His work involves consideration of the peculiar fact that plants (or at least some of them) will absorb fixed N whether they need it for growth purposes, or not. It is therefore important to understand how and when fixed N is used by a plant, and to develop a model of this activity. The model might then be used, in conjunction with continuous measurement of fixed N removal from the nutrient solution, to regulate N availability to a crop.

The role of fixed N in plant metabolism is primarily to allow the synthesis of nucleic and amino acids by the plants, which are subsequently used in the synthesis of protein, DNA and RNA. Plant protein is adequate in quality and quantity for human diets. Growth of wheat in controlled environments actually increases the protein content of wheat from about 14% to between 18 to 22% (Bugbee, 1988). In contrast, soybeans can provide as much as 35% of their mass as protein. Other major human nutrients that can be supplied by plants are lipids and carbohydrates.

Carbohydrates are the major nutrient of the human diet and can be obtained from many diverse plants, including potato. The growth of Irish potato is being studied by T. Tibbitts (Univ. of Wisconsin). Tibbitts' studies (Tibbitts et al., 1988) have concerned the effect of environmental parameters on the production of edible potato tubers. The parameters examined have been CO_2 concentrations, photoperiod, and light irradiance (Wheeler & Tibbitts, 1986; Wheeler et al., 1986). His data suggest that potato plants, in contrast to wheat, produce more tuber mass when the CO_2 concentration is raised if the light irradiance is low (400 μmol m^{-2} s^{-1}). Again in contrast to the situation that is obtained with wheat, high CO_2 concentrations (1000 μmol mol^{-1}) and continuous high-light irradiance (800 μmol m^{-2} s^{-1}) cause plant damage. Simulated orbital photoperiods cause a decrease in food yield and in total growth.

Future studies will ascertain the effects of temperature and the growth rates and productivity of new potato cultivars. The new studies will also evaluate the use of "light pipes" for supplying appropriate light levels, and in-canopy lighting. Tibbitts' studies, and those conducted at Kennedy Space Center, suggest that tuberization will proceed in hydroponic solution (Wheeler et al., 1988).

In general, the efficiency with which growing potato plants (*Solanum tuberosum* L.) use light and volume is less than wheat; however, the productivity data strongly suggest that comparisons must be conducted using many parameters, including nutritive value, and the use of other resources. Neither potato plants nor wheat (*Triticum aestivum* L.) produce significant quantities of lipids or oils, which are required in the human diet.

The production of edible oils by plants will be studied during the coming year by C. Mitchell (Purdue Univ.). Bubenheim (Ames Research Center, ARC) and Mitchell (1987) have identified a group of candidate species, including peanut (*Arachis hypogaea* L.), rapeseed (*Brassica napus* L.), and safflower (*Carthamus tinctorius* L.) as potential oil producers. Evaluations made under greenhouse conditions have suggested that several of these species will be good candidates for intensive study in controlled environments. The utility of cowpea (*Vigna unguiculata* L.) seed and edible foliage has been evaluated (Bubenheim et al., 1988, 1989).

Oils are also produced by blue-green algae. During the coming year L. Packer (Univ. of California, Berkeley) will investigate the use of stress conditions to shift the metabolism of blue greens (Cyanobacteria) to overproduce lipids. The major thrust of this work (Lefort-Tran et al., 1988), however, is the direct fixation of N_2 by the organisms. The data gathered to date suggest that when grown at high cell densities this single-celled organism is a viable candidate for consideration as a N_2 fixer in a CELSS system.

RESEARCH ACTIVITIES AT AMES RESEARCH CENTER

The ability to manipulate environmental conditions within controlled environment chambers raises the possibility that such manipulations might be used to regulate the allocation of resources within the plant. Studies directed at this objective have been conducted by J. Gale (National Res. Counc. Assoc., ARC). The data suggest that by changing dark/light periods, temperature and CO_2 concentration, a model plant (*Lemna gibba*, the common duckweed) can be induced to shift its allocation patterns to produce different amounts of protein, carbohydrate, and lipids (Gale et al., 1988).

While duckweed is regarded as a weed, it rapidly and efficiently produces biomass. The quality of its protein is good, and the quantity is of the order of 40% of the total biomass. Gale (1988) has suggested that duckweed, which has a taste similar to alfalfa (*Medicago sativa* L.) sprouts, would be an acceptable human food, and quite nutritious. Further, calculations suggest that the edible productivity of the plant (it is totally edible), based on volume and light considerations, place it on a par with wheat. While it is not likely that a crew would want to subsist on duckweed, the tiny plant's potential as a food source has probably been underestimated.

The tolerance of duckweed to unusual concentrations of salts has been explored by B. Macler (National Res. Counc. Assoc., ARC). The rationale for this study has been to evaluate the possibility that plant growth in increased concentrations of salts may be of practical consideration. The methods of waste processing for a CELSS have not been thoroughly explored, and one aspect of that problem is the removal or separation of salts from waste streams. It is conceivable that there may be system advantages to using water with high salt concentrations.

Waste processing involves the conversion of inedible materials to the inorganic chemicals that are appropriate for plant growth. However, before waste biomass (mostly celluloses) is oxidized certain materials may be profitably extracted. Of particular interest in this regard is the possible hydrolysis of cellulose to simple sugars.

Cellulose sugar recovery requires that the celluloses first be treated chemically or physically to allow hydrolysis to occur (splitting of polysaccharides to component sugars). D. Wang (ARC) and D. Strayer (Bionetics, Kennedy Space Center) have conducted studies to evaluate the processes that will be required to prepare celluloses for hydrolysis. They have investigated the use of purified enzymes (MacElroy & Wang, 1988) and specific microorganisms (Garland & Strayer, 1987; MacElroy & Wang, 1988) to effect the hydrolysis. Following hydrolysis simple sugars would be collected and used to supply C and energy to edible organisms, such as yeasts, or to produce edible materials, such as cultured plant cells.

After all usable organic materials have been recovered a CELSS will oxidize the remainder and recover CO_2 and mineral salts for use by plants. The process of waste oxidation is being studied by T. Wydeven (ARC) and Y. Takahashi (Niigata Univ.). The focus of the work is the use of super-critical water in which to oxidize organics (Takahashi et al., 1988). A particular interest is to determine whether conditions exist under these high temperature, high pressure regimes that will minimize the destruction of fixed N.

Studies have been conducted on model waste materials, and on real wastes (human wastes and plant materials) (Wydeven, 1983; Wydeven et al., 1988). The efficiency of the process has been evaluated, and the conditions necessary for complete oxidation have been identified. The data also suggest that, at temperatures close to the critical point of water, the destruction of reduced N compounds (amines, $-NH_2$, and ammonia, NH_3) is minimal.

CONCLUSION

Work conducted to date under the sponsorship of the CELSS program clearly indicates that bioregenerative life support systems have a role to play in supporting humans in space. While their initial cost (in terms of mass lifted into orbit) is high, they will function indefinitely, substantially reducing the cost of life support over the lifetime of a mission. For example, in the case of a low Earth orbiting (LEO) vehicle, such as space station, a bioregenerative life support system will "pay for itself", compared to incomplete physical-chemical regenerative methods, in a period of 5 to 7 yr. As the efficiency of plant productivity is increased through research, and the volume and power demands of a bioregenerative system decrease, the "pay-off time" also decreases. To date, the results of the CELSS research program have substantially improved the competitive position of bioregenerative life support systems, compared to resupply methods or partial regeneration by physical-chemical techniques.

Future research and technology development in the area of bioregenerative systems will likely focus on (i) recycling issues, such as recovery of usable waste materials, oxidation of irrecoverable organics, and recovery of minerals; (ii) system stability issues, such as sensing, monitoring, and interpreting changes in the flows and states within the system, control schemes to maintain selected system states, and accommodating to such events as crew changes, variations in crew activity, and outbreaks of plant disease; and (iii) converting edible biomass into acceptable food.

Bioregenerative systems will probably be required for economy of life support, and to enable the development of colonies on the Moon or Mars. It is also likely that they will be in demand for the psychological support of crews, who will be isolated from the Earth for a long time. In this case, supplies of fresh food will be of inestimable value to crew morale. The CELSS program's goals are to provide for all aspects of crew life support, and to provide the technology required for the Agency's future missions involving long duration space flight.

REFERENCES

Bubenheim, D.L., and C.A. Mitchell. 1988. Cowpea harvest strategies and yield efficiency for space food production. HortScience 23:106.

Bubenheim, D.L., C.A. Mitchell, and S. Nielsen. 1989. Utility of cowpea foliage in a crop production system for space. First National Symp. on New Crops: Research, Development, and Marketing. Indianapolis, IN. 24–27 Oct. (In press.)

Bubenheim, D.L., and C.A. Mitchell. 1987. Evaluation of new candidate crop species for CELSS. p. 27. *In* Space Life Sciences Symposium: Three Decades of Life Science Research in Space. Washington, DC. June.

Bugbee, B. 1988. Exploring the limits of crop productivity: I. Photosynthetic efficiency of wheat in high irradiance environments. Plant Physiol. 88:869–878.

Bugbee, B., and F.B. Salisbury. 1985. Wheat production in the controlled environments of space. Utah Sci. 46:145–151.

Bugbee, B., and F.B. Salisbury. 1988. Current and potential productivity of wheat for CELSS. Controlled Ecological Life Support Systems (2). Adv. Space Res., Pergamon Press, New York. (In press.)

Gale, J., D. Smernoff, B. Macler, and B. MacElroy. 1988. The carbon balance and productivity of Lemna gibba: A candidate plant for CELSS. Controlled Ecological Life Support Systems (2). Adv. Space Res., Pergamon Press, New York. (In press.)

Goyal, S., and R.C. Huffaker. 1986a. A novel approach and a fully automated micro-computer-based system to study kinetics of NO_3, NO_2, and NH_4 transport simultaneously by intact wheat seedlings. Plant, Cell Environ. 9:209–215.

Goyal, S., and R.C. Huffaker. 1986b. The uptake of NO_3, NO_2, and NH_4 by intact wheat (*Triticum aestivum*) seedlings. I. Induction and kinetics of transport systems. Plant Physiol. 82:1051–1056.

Lefort-Tran, M., M. Pouphile, S. Spath, and L. Packer. 1988. Cytoplasmic membrane changes during adaptation of the fresh water cyanobacterium synechococcus 6311 to salinity. Plant Physiol. 87:767–775.

MacElroy, R.D., and J. Bredt. 1984. Life sciences and space research XXI(2), Adv. Space Res. 4(12):1–11. Pergamon Press, New York.

MacElroy, R.D., and D. Wang. 1988. Waste recycling issues in bioregenerative life support, in controlled ecological life support systems (2). Adv. Space Res., Pergamon Press, New York. (In press.)

MacElroy, R.D., and D. T. Smernoff (ed.). 1987. Controlled ecological life support systems. Adv. Space Res. 7(4). Pergamon Press, New York.

MacElroy, R.D., T.Tibbitts, T. Volk, and B. Thompson (ed.). 1988. Controlled ecological life support systems (2). Adv. Space Res., Pergamon Press, New York. (In press.)

MacElroy, R.D., N.V. Martello, and D.T. Smernoff (ed.). 1986. Controlled Ecological Life Support Systems: CELSS '85 Workshop. NASA TM 88215. Ames Res. Ctr., Moffett Field, CA.

Oser, H., J. Oro, R.D. MacElroy, H.P. Klein, D.L. DeVincenzi, and R.S. Young (ed.). 1984. Life sciences and space research XXI(2). Adv. Space Res. 4(12). Pergamon Press, New York.

Takahashi, Y., T. Wydeven, and C. Koo. 1988. Subcritical and supercritical water oxidation of CELSS model wastes. Controlled Ecological Life Support Systems (2). Adv. Space Res., Pergamon Press, New York. (In press.)

Tibbitts, T.W., S.M. Bennett, R.C. Morrow, and R.J. Bula. 1988. Utilization of white potatoes in CELSS. Controlled Ecological Life Support Systems (2), Adv. Space Res., Pergamon Press, New York. (In press.)

Tolley, L.C., and C.D. Raper, Jr. 1986a. Cyclic variations in nitrogen uptake rate in soybean plants. Plant Physiol. 78:320–322.

Tolley-Henry, L., and C.D. Raper, Jr. 1986b. Utilization of ammonium as a nitrogen source: Effects of ambient acidity on growth and nitrogen accumulation by soybean. Plant Physiol. 8:54–60.

14 Life Support Systems Research at the Johnson Space Center

D. L. Henninger

*Johnson Space Center
Houston, Texas*

Life support systems research involving biological systems is just beginning at the Johnson Space Center and is directed at the development of a lunar-derived agricultural soil capable of supporting plant growth in a manner similar to the way soils support plant growth on Earth. Upcoming research will be directed at a series of experiments directed at the response of lunar regolith to a variety of solvents, environmental conditions, and other induced weathering parameters. Transformation of lunar regolith into other minerals such as zeolites is under investigation. These experiments are directed at developing a long-term productive, resilient soil for plant growth that will be able to supply food to lunar base crews for years.

Other research that has been going on for many years at the Johnson Space Center has centered around what is known as the environmental control/life support systems (ECLSS) program. The goal of the ECLSS work has been to conduct the research and develop the technology to minimize the dependence of the life support systems on expendables that must be re-supplied from Earth. These regenerative systems will be the mainstay of Space Station Freedom's life support systems.

Regenerative and the more complex bioregenerative life support systems of the controlled ecological life support system (CELSS) program will be combined to form a hybrid that will then represent a true CELSS. Development of such a life support system is critical to NASA's future long-duration missions. Space Station Freedom crew tours of duty will extend up to 24 wk and perhaps longer. Lunar base crews might spend years on the lunar surface without return to Earth and trips to Mars will involve several years. Life support systems will have to operate for a long time with minimum maintenance.

LUNAR BIOREGENERATIVE LIFE SUPPORT SYSTEMS

Bioregenerative life support systems research at the Johnson Space Center is focused on the use of lunar regolith as a productive, resilient plant growth medium. Three major activities are currently underway: (i) dissolution experiments to ascertain the response of lunar regolith to various solvents and weathering environments, (ii) transformation of lunar minerals into minerals such as zeolites which would be more conducive to plant growth, and (iii) development and characterization of simulated lunar regolith for preliminary experimentation (leading to experiments with actual returned lunar sample material) and experiments requiring relatively large amounts of starting material.

Dissolution Experiments

Mechanical forces of impact events were once active on the Moon, but such events are of little importance in alteration of lunar regolith now. Since the Moon is essentially devoid of water, the chemical processes of terrestrial weathering are not at work. However, when we establish an Earth-like environment we can introduce and control an environment of our choosing. This presents a host of questions concerning the lunar regolith's response to such Earth-like environments that we can impose, control, and change. Parameters such as temperature, atmospheric composition and pressure, and weathering solvents can be controlled as a function of time. Specific environments can be introduced to perform certain functions and then changed to perform another function; eventually getting to the point of having a true lunar soil capable of growing plants.

This series of experiments will examine the effects of various solvents on simulated lunar regolith as a function of time. Solvents will range from water to solvents representing a range of complexing abilities and will include humic and fulvic acids extracted from terrestrial soil samples. Humic and fulvic acids are complex organic acids common in the root zone environment of plants and promote good soil structure, increase cation exchange capacity (CEC), improve pH buffering, and increase water-holding capacity (Bohn et al., 1979).

Dissolution of lunar regolith was examined by Keller and Huang (1971) where they subjected Apollo 12 lunar sample material to deionized water, CO_2-charged water, 0.01 M acetic acid and 0.01 M salicylic acid. A terrestrial basalt similar to lunar regolith was tested for comparison. They found that dissolution approached equilibrium after 81 d at which time the amount of lunar material dissolved in organic acids was about 2.5 times that of basalt, whereas in water and CO_2-charged water, basalt was slightly more soluble than lunar material. In salicyclic acid, Al was dissolved differentially from lunar material in a higher proportion than were Ca, Fe, Na, Mg, Si, and K in decreasing order; from basalt, Mg, was dissolved preferentially to Fe, Al, Ca, Na, Si, and K in decreasing order. The final conclusion of Keller and Huang (1971) was that the higher solubility of the lunar regolith might yield more plant essential nutrients to the soil solution than a terrestrial basalt, but also more toxic trace metals, if subjected to the pedologic weathering environment on Earth.

Information from dissolution studies will aid in determining the rates of release of essential plant growth nutrients and also materials that are potentially toxic to plant growth. It is expected that a large amount of essential plant growth nutrients will have to be added to the lunar regolith to have a productive plant growth medium. It might be necessary to develop methods to remove or otherwise tie-up toxic elements and compounds.

Transformation of Lunar Minerals

Zeolites are crystalline, hydrated aluminosilicates of alkali and alkaline earth cations that possess infinite, three-dimensional crystal structures. They are further characterized by an ability to lose and gain water of hydration reversibly and to exchange some or all of their constituent cations, both without change of structure. Based on their unique adsorption and cation exchange properties, zeolites might prove to be an excellent material upon which to base the development of a solid plant growth medium or soil. About 50 natural species have been recognized, and at least 150 species having no natural counterparts have been synthesized in the laboratory. Natural zeolites have CEC of 200 to 300 $cmol_c$ kg^{-1}, whereas, synthetic zeolites may have a CEC as high as 500 to 600 $cmol_c$ kg^{-1}. By comparison, a predominantly montmorillonite clay soil in the Midwest can potentially have a CEC as high as 100 $cmol_c$ kg^{-1} (Brady, 1974).

Returned lunar samples did not contain any zeolites. However, zeolites can be relatively easily synthesized from simulated lunar glass under mild hydrothermal conditions (Ming & Lofgren, 1989). Ming and Lofgren (1989) synthesized zeolites (analcime, ZK-19, sodalite hydrate, zeolite A, zeolite P_t), feldspathoids (cancrinite), tobermorite (double-chained silicate), calcite, and several unnamed synthetic minerals with no known natural analogs. Glass is a major constituent of the lunar regolith resulting from numerous impact events during the lunar history (Williams & Jadwick, 1980).

A productive extraterrestrial soil may consist of zeolites saturated with essential plant nutrients (e.g., N in the form of NH_4, K, Ca, Mg, Zn, and possibly Fe) (see Chapter 7 in this book). The few essential plant nutrients that are not readily adsorbed on zeolite exchange sites may be added directly as a fertilizer mixed with the synthetic soil or in a solution phase through the irrigation or watering systems. Granulated zeolites will provide ideal physical conditions for an extraterrestrial soil and enhance water movement and aeration.

A study is currently underway to examine the ability of zeolite/apatite mixtures to provide N, P, and K through dissolution and ion exchange (Allen & Hossner, 1988). The primary objective is to develop a clinoptilolite-based substrate (highly siliceous zeolite) that is capable of slowly releasing N, P, and K into the soil solution at concentrations ideal for plant growth. In this study, N and K will be provided by direct exchange of NH_4 and K from the zeolite into solution. Phosphorus will be provided by the dissolution of apatite. The clinoptilolite will serve as a sink for Ca dissolved in solution from apatite, increasing the solubility of the apatite until the desired concentration of phosphate is obtained in solution. The clinoptilolite and apatite have been fully characterized and the selectivity coefficients of clinoptilolite for

NH_4, K, and Ca are being determined. Then the best substrate mixture will be determined through leaching and batch equilibrium experiments using various proportions of clinoptilolite and apatite. A computer model will be developed to predict the equilibrium concentrations of cations in solution, the relative amounts of cations occupying the exchange sites and the cation-buffering capacity of the clinoptilolite. Finally, plant growth will be monitored along with nutrient availability as a function of time in a series of plant growth trails.

Simulated Lunar Regolith

The lunar samples returned during the Apollo program represent an extremely valuable resource and are made available to researchers only in small quantities. Even then, researchers are usually required to fully verify their experimental procedures with some substitute material before using actual lunar material. Typical allocations of lunar sample are in the tens of milligram quantities. Thus experiments requiring sample material of >1 g are forced to resort to use of some sort of simulated material. At this time, only one source of simulated lunar material is known and is terrestrial basalt of similar composition to lunar basalts (Weiblen & Gordon, 1988). Relatively large amounts of lunar soil simulant have been prepared from a fine-grained basaltic rock that closely resembles the Apollo 11 high-Ti basalts in mineralogy and bulk chemistry. The crushed and ground material has the size distribution of the Apollo 11 soil samples. They also have successfully produced agglutinate-like material using a plasma arc melting technique.

There are limitations to the degree of fidelity any simulant can achieve. Even the returned Apollo samples were altered when they were removed from the lunar surface; most notably, in terms of the in-situ characteristics of bulk density and stratigraphy. The least physically disturbing method of sampling the lunar soil was with the large diameter core tubes used on Apollo 15, 16, and 17 (Carrier et al., 1971). The lunar samples brought back to Earth have undergone further changes in the laboratory because the lunar conditions of hard vacuum and freedom of water molecules and other atmospheric gases cannot be maintained on Earth. In the lunar sample curatorial facility, "pristine" samples are stored and handled only under dry N_2. Despite this, small amounts of water and other gases are probably adsorbed on the highly reactive surfaces of lunar soil grains.

Simulating the lunar soil for laboratory experimentation is approached from three aspects: soil grain distribution, soil particle-type distribution, and particle chemistry (Allton et al., 1985). Grain-size distribution curves have been determined for most Apollo soils (Carrier, 1973a). The grain-size distribution of simulants should be prepared with the fewest sieve sizes that adequately characterize the grain-size distribution curve and yet are practical to use. For example, simulant composition should be defined as 90% finer than 1 mm, 75% finer than 0.25 mm, and 50% finer than 0.075 mm.

Most particles comprising the lunar soil are igneous or breccia lithic grains, mineral grains, glass fragments, and agglutinates (mixtures of glass and mineral fragments fused together by impact events) (see Chapter 4 in this book). A simulant could be prepared with crushed basalt or minerals to substitute for the lithic and mineral fragments and by using crushed glass to substitute for the glass fragments and agglutinates. Glasseous basaltic melts have been prepared in small quantities in the laboratory (D.W. Ming, 1986, personal communication) using a formula or recipe developed by Williams (R.J. Williams, 1986, personal communication). These glasses have been used in zeolite synthesis experiments described earlier.

Simulants must be prepared with great care and closely mimic the characteristics important to the line of experimentation to be conducted. A simulant prepared for one purpose will often be totally inadequate for another purpose. For research on use of lunar regolith as a solid plant growth medium or soil, one of the most important components of a lunar simulant is the glasseous component. Because this regolith component is chemically the most reactive, special early emphasis will be placed on developing lunar glass simulants for use in the dissolution areas of research.

The Future

Reliable and efficient methods of life support must be developed to support human crews on long-duration space missions sometimes at great distances from the Earth. The hybrid life-support system derived from the regenerable physical/chemical systems and the biological systems must be developed; the so-called controlled ecological life-support system (CELSS). Resupply of consumables is limited to short-duration missions with small crew sizes. Missions to the Moon or Mars and beyond require bioregeneration of materials to supply food, O_2, and process all wastes from the crew environment.

One scenario to the exploration of the solar system places a human-inhabited lunar base near the beginning of our exploration of the solar system. A lunar base CELSS would not only supply food to lunar base crews but could resupply food to human crews at other space locations such as space stations. Such resupply through lunar base agricultural production will reduce the high transportation costs from Earth. Furthermore, industrial and human wastes from space stations could be transported to the lunar surface for reclamation and use. Interplanetary spacecraft with human crews will be extremely large vehicles requiring on-orbit assembly and supply. Initial supplies of food could be provided from lunar base agricultural production eliminating the high cost of transportation from Earth because of the high-energy requirements to achieve Earth orbit. Likewise, wastes from returning spacecraft could be sent to the lunar surface for reclamation and use rather that burdening Earth-return vehicles with the task.

A true CELSS will likely have its greatest application at a lunar base rather than a space station or other space vehicles simply because of the tight mass, volume, power, and thermal restrictions associated with space vehicles. The Moon also offers a gravity gradient that will make the engineering of systems involving fluid flow much easier. Assuming the growth of plants at a lunar base is inevitable, the question of required technology arises. Terrestrial agricultural production practices are well developed, and it seems fitting that as humans seek to establish a colony on the Moon that they take a familiar practice with them.

MERCURY, GEMINI, APOLLO, AND SHUTTLE

The life support technology used on the Mercury, Gemini, and Apollo space missions used expendables. Expendable lithium hydroxide canisters were used for removal of CO_2 from the cabin atmosphere; stored O_2 was used for breathing; and activated charcoal was used for trace contaminant removal. Stored water was used for drinking and personal hygiene while the menu consisted of stored, freeze-dehydrated food. Urine and wash water were collected and stored with solid wastes stabilized through treatment, stored, and returned to Earth.

Space missions prior to Skylab used a 34.5 kPa (5 lb in^{-2} absolute) O_2 atmosphere with no inert diluent gas; Skylab used a two-gas atmosphere with N at 10.35 kPa (1.5 psia) as the diluent. The total pressure was 34.5 kPa (5 psia). Skylab used a regenerable CO_2 removal subsystem using molecular sieves for removing the CO_2. However, silica gel was needed to remove water vapor from the air prior to processing to prevent contamination of the molecular sieves. The CO_2 was discharged overboard rather than processed as a source of O_2.

The Shuttle is a resuable spacecraft and uses a cabin pressure of 101.43 kPa (14.7 psia) or the equivalent of one Earth atmosphere. The gas composition is approximately 20% O_2 and 80% N_2. Since Shuttle flights are of short duration, the life support system continues to use expendables.

SPACE STATION

Space Station Freedom will be a vehicle with a projected design life of decades and crew sizes will be eight or more. Furthermore, Freedom is being designed to allow it to evolve into a large and more complex vehicle so that it can conduct progressively more difficult missions. Freedom's life support systems will be based on a history of research and technology development programs designed to develop processes and techniques to minimize resupply costs.

REGENERATIVE LIFE SUPPORT SYSTEMS TECHNOLOGY

Life support systems technology can be divided into three major groups: (i) atmosphere management, (ii) water and waste management, and (iii) food management. Other areas are sometimes also considered part of life support systems and include crew habitability (e.g., furnishings, housekeeping equipment), crew support (e.g., clothing, exercise equipment), and extra-vehicular/intervehicular activity. These latter areas will not be discussed here.

Atmosphere Management

The atmosphere management area consists of an air revitalization system, atmosphere pressure and composition control system, and module temperature and humidity control system. The air revitalization system uses regenerative processes and consists of the following subsystems:

1. Carbon dioxide concentrator.
2. Carbon dioxide reduction.
3. Oxygen regeneration.
4. Trace contaminant control.
5. Atmosphere quality monitoring.

The atmosphere quality monitoring subsystem does not use regenerative techniques and will not be discussed here.

Carbon Dioxide Concentration

Carbon dioxide concentration has two candidate processes that have emerged as leading contenders (Samonski, 1984); the electrochemical CO_2 concentrator (EDC) and the solid amine water desorbed (SAWD) methods (Fig. 14–1). The EDC is an electrochemical method that continuously removes

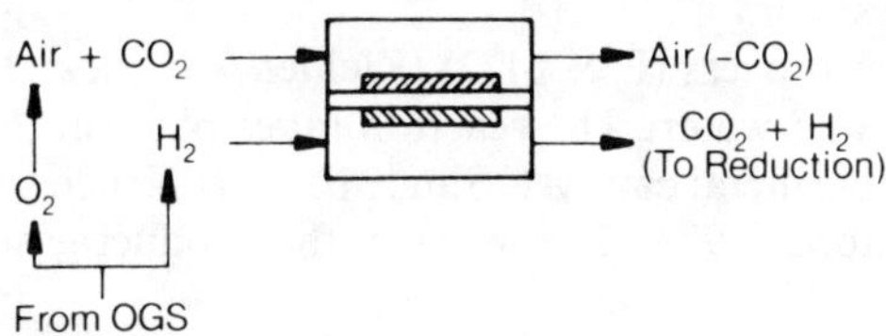

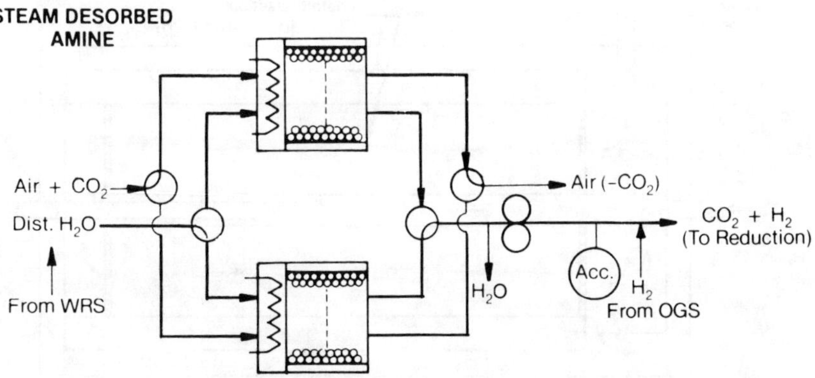

Fig. 14–1. Conceptual comparison of alternative regenerative CO_2 concentrators (Samonski, 1984).

CO_2 from the atmosphere and concentrates it for subsequent recovery of the O_2. The CO_2 removal and concentration takes place in series of electrochemical cells consisting of porous electrodes separated by a matrix containing an aqueous carbonate electrolyte such as cesium carbonate. The EDC technology has a high process efficiency (approximately 91%) and demonstrated long-term performance (more than 20 000 h on a single unit) (Lance & Schubert, 1981; Heppner & Schubert, 1983).

The solid amine water desorbed (SAWD) method (Heppner & Schubert, 1983) of CO_2 concentration uses a weak base ion exchange resin whose active ingredient is diethylene triamine which is polymerized into a porous polystyrene divinylbenzene co-polymer substrate. The amine resin absorbs CO_2 by first combining with water to form a hydrated amine. The CO_2 then reacts with the hydrated amine to form a bicarbonate. The amine is regenerated by applying heat in the form of steam to break the bicarbonate bond and releases the CO_2. Thus, this system must be run in a cyclic mode and continuous CO_2 removal can only be achieved with multiple units.

Carbon Dioxide Reduction

The two primary candidate processes for reducing the concentrated CO_2 with H_2 to form water for subsequent removal of O_2 are the Sabatier and Bosch processes. The major difference between the two processes is the final form of the C; the Sabatier process produces CH_4 while the Bosch process produces solid C.

The Sabatier process (Fig. 14–2) (Kleiner & Cusick, 1974) reduces CO_2 with H_2 to CH_4 and water. The reaction takes place in the range of 175 to 620 °C with a noble metal catalyst. Single pass efficiencies through the Sabatier reactor are about 99%. The water in the product gases is condensed in

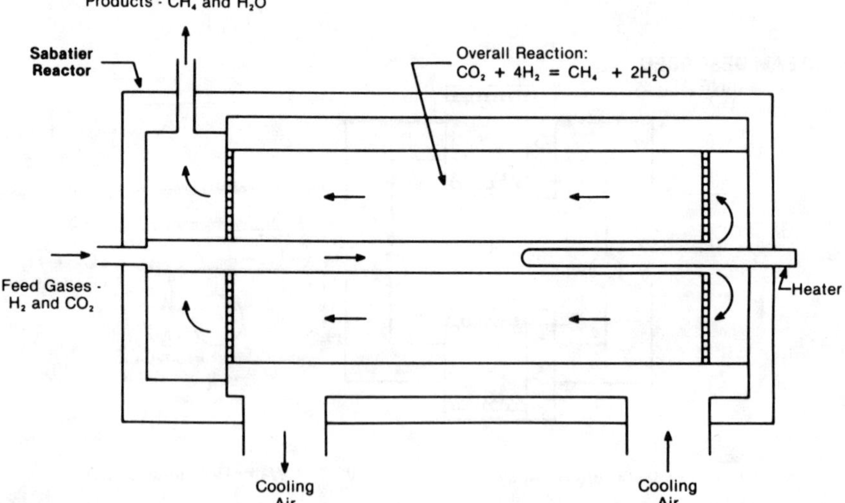

Fig. 14–2. The Sabatier process (Samonski, 1984).

a liquid-cooled porous plaque condenser/separator. The gases from this separator, primarily CH_4 and unreacted CO_2, are then vented overboard or could be stored.

The Bosch process (Fig. 14–3) (Heppner et al., 1979) reduces CO_2 with H_2 to form solid C and water. The reaction takes place in the range of 525 to 727°C in the presence of an Fe catalyst. Single pass efficiencies through a Bosch reactor are $<10\%$. To achieve complete conversion, the process gases must be recycled with continuous deposition of C and removal of water vapor. The C is collected in an expendable cartridge.

Oxygen Regeneration

Oxygen regeneration must extract the O_2 contained in water recovered through the CO_2 reduction process. The O_2 is used for rebreathing and the recovered H_2 is used in the CO_2 reduction process. Numerous electrolysis approaches have been investigated and three have emerged as viable candidates:

1. Static feed electrolyzer using an alkaline electrolyte.
2. Liquid feed electrolyzer using an acid electrolyte in the form of a solid polymer.
3. Vapor feed electrolyzer using an acid electrolyzer.

Aqueous KOH is the electrolyte used in the static electrolyzer (Fig. 14–4) (Schubert et al., 1981). The electrolyte retaining matrix is about 0.03-cm thick and there is no free electrolyte external to the module. The system operates at pressures up to and exceeding 3.45 MPa (500 psia). The static feed elec-

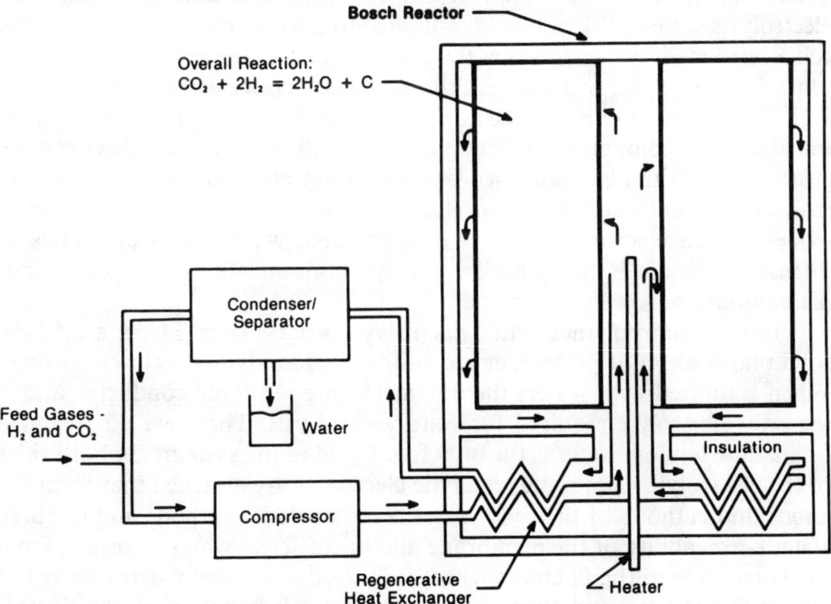

Fig. 14–3. The Bosch process (Samonski, 1984).

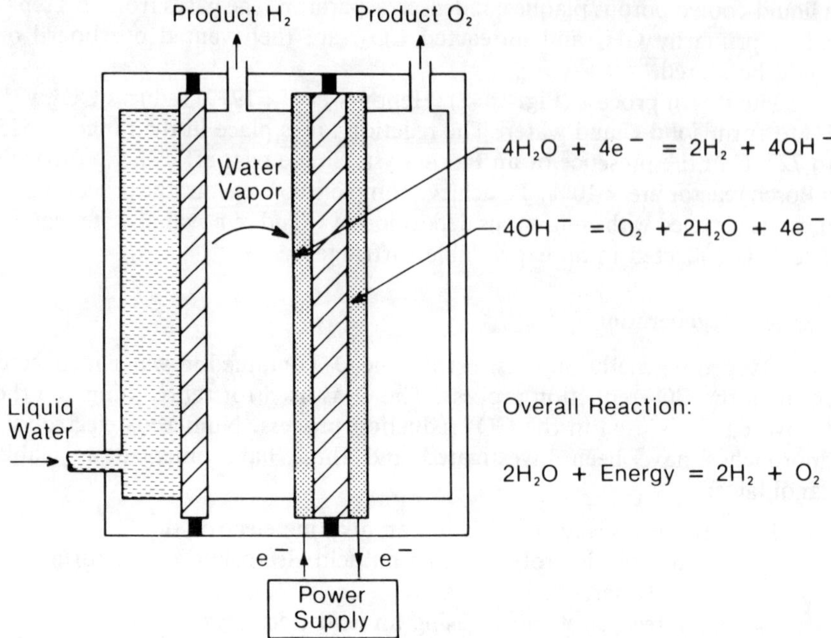

Fig. 14-4. Static feed electrolysis O_2 regeneration process (Samonski, 1984).

trolysis system operates at low cell current densities (160 mA cm^{-2}) and at modest temperatures (80 °C), as a result negligible waste heat is generated. It does not require a gas/liquid separation process as does the liquid feed electrolysis system. When power is applied to the electrodes, water from the cell electrolyte is decomposed at the H electrode and the cell electrolyte concentration increases, with the resulting decrease in cell electrolyte vapor pressure. Vapor pressure in the feed matrix causes water vapor to diffuse from the water feed compartment through the H cavity into the cathode electrode. A new equilibrium is established based on the water requirements for electrolysis and continues as long as electrical power is applied to the cell electrodes. When electrical power is discontinued, water vapor continues to diffuse across the H cavity until electrolyte concentrations within the module equilibrate.

In the solid polymer water electrolysis process (Fig. 14-5), a 0.03-cm solid plastic sheet of perfluorinated sulfonic acid polymer is the electrolyte. When saturated with water, the polymer is a good ionic conductor and is the only electrolyte required for water electrolysis. There are no free acid or alkaline liquids; water is the only free liquid in the system. On both sides of the electrolyte membrane, catalytic electrodes are attached that form the anode and cathode of the cell. Water is fed to the H cathode and the high water permeability of the membrane allows sufficient water transport from the H cathode to the O anode where it is electrochemically decomposed to release O_2, H ions, and electrons. The H ions migrate through the electrolyte while the electrons travel through the external electrical circuit to electro-

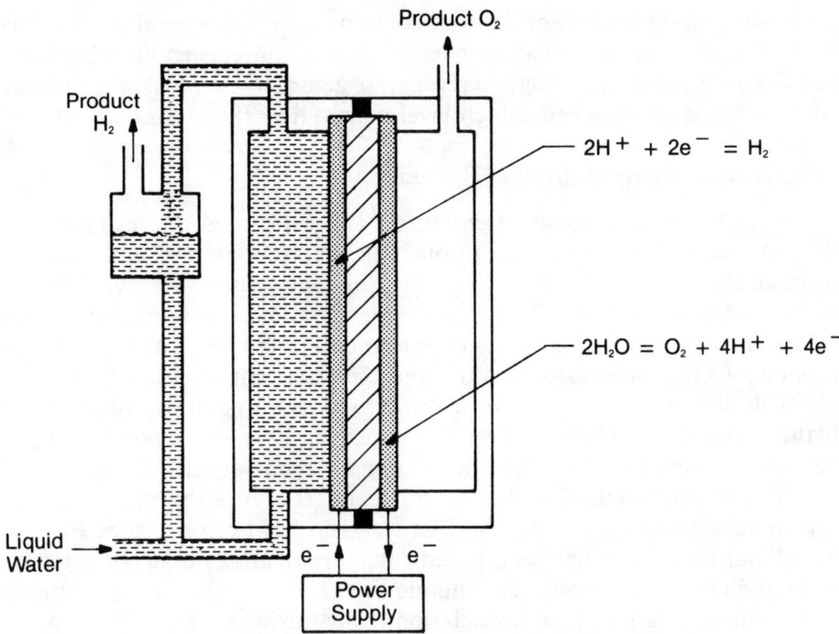

Fig. 14–5. Solid polymer water electrolysis O_2 regeneration process (Samonski, 1984).

chemically recombine H ions to form H_2 at the cathode. The H_2 is then flushed out of the cell with the recycling flow of the process feed water. The generated O_2 is exhausted from the cell without free water, although it is saturated with water vapor at the O_2 outlet temperature and pressure conditions.

The third water electrolysis approach is the vapor feed electrolyzer that uses a hygroscopic electrolyte (such as sulfuric or phosphoric acid) that absorbs water vapor from the air stream (Marshall et al., 1977). Within the unit, water vapor is absorbed at the anode-electrolyte interface and the O_2 generated by electrolysis is injected into the air flowing into the anode compartment. Hydrogen is generated at the cathode and is then used at the anode of the electrochemical CO_2 concentrator.

Trace Contaminant Control

Trace contaminants arise from a variety of sources including spacecraft components as well as payloads. As the number and diversity of payloads increase, the variety and quantity of trace contaminants also increase. Additionally, as mission durations increase dramatically, crew exposures increase and trace contaminant concentration tolerances will generally be considerably lower than with shorter duration missions.

The primary methods to handle trace contaminants aboard space vehicles include catalytic oxidation, charcoal adsorbers, and chemical adsorbers. Chemical adsorbers are needed to remove acidic gases generated as by-products in the oxidation of trace contaminants. Considerable work has been done to characterize catalytic oxidizers and nonregenerable adsorbers. For the most part, technology to regenerate activated charcoal and acid gas adsorbers has not reached a high level of maturity (Samonski, 1984).

Independent Air Revitalization System

NASA has focused its attention on two integrated air revitalization systems, one has been developed through the preprototype level and the other through the breadboard level. The first, the independent air revitalization system (IARS) provides CO_2 removal, O_2 generation, and humidity control (Marshall et al., 1977). The two major elements of the IARS are the electrochemical CO_2 concentrator module and the water vapor electrolysis module. The electrochemical CO_2 concentrator is similar to that described earlier in the section Carbon Dioxide Concentration while the water vapor electrolysis system was described in the section Oxygen Regeneration.

The second method is the integrated air revitalization system (ARS) shown schematically in Fig. 14–6. This system has been developed to the breadboard level and has been tested for more than 125 d at both steady-state and cyclic conditions. The unique aspect of the ARS is the N-supply subsystem that is based on dissociation of hydrazine (N_2H_4). The two gas atmosphere of Space Station Freedom using N_2 as the diluent gas will require supply and storage of N for module leakage make-up. Nitrogen can be stored as a cryogenic liquid, as a high pressure gas, or as part of a chemical that is readily converted. Such a method is to catalytically decompose N_2H_4 into N_2 and H_2 and then to separate the gases. This approach is more

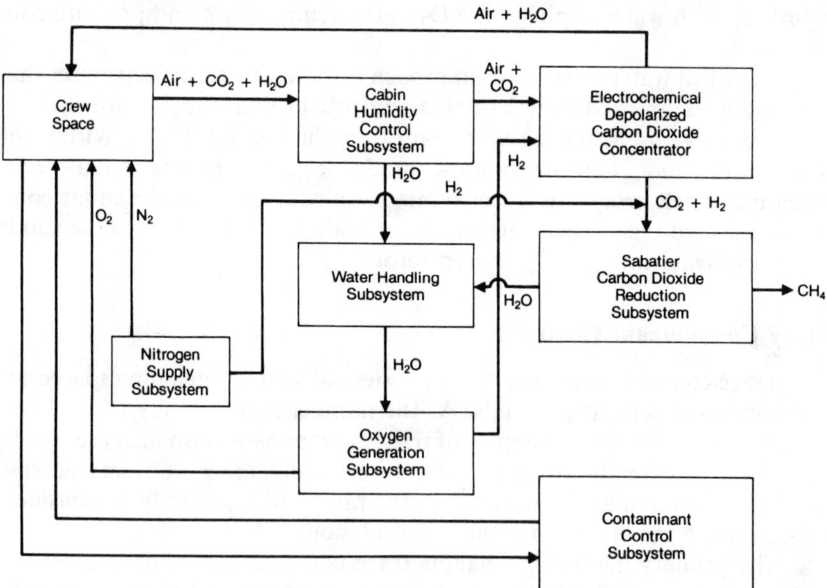

Fig. 14–6. Air revitalization system block diagram (Samonski, 1984).

attractive if the hydrazine is used as a propellant and the Sabatier CO_2 process is used. The Sabatier process has a high mass balance with insufficient H_2 to reduce all the CO_2. The N_2H_2 thus can provide the extra H_2 needed to recover all the metabolically consumed O_2 from the CO_2 produced by the crew.

The N generation system based on dissociation of N_2H_2 (Fig. 14–7) operates at a temperature of approximately 730 °C and a pressure of 1720 kPa. Note that this is not a regenerable technology since hydrazine is expended to generate the N_2 and H_2. However, its attractiveness resides in the fact that hydrazine is often used as a propellant.

Water and Solid Waste Management

The water and waste management area consists of the water reclamation system and solid wastes management. The mechanical devices such as the hand wash unit, full-body shower, laundry washer and dryer, toilet, urinal, emesis unit, and trash compactor are interrelated with the waste processing systems. The mechanical devices will not be discussed here.

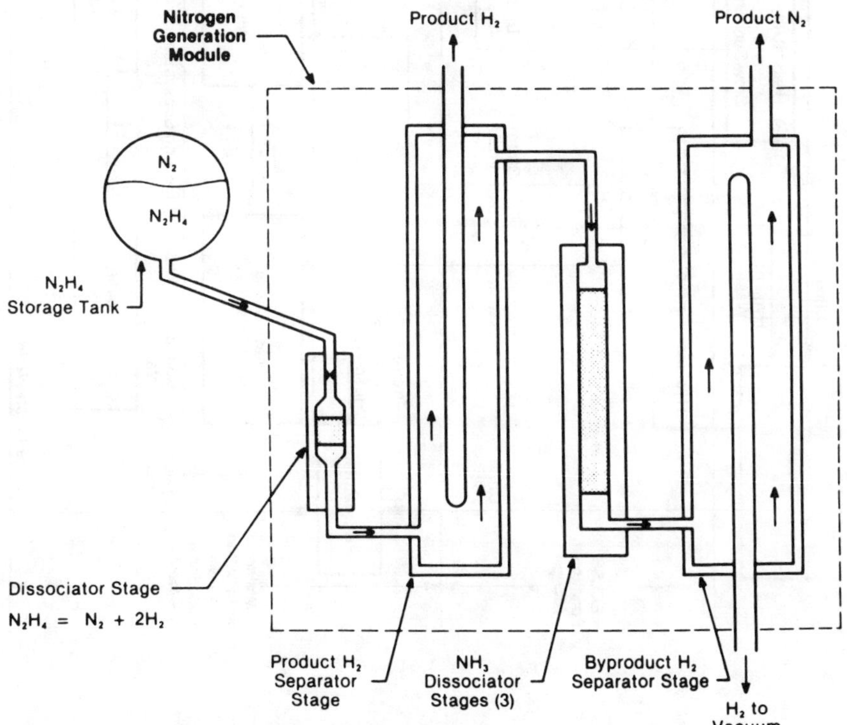

Fig. 14–7. Hydrazine dissociation N generation process (Samonski, 1984).

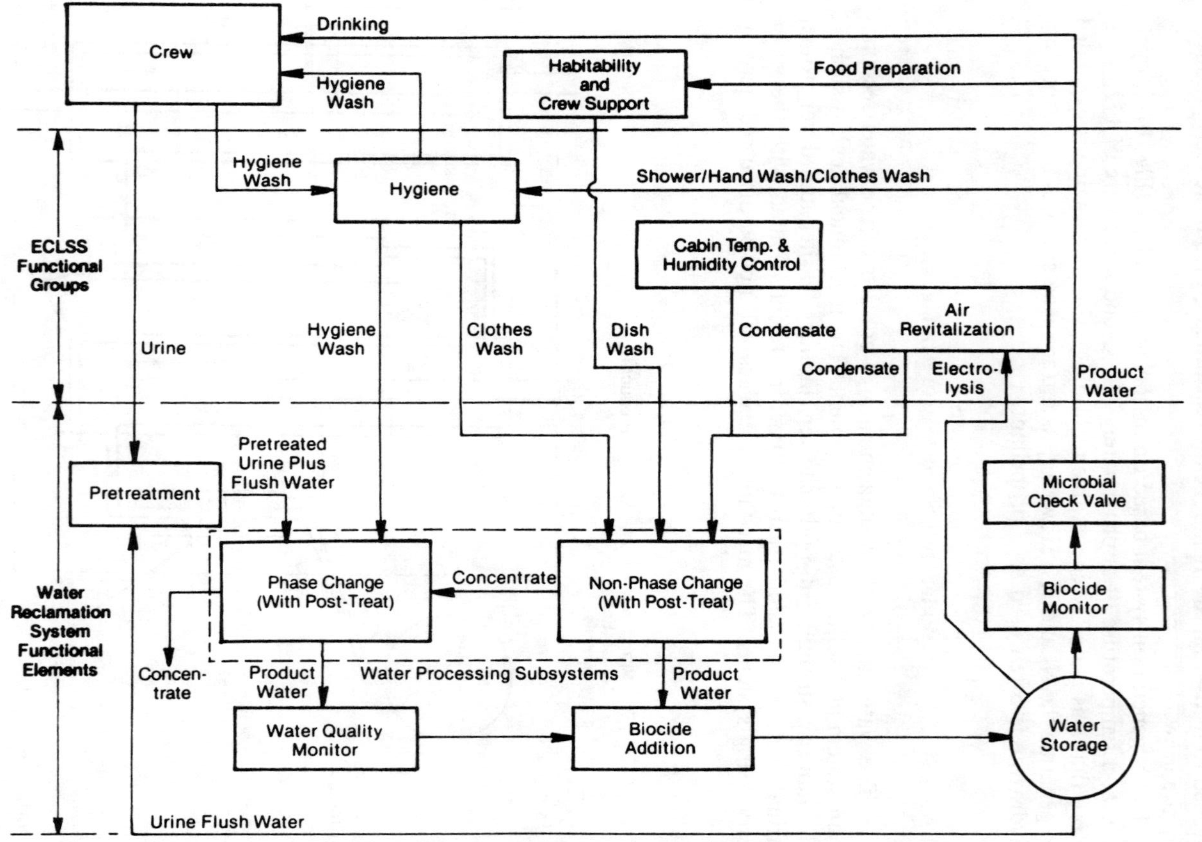

Fig. 14–8. Water reclamation system (Samonski, 1984).

Water Reclamation System

The water reclamation system consists of the following subsystems:

1. Phase change water recovery and/or filtration water recovery.
2. Potable and/or reuse water posttreatment.
3. Water quality monitor.
4. Biocide addition and monitoring of stored water.
5. Water storage.
6. Biocide monitor at points of usage.
7. Waste water storage.

Figure 14–8 depicts the water reclamation system including the sources of the waste water and the uses for the reclaimed water. Two primary methods are used to reclaim waste water: phase change processes and membrane or filtration processes.

Phase Change Processes. Phase change processes are basically distillation and condensation. These processes have been developed to recover the energy expended in the distillation phase during the condensation phase. Two methods are undergoing development; the vapor compression distillation process and the thermoelectric/membrane evaporation process.

The vapor compression distillation process (Fig. 14–9) (Ellis et al., 1979; Schubert, 1983) recovers latent heat by slightly compressing the vapor to raise its saturation temperature and condensing the vapor on the surface in direct thermal contact with the evaporator. The heat flux from the condenser to the evaporator is sufficient to evaporate a mass of water equal to that being condensed. The latent heat of condensation is thus recovered for the evaporation process. A small amount of energy is required, however, to compress the vapor and overcome thermal and mechanical losses. The amount of energy that must be supplied is approximately 95 W-h kg^{-1} of water produced at an operating temperature of 32 °C. A preprototype has been built and operates at a capacity of 33 kg d^{-1} of water. This preprototype has been designed for microgravity phase separation wherein the evaporator, condenser, and

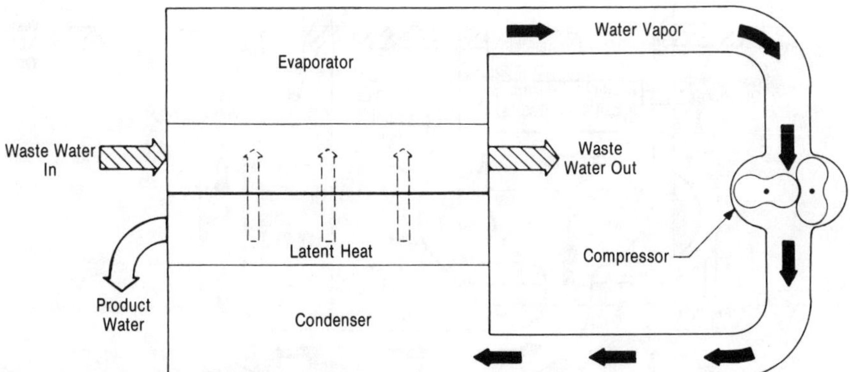

Fig. 14–9. Vapor compression distillation concept (Samonski, 1984).

condensation collector are rotated. The solids concentration in the recycle loop slowly increases until more than 96% of the original water contaminated in the waste has been removed and a solids concentration of approximately 56% has been achieved.

The thermoelectric/membrane evaporator concept is shown in Fig. 14–10. This system uses a vacuum distillation process for reclaiming water from waste water. The waste water is first chemically pretreated to inhibit urea breakdown and to control bacterial growth. This mixture is heated to 65°C by a heat exchanger in contact with the hot junction surfaces of a thermoelectric heat pump. The heated waste water passes through a tubular hollow fiber membrane evaporator module. The exterior surfaces of the membrane tubes are exposed to reduced pressure (17.2 kPa or 2.5 psia) enabling water to diffuse through the tube walls and to evaporate, forming steam at the tube outer surfaces. Other liquid and dissolved solid constituents of the waste water remain in the recycle loop where it is passed through a 25-μm filter to remove solids before it is re-introduced to the thermoelectric heat exchanger. The solids concentration in the recycle loop increases until about 94% of the original water content has been recovered and a 38% solids concentration is achieved. Then the process is halted, the recycle tank and filter are replaced and the process is continued.

Evaporated steam condenses to water on a chilled porous condenser plate that is in thermal contact with the cold junction surfaces of the thermoelectric heat pump. The chilled porous plate acts as a wick and a liquid/vapor separator. The condensate goes through the porous plate, is collected in an accumulator, and is then pumped to a posttreatment module.

The posttreatment module contains activated C and ion exchange resin beds to filter the product water to remove odors and trace amounts of dissolved NH_3 and organic molecules. After this treatment, the polished water is pumped to the water storage tank for subsequent use.

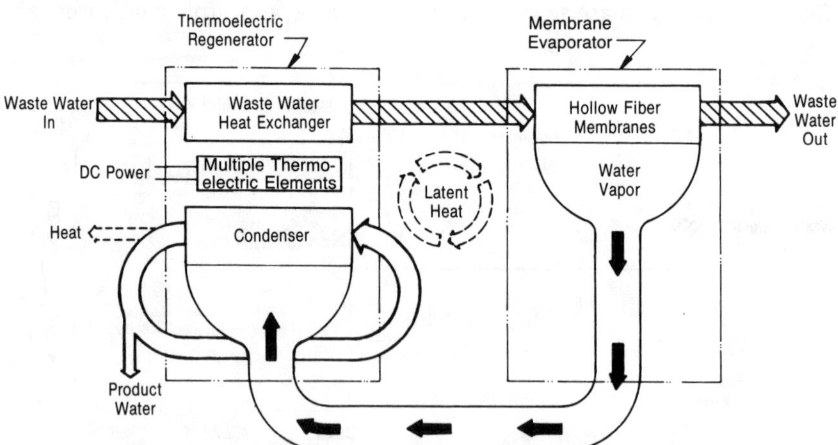

Fig. 14–10. Thermoelectric/membrane evaporator concept (Samonski, 1984).

A preprototype thermoelectric/membrane evaporator system has been built and tested. It has a capacity of 22 kg d^{-1} of water and operates at 65 °C (Roebelen et al., 1982).

Membrane Processes

The NASA has evaluated numerous membrane processes including reverse osmosis, hyperfiltration, ultrafiltration, and electrodialysis. Hyperfiltration and reverse osmosis are processes in which water and contaminants are separated by forcing the water across a semipermeable membrane under pressure. The membrane allows water to pass selectively through the membrane while rejecting dissolved and suspended solids (Reysa et al., 1983). Membranes used in water purification processes are good at filtering out ionic species and larger organic species. Smaller organic molecules (such as urea) and non-ionized acids and bases are not rejected well.

Water Quality Monitoring, Storage, and Maintenance. Water quality provides for monitoring of pH, conductivity, and organic content. Organic content can be expressed as total organic C (Lantz et al., 1981) or as electrochemical organic content (Davenport & Lantz, 1980). Such devices have been tested in the laboratory on several water reclamation systems.

Product water will be stored in a series of three stainless steel bellows tanks; one being filled, the second being tested for microbial content, and the third for current use. There could also be two water loops. One would be for potable water and the other for reuse water for such purposes as hand washing, showering, and laundry. In this case, more than three storage tanks would be required.

A biocide, such as I, is added to the recovered water to prevent subsequent microbial contamination. In addition, a biocide monitor is used at the final points of water use to ensure that the biocide is still present. Thereby, the need for a microbiological analyzer, which is a more complex instrument, is eliminated.

Solid Waste Management System

Solid waste management systems have historically consisted of using chemical germicides to stabilize fecal matter and more recently vacuum drying. At the other end of the spectrum is partial and complete oxidation of organic wastes. Oxidation methods such as wet oxidation, supercritical water oxidation, and dry oxidation all require O_2 and produce CO_2 and other gases which must then be handled by the air revitalization system. Figure 14-11 is a schematic of the Shuttle solid waste collection and treatment system. In general, the technology for regenerative solid waste treatment is not as well developed as that for air and water.

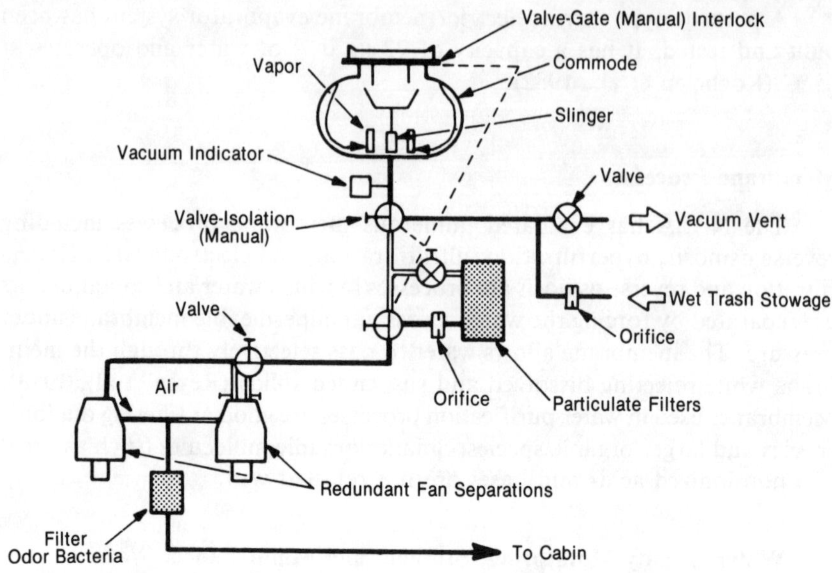

Fig. 14–11. Shuttle solid waste collection and treatment subsystem (Samonski, 1984).

Control and Monitoring

The goal of control and monitoring is to minimize servicing and maintenance time required by the life support systems while maintaining a high level of reliability. The monitoring instrumentation and associated computer hardware and software provides for fault detection, fault isolation, fault prediction, and then provides for fault correction. Control instrumentation maintains process parameters within prescribed limits, controls operating modes and changes in modes, and controls interactions between elements of the life support system.

A great deal of work is underway in the area of control and monitoring with much more of this function being handled by computer software rather than manually or with hardwired controllers.

The Future

The goal of the advanced life support research has been to conduct the research and develop the technology with which NASA can conduct future missions. Space Station Freedom is now the center of attention and will continue to require concentrated efforts to bring such systems to the flight readiness stage.

The regenerative air revitalization technology has reached the highest level of maturity. Water recovery technology must be developed to this same level and then solid waste and finally food management must follow.

Future NASA research and development technology efforts will be at multiple levels (Samonski, 1984):

1. Carrying out scientific and engineering work on basic technology.
2. Completing engineering assessments to aid in identification of high leverage technology.
3. Designing, building, and testing major hardware components.
4. Developing complete subsystems based on new technology approaches.
5. Integrating function components to provide cost-effective and reliable systems for air revitalization, water reclamation, and possibly waste management.

REFERENCES

Allen, E.R., and L.R. Hossner. 1988. Supplying nitrogen, phosphorus, and potassium through ion exchange using a zeolite:apatite substrate. p. 193. *In* Agronomy abstracts. ASA, Madison, WI.

Allton, J.H., C. Galindo, and L.A. Watts. 1985. Guide to using lunar soil and simulants for experimentation. *In* W.W. Mendell (ed.) Lunar bases and space activities of the 21st century. Lunar and Planetary Inst., Houston.

Bohn, H.L., G.L. McNeal, and G.A. O'Conner. 1979. Soil chemistry. John Wiley and Sons, New York.

Brady, N.C. 1974. The nature and properties of soils. 8th ed. Macmillan Publ. Co., New York.

Carrier, W.D. III, S.W. Johnson, R.A. Werner, and R. Schmidt. 1971. Disturbance in samples recovered with the Apollo core tubes. Proc. Lunar Sci. Conf. 2nd, 1971 3:1959–1972.

Davenport, R.J., and J.B. Lantz. 1980. Evaluation of an electrochemical organic content analyzer for Army water supply technology. Final report. Contract DAMD17-79-C-9058, TR-391-4. Life Systems, Cleveland.

Ellis, G.S., R.A. Wynveen, and F.H. Schubert. 1979. Development of a preprototype vapor compression distillation water recovery subsystem. NASA-CR-160332. Final report, Contract NAS9-15267, ER-312-4. Life Systems, Cleveland.

Heppner, D.B., T.M. Hallick, D.C. Clark, and P.D. Quattrone. 1979. Bosch: an alternative CO_2 reduction technology. ASME Paper 79-ENAs-32. ASME, San Francisco.

Heppner, D.B., and F.H. Schubert. 1983. Electrochemical and steam desorbed amine CO_2 concentration: Subsystem comparison. SAE Paper 831120. SAE, San Francisco.

Keller, W.D., and W.H. Huang. 1971. Response of Apollo 12 lunar dust to reagents simulative of those in the weathering environment of Earth. Proc. Lunar Sci. Conf. 2nd, 1:973–981.

Kleiner, G.N., and R.J. Cusick. 1974. Development of an advanced Sabatier CO_2 reduction subsystem. ASME Paper 81-ENAs-11. ASME, San Francisco.

Lantz, J.B., R.J. Davenport, R.A. Wynveen, and W.J. Cooper. 1981. Development of a TOC/COD analyzer for process applications. *In* W.J. Cooper (ed.) Chemistry in water reuse. Vol. 1. Ann Arbor Sci. Publ. Ann Arbor, MI.

Lance, N., Jr., and F.H. Schubert. 1981. Regenerative CO_2 collection for spacecraft application. ASME Paper 81-ENAs-28. ASME, San Francisco, CA.

Marshall, R.D., R.R. Woods, T.M. Hallick, and F.H. Schubert. 1977. Development of a three-man preprototype independent air revitalization subsystem. ASME Paper 77-ENAs-31. ASME, San Francisco.

Ming, D.W., and G.E. Lofgren. 1989. Crystal morphologies of minerals formed by hydrothermal alteration of synthetic lunar basaltic glass. *In* L.A. Douglas (ed.) Proceedings of the Soil Micromorphology Workshop. Elsevier Publ., Amsterdam. (In press.)

Reysa, R.P., D.F. Price, T. Olcoh, and J.L. Gaddis. 1983. Hyperfiltration wash water recovery subsystem—design and test results. SAE Paper 831112. SAE, San Francisco.

Roebelen, G.J., Jr., G.G. Dehner, and H.E. Winkler. 1982. Thermoelectric integrated membrane evaporation water recovery technology. SAE Paper 820849. SAE, San Diego.

Samonski, F.H., Jr. 1984. The development status of candidate life support technology for a space station. *In* 35th Congress of the International Astronautical Federation. Lausanne, Switzerland. 7–13 Oct. Am. Inst. of Aeronautics and Astronautics, New York.

Schubert, F.H. 1983. Phase change water recovery techniques: Vapor compression distillation and thermoelectric/membrane concepts. SAE Paper 831122. SAE, San Francisco.

Schubert, F.H., K.A. Burke, and P.D. Quattrone. 1981. Oxygen generation subsystem for spacecraft. ASME Paper 81-ENAs-40. ASME, San Francisco.

Weiblen, P.W., and K. Gordon. 1988. Characteristics of a simulant for lunar surface materials. p. 254. *In* Symposium on lunar bases and space activities in the 21st century. 5–7 Apr. Abstracts, LPI Contribution 652. Lunar and Planetary Inst., Houston.

Williams, R.J., and J.J. Jadwick (ed.) 1980. Handbook of lunar materials. NASA Rep. RP-1057. NASA Sci. and Tech. Info. Office, Washington, DC.

15 Physical and Chemical Considerations for the Development of Lunar-Derived Soils

P. A. Helmke and R. B. Corey

University of Wisconsin
Madison, Wisconsin

The conversion of pristine geological materials to soils that are capable of supporting growth of higher plants offers exciting opportunities and challenges to scientists. For millions of years, soil-forming processes have been occurring on Earth, but analogous processes have not been active on the Moon because of the absence of water and life. Humankind has long been interested in soils and their ability to support plants, ranging from the ancient Egyptian understanding of the role the annual flooding of the Nile River played in ehancing soil fertility to the latest attempts at altering plant-soil interactions by using techniques of genetic engineering. The relative abundance of food and fiber in the industrialized countries is clear evidence of the success we have had in applying the knowledge of soils and plants accumulated during the last several centuries.

Food production under lunar conditions presents unprecedented challenges. Because of the hostile surface environment (e.g., no atmosphere and high levels of radiation), the plant-growing area must be enclosed underground with an atmosphere containing adequate levels of O_2 and CO_2. Suitable temperature and adequate lighting also are essential. Once these environmental requirements have been established, a rooting medium must be developed that maintains aerobic conditions for root respiration and provides water and nutrients at optimum levels for plant growth. Lunar materials are logical sources from which to extract essential plant nutrients, and to possibly use as a rooting medium. Maximum production of plants per unit volume of growth chamber is a likely goal driven by cost considerations. The relative crop productivity in different rooting media and the cost to prepare and maintain each medium must be determined by research.

Never before have we been faced with the prospect of creating an ecosystem from pristine materials. Such a creation will require a major increase in our basic understanding of soil processes. The scarcity of lunar materials for experiments and the time needed for the empirical research approach to work requires the application of basic research to develop lunar soils for crop production. This chapter outlines the physical and chemical needs that a plant-growing system must provide and how the lunar regolith might be used in providing those needs. Areas where more basic research is needed also are indicated.

FUNCTIONS OF ROOTING MEDIA

Types of Rooting Media and Nutrient Delivery Systems

Using porous media such as soils is one way to meet the needs of plants for: root anchorage, water retention and supply, aeration, pH and ionic strength control, and nutrient buffering and supply. These needs can also be met by hydroponic methods collectively known as *nutriculture*. An obvious advantage of using a porous medium, particularly one with ion-exchange properties, is that it buffers many of the critical reactions in plant growth systems.

The types of nutrient delivery systems range from those that use nutrient solutions without solid media (systems A, B, and C below) to those that use solid media components for one or more of the following purposes: root anchorage, water retention, aeration, nutrient buffering, pH buffering, and ionic strength control (systems D, E, F, and G below). These nutrient delivery systems include:

A. Flowing solution cultures (hydroponics) (Hewitt, 1966).
B. Porous membrane systems in which roots contact a porous membrane that separates the root system from a flowing nutrient solution (nutrient film techniques) (Dreschel et al., 1987).
C. Mist solution culture in which a mist of nutrient solution is applied continuously or intermittently to the plant roots (aeroponics) (Hewitt, 1966).
D. Sand or gravel cultures with drop or pulsed nutrient solution additions (Hewitt, 1966).
E. Media providing varying capacities for capillary water retention and for nutrient retention on ion exchange and/or specific adsorption sites (with or without internal ionic strength control) (Hewitt, 1966).
F. Media similar to media E but with more or less continuous, internally generated nutrient release (slow release fertilizer or organic matter decomposition) (Hewitt, 1966).
G. Combinations of B and E or B and F. If the flowing nutrient solution beneath the membrane is maintained at a negative pressure with respect to the atmosphere to which the rooting medium is exposed, the rooting medium will remain unsaturated (well aerated) so long as the negative pressure is sufficient to drain the large pores (Morrow et al., 1988).

Systems A through D require constant or frequent intermittent additions of nutrient solution in excess of plant requirements to assure adequate nutrient availability and aeration and to flush accumulated salts from the system if evapotranspiration is high. Nutrient solutions are usually recirculated and nutrient concentrations, ionic strength, and pH are periodically readjusted. Microbial growth must be controlled. Systems E through G, depending on the media components and amounts of those components used per unit area, could have sufficient nutrient buffering to extend periods between maintenance nutrient additions to weeks, months, or even years. Replacement of water lost by evapotranspiration is essential for all systems.

Plant Growth Media from Lunar Soil

The most expedient approach to producing productive rooting media from lunar material might involve only three major steps: (i) selecting and pretreating the parent material to obtain proper particle size; removing deleterious phases such as free Fe, Ni, Cr, and Co; and leaching excess soluble salts; (ii) an additional incubation period with supervised water and chemical management to promote initial formation of clays and hydrous oxides; and (iii) a final period that evolves into productive agriculture and which includes inoculation with desirable microorganisms and introduction of plants along with management of water and fertility.

If the lunar regolith can be converted to a zeolite-like mineral as Ming (see Chapter 7 in this book) and Ming and Lofgren (1989) suggested, system E could be used with zeolite as the rooting medium. Because of the capillary water retention and the nutrient and pH buffering in this system, it is much more forgiving of minor to even major errors in nutrient addition or water scheduling. Addition of an excess of a sparingly soluble compound such as gypsum ($CaSO_4 \cdot 2H_2O$) would establish the minimum ionic strength in solution and anion exchange resins could be added to fix the ratios of the soluble anions if desired. Otherwise ionic strength and anion ratios would have to be maintained by periodic additions of nutrient solutions. If organic matter or slow release fertilizers are added to the version of system E described above we have system F. If nutrient release rates from these nutrient sources are closely matched to uptake rates, this system has the potential for operating for long periods with only the addition of water which could be supplied automatically. Such systems would be attractive for use on a lunar base because it would permit efficient recycling of plant nutrients by using composted organic wastes.

PHYSICAL CONSIDERATIONS

The size of the particles is the most important physical property of a rooting medium. Water retention and root aeration are critically dependent upon particle size. A dense growth of plants with extensive root systems requires a rooting media that supports a reservoir of plant-available water to meet evapotranspiration demands and supplies O_2 to the roots by either diffusion through air-filled pores or by continuous influx of oxygenated nutrient solution. Except in a flow-through system, a rooting medium saturated with water is not desirable because of the limited influx of O_2. The pores between the particles in a soil must be of sufficient size that the soil solution drains from enough of the pores to provide an unrestricted diffusion path from the atmosphere to the roots.

Water in soil pores is held against the force of gravity by the surface tension of the soil solution. The equation used to calculate the rise of a liquid in a capillary tube can be used to approximate the pore size needed for drainage of a lunar soil.

$$h = \frac{\gamma \, 2\cos\theta}{\rho \, g \, r} \tag{1}$$

where γ is the surface tension of the soil solution, θ is the contact angle, ρ is the density of the soil solution, g is the acceleration of gravity, and r is the radius of the pore (Baver et al., 1972). The use of this equation assumes that the vertical pore spaces are continuous and that the bottom of the soil is saturated with water.

The lunar regolith has been described as having a texture of a silty sand or sandy silt with mean particle sizes between 0.04 and 0.13 mm with few particles larger than 1 mm (Carrier et al., 1973). A soil made from this material would remain saturated and be unsuitable for plant growth under lunar gravity. A lunar soil 10 or 100 cm deep would have to have pore diameters of 1.8 to 0.18 mm, respectively, in order to drain the surface layer. This compares to pore diameters of 0.3 to 0.03 mm for soils experiencing 1 g of gravity. Smaller particles could be used if a partial vacuum were applied to the base of the medium, but the effective continuity of the pore space will limit the thickness of the soil that can be drained in this fashion.

The particle size of the lunar soil will have to be larger than the pore sizes calculated above because the pores between particles are smaller than the size of the particles. For proper drainage to occur at one-sixth gravity, the texture of a lunar soil will have to range from a medium sand to fine gravel (0.25 to >2.0 mm). The content of fine material would have to be low enough as to not block the larger pores in the soil. Consideration of the drainage characteristics implies that the lunar regolith must undergo some processing to develop the proper range of particle sizes.

One way of meeting this requirement is to produce a totally synthetic soil, such as zeolites and other minerals synthesized from lunar materials (see Chapter 7 in this book; Ming & Lofgren, 1989). These materials would need additional processing to produce the proper particle sizes. Another approach is to sinter the lunar regolith and crush and sieve the product to the proper size range. The proposed extraction of He-3 from the lunar regolith will heat the regolith to 600 to 700 °C (Kulcinski, 1989), and the sintering process may be linked to the extraction of lunar volatiles. The refuse from other resource recovery schemes may also be potential sources of material for a solid plant-growth medium. Unlike crushed rocks, sintered materials are likely to retain some porosity and have a relatively large surface area. These properties will enhance the rate of surface dependent reactions such as mineral weathering and nutrient release.

Either the synthetic or sintered material can produce the properly sized material to satisfy the physical constraints, but the chemical behavior of the two systems will be quite different. Each must provide means to supply and buffer the essential nutrients and provide suitable habitat for microorganisms. The following discussion emphasizes the chemistry of a lunar soil developed from the lunar regolith, although many of the principles can be applied to a synthetic soil.

CHEMICAL CONSIDERATIONS

The elements considered as nutrients essential for plant growth and their bioavailable forms are given in Table 15–1. The approximate ranges of concentrations of the major nutrients in the soil solution are given in Table 15–2.

Table 15–1. Essential nutrient elements and common chemical forms absorbed by plant roots from soils (Barber, 1984).

Nutrient	Forms commonly absorbed by plants
Carbon	CO_2
Hydrogen	H_2O
Oxygen	H_2O, O_2
Nitrogen	NO_3^-, NH_4^+
Phosphorus	$H_2PO_4^-$, HPO_4^{2-}
Potassium	K^+
Calcium	Ca^{2+}
Magnesium	Mg^{2+}
Sulfur	SO_4^{2-}
Chlorine	Cl^-
Manganese	Mn^{2+}
Iron	Fe^{2+}
Boron	H_3BO_3
Zinc	Zn^{2+}
Copper	Cu^{2+}
Molybdenum	MoO_4^{2-}

Table 15–2. Concentrations of major nutrients in the soil solution (Barber, 1984).

Nutrient	Amount in solution
	—— μmol L^{-1} ——
NO_3^-	100–20 000
NH_4^+	100–2 000
$H_2PO_4^-$, HPO_4^{2-}	1–20
K^+	100–1 000
Ca^{2+}	100–5 000
Mg^{2+}	100–5 000
SO_4^{2-}	100–10 000

The concentrations of the micronutrients are in the micro- to nanomolar range, except for chlorine which can be much higher (Mortvedt et al., 1972; Lindsay, 1979). Few data are available for many of the trace elements. Elements, especially metals in the solution phase may exist as dissolved species complexed with organic or inorganic ligands (Fig. 15–1), which may then have positive, negative, or neutral charge. The complexed species and the free cations interact with solid phases by precipitation-dissolution or by sorption-desorption (ionic or coordinate covalent bonding to surface sites). Few data exist as to which mechanism predominates for trace elements (Lindsay, 1979; Stites & Helmke, 1989). For most metals, the free hydrated cation is the form thought to be assimilated by plant roots (Chaney et al., 1972; Allen et al., 1980; Checkai et al., 1987).

Central to the thesis of understanding the behavior of elements in the soil-water-plant system, and therefore the chemical function of soils in supporting plants, is knowledge of the factors affecting concentrations of elements dissolved in the solution phase. The following discussion focuses on the equilibria and reaction kinetics that are thought to control the composition of the terrestrial soil-water system. The potential roles of the minerals found in lunar materials will be explained in terms of their terrestrial analogs. Most of the reactions will be discussed in terms of thermodynamic equilibria, but one should be aware that equilibrium is seldom if ever reached in soil-water-plant systems. In fact, it is impossible for a plant to grow in a system at thermodynamic equilibrium. However, the thermodynamic model can be profitably used to predict the approximate composition of reacting systems if the disturbance from equilibrium is acceptably small.

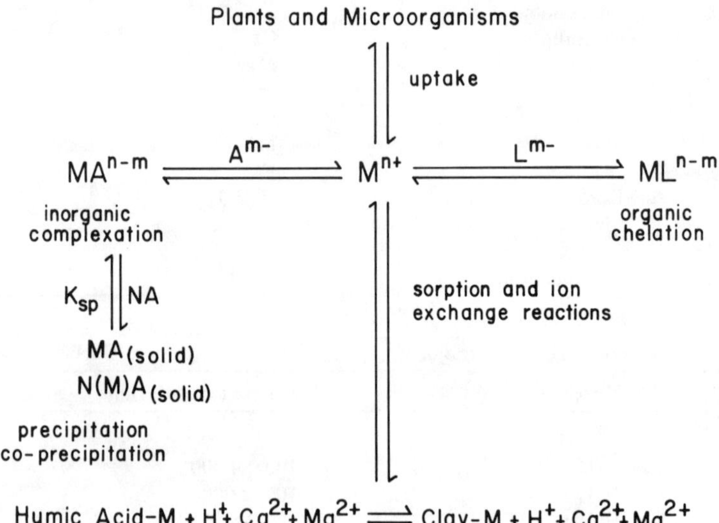

Fig. 15–1. Potential reactions of a dissolved cationic nutrient (M^{n+}) in the soil-water-plant system. L^{m-} and A^{m-} are organic and inorganic species, respectively, that form soluble species with M^{n+}. MA and N(M)A are pure and mixed solid phases, respectively, that are reacting with the soil solution.

Ionic Strength and pH

Plants are tolerant of a relatively narrow range of values of ionic strength and pH. The ionic strength of the solution in typical soils is generally < 0.01 M. The growth and yield of most plants will be restricted if the ionic strength is greater than about 0.04 M because of osmotic effects and/or interference with nutrient uptake. Excessive Na^+ on cation exchange sites caused by a high ratio of Na^+ to divalent cations in solution can disperse clays causing low water permeability and poor aeration in medium- to fine-textured soils. Appropriate values of soil pH for most plants range from about 5 to 8, with 6 to 6.5 being optimum for many species.

The dominant cations in terrestrial soil solutions are Ca^{2+}, Mg^{2+}, Na^+, and K^+. The dominant anions are HCO_3^-, Cl^-, NO_3^-, and SO_4^{2-}. The sources of Na, K, and Ca are primarily feldspars and those for Mg are pyroxenes and olivines. Lunar materials are generally depleted in volatile elements, including Na and K, compared to terrestrial materials. Lunar feldspars are composed dominantly of anorthite plagioclase with 4 to 25% albite and 0 to 3% orthoclase (Papike et al., 1976). They are common in the basalts of the mare and abundant in the lunar highlands. These minerals are unstable under the conditions found for terrestrial soils, and they tend to weather quite rapidly in such environments.

Ionic strength and pH in soils are buffered by precipitation-dissolution phenomena and sorption-desorption reactions (Fig. 15–1). Sorption-desorption reactions are generally fast, minutes to hours, while precipitation-dissolution is generally slow, months to millennia. The ionic strength of the solution in soils supporting growing plants is always in a state of flux. Addition of water to soil dilutes the soil solution while evapotranspiration concentrates it. Drainage removes some of the dissolved elements. All of the chemical elements in the soil-water-plant system are involved in the simultaneous reactions of precipitation-dissolution, sorption-desorption, complexation, and biouptake/release. A state of nonequilibrium is the common condition because of rate-limiting solubility steps, diffusion constraints, and biomediated reactions.

Sufficient basic studies on mineral weathering have been done to indicate that a chemical environment conducive to plant growth using lunar materials under proper management is a reasonable goal. When fresh feldspars are mixed with water, the dissolved concentrations of Na, K, Ca, Al, and Si increase quickly (Carroll, 1970). This is accompanied by an increase in pH because the dissolution of silicates consumes protons and water to produce cations and silicic acid (Bohn et al., 1985). The reaction occurs preferentially at discrete sites of excess surface energy, that is, at crystalline defects (Berner & Holdren, 1979; Holdren & Speyer, 1985). The Al concentration is initially controlled by the solubility of microcrystalline gibbsite, but the solution composition evolves to one controlled by microcrystalline halloysite (Busenberg & Clemency, 1976; Busenberg, 1978). When the solutions contain Ca, Mg, and K, the solution composition evolves along the halloysite-montmorillonite boundary for a considerable length of time, and then moves into the montmorillonite stability field (Busenberg, 1978). Pyroxenes dissolve incongruently in the presence of O_2 to give dissolved Si, Ca, and Mg and solid ferric oxides (Schott et al., 1981; Schott & Berner, 1983). A mixture of the feldspars and pyroxenes should produce the cations and silicic acid needed to form kaolinite and montmorillonite.

The reactions of feldspar and pyroxene described above are apparent within a few months in laboratory dissolution studies (Busenberg & Clemency, 1976; Busenberg, 1978). Compared to the parent material of terrestrial soils, a significant fraction of the lunar materials should be more reactive because of the abundance of very small particles, significant glassy material, and crystal defects from impact events (see Chapter 4 in this book). The initial solution upon wetting of lunar material will probably have a high pH and have relatively high concentrations of alkali and alkaline earth metals and high concentrations of Al and Si because of the hydrolysis of silicates. These conditions may be extreme enough to promote pozzolanic (cementitious reactions). The initial alkalinity will need to be reduced by leaching with water. Thereafter, managing the water budget and acid-base additions to maintain a slightly acid pH, to keep soluble Si levels near saturation with amprphous silica, and to maintain relatively high concentrations of Mg andO_2, should lead to conditions conducive to the formation of montmorillonites and hydrous oxides of Al and Fe. Such studies have not been reported for lunar materials. More research is obviously needed to find the best conditions to promote these weathering reactions. In summary, the initial conditioning process that converts lunar materials to soils must create new solid phases that can buffer the chemical composition of the solution phase.

NUTRIENT ELEMENTS

The ability of soils to supply nutrients to plants and to buffer nutrient concentrations has been and continues to be a major objective of research in soil science (Mortvedt et al., 1972; Barber, 1984; Marschner, 1986). The concentrations of total nutrients in soils are well known and many procedures have been developed in attempts to quantify the bioavailability of nutrients (Page et al., 1982). Most attempts at quantifying the bioavailability of nutrients in soil involve empirical correlations between the concentration of the selected nutrient in the soil as measured by a selective extraction procedure and the corresponding concentration of the nutrient in various plants grown on the soil (each state has their set of analytical procedures and correlations for their important crops and soil types). This approach has been enormously successful in modern agriculture, but it has not been equally successful in increasing our understanding of the mechanisms that control the bioavailability of nutrients and their uptake by plants (Barber, 1984). Some elements such as N, are not amendable to a quantitative approach. Their bioavailability is controlled by microorganisms and so many other soil reactions, that all of the variables affecting such elements have not been successfully integrated into an analytical procedure.

It is natural to assess the potential fertility of the lunar regolith by comparing the concentrations of elements in lunar material to terrestrial soils, but this comparison demands caution. Within a single soil, different plants vary in their ability to take up nutrients, and slight variabilities in the physical and chemical properties of the soil can also have an effect on bioavailability (Barber, 1984). Although the concentrations of nutrients in terrestrial soils are known, it is not necessarily known whether these are the optimum levels for all crops.

Potassium

The concentration of K in terrestrial soils ranges from 0.4 to 30 g kg^{-1} with a selected average concentration of 8.3 g kg^{-1} (Lindsay, 1979). This compares to 0.1 to 3.5 g kg^{-1} for lunar basalts, with most lunar crystalline material containing 0.2 to 0.8 g kg^{-1} (Williams & Jadwick, 1980). Micas, feldspars (orthoclase and microcline), and illite are the dominant K-bearing minerals in terrestrial soils, but of these, only the feldspars are found in lunar materials. Significant amounts of K are found in KREEP (K, Rare-Earth Element, P) components of lunar materials, but this is a relatively minor solid phase (Hubbard & Gast, 1971).

The amount of K abosrbed by crops is often greater than any other mineral nutrient (Barber, 1984). Nitrogen uptake may exceed K for some crops. The form of K that is usually assumed to be absorbed is K$^+$. Considerable amounts of K are recycled back to the soil depending upon how much of the crop is harvested. The concentration of K in soil solutions is buffered by exchange reactions with Ca^{2+}, Mg^{2+}, and other cations. The lunar soils will not contain micas and muscovite, which is a natural but slowly released source of K in terrestrial soils. The high nutrient demand for K indicates that dissolution of the lunar feldspars and glassy phases will supply insufficient K for most crops. This will require chemical extraction of K from lunar material for use as a fertilizer.

Calcium

Calcium requirements for plants are usually less than requirements for K (Wallace et al., 1966). Calcium appears to have its greatest significance in maintaining conditions for the appropriate balance of other nutrient concentrations within the plant. It is an important cation in soil exchange reactions and also for the function of the plasma membrane during ion uptake by plants.

Calcium is abundant in lunar materials (7.9–10.9% Ca) as a constituent of anorthite and pyroxene (see Chapter 4 in this book). Dissolution of lunar materials should provide sufficient Ca in lunar soils for plant growth. Additional Ca may occasionally be added to maintain a desirable mix of exchangeable cations and, as a constituent of liming materials, to maintain soil pH.

Phosphorus

The concentrations of P in terrestrial soil solutions are low compared to those of the other major plant nutrients such as N, K, Ca, and Mg. Controversy exists over the amount of P that may be adsorbed on the surfaces of such soil constituents as Fe and Al oxides vs. the amount that is precipitated as discrete mineral forms, such as the phosphates of Ca, Al, or Fe (Lindsay, 1979). More than one-half of the total P in the surface horizon of some terrestrial soils may be associated with the organic matter content of the soil. A large fraction of the P in terrestrial soils that is in mineral form is not readily available for adsorption by plants.

Concentrations of P in the lunar regolith range from 0.01 to 0.25% (Rose et al., 1975), which are similar to slightly less than those found for terrestrial soils (0.02–0.5%) (Lindsay, 1979). The dominant mineral forms of P in lunar materials are the calcium phosphates such as apatite and whitlocktite, which occur as trace minerals. These have a low solubility under normal soil conditions, and they are not expected to supply significant plant-available P in the lunar soils. Additional P will have to be extracted from lunar materials, such as KREEP-rich components, and used as a fertilizer to maintain adequate levels of soil P. The management of P may be initially difficult in the lunar soils because of the low content of organic matter and the potential abundance of freshly precipitated oxides of Fe and Al, which react with phosphate to form insoluble compounds (Lindsay, 1979). Research is clearly needed on how P will behave in these newly formed soils.

Magnesium

In terrestrial soils, Mg is a constituent of many minerals, and it occurs in plant available forms on the cation-exchange complex and as Mg^{2+} ion in the soil solution. Magnesium is an important constituent of several secondary minerals significant to terrestrial soils, such as vermiculite, illite, smectite, and chlorite. Magnesium is held less tightly than Ca by the exchange complex of soils. The ratio of concentrations of Ca to Mg on the exchange sites is usually about twice that of the soil solution. Increasing K or Ca concentrations in the soil solution reduces the plant uptake of Mg (Barber, 1984). Like Ca, little Mg is associated with the organic matter content of soils.

The important sources of Mg in lunar materials are pyroxenes and olivine. These minerals are unstable in soil environments and some of the Mg released by their dissolution reacts with Al and Si to form clay minerals, such as smectites. Plants take up roughly similar quantities of Mg and Ca. Because Mg is held less tightly by the exchange sites in the soil, however, less Mg is needed in the soil to maintain equal rates of supply (Barber, 1984). The lunar regolith contains up to 11% MgO (see Chapter 4 in this book), which should result in adequate supplies of plant available Mg in the lunar soils.

Sulfur

Concentrations of S in terrestrial soils vary widely (30–10 000 mg kg^{-1}) (Lindsay, 1979). In most soils, almost all of the total S is chemically combined with soil organic matter. It also occurs as sulfate in solution, sulfate adsorbed on minerals, and sulfate- and sulfide-bearing minerals. Metal sulfides are unstable in well-aerated soils because O_2 oxidizes the sulfide to sulfate. Terrestrial soils with a low concentration of Fe oxides have a small capacity to adsorb sulfate. The amount of adsorption increases as the concentrations of allophane or amorphous aluminum and iron oxides increase and pH decreases. High concentrations of specifically adsorbed sulfate can occur because of the pH-dependent charge of Fe and Al oxides (Bohn et al., 1985). Decomposition of soil organic matter releases S in plant-available forms. Sulfur-oxidizing microorganisms such as *Thiobacillus thiooxidans* convert reduced forms of S to sulfate, but these will not be in the lunar soil unless they are introduced. Important additional sources of S in terrestrial systems are aerosolic inputs. Plants can take up SO_2 directly through their leaves, and rainfall can add significant amounts of soluble S compounds to the soil.

Sulfur in lunar regolith materials occurs almost entirely as the iron sulfide, troilite, although trace amounts of mackinawite and chalcopyrite, and other minerals have been reported (see Chapter 4 in this book). Concentrations of S range from 300 to 1200 mg kg^{-1} in the lunar regolith, with somewhat higher values in basalts (Kaplan & Petrowski, 1971; Gibson & Moore, 1973; Petrowski et al., 1974). Troilite is expected to alter rapidly to iron oxides and soluble sulfides when the lunar materials are exposed to water. The sulfides will be oxidized to sulfates in the presence of O_2, but the reaction is slow in the absence of microorganisms and at high values of pH (Singer & Stumm, 1970; Goldhaber, 1983). The initial sulfate in lunar soils will probably be highly leachable if the pH is high.

Maintaining optimum amounts of S in lunar soils will require intensive management. This is expecially so during the initial development and cropping of the lunar soils. The expected accumulation of organic matter in the soils from plant and animal wastes after a lunar habitat is functional will result in continuous recycling or organic S to plant-available forms. However, secondary sources of S will need to be developed. The status of plant-available S will be most easily managed if the release of S from the fertilizer is controlled.

Chlorine

The chlorine requirement of crops is small, even though Cl may be one of the most abundant anions in plants. Its role in plants is not well established but its functions include those of a regulator of osmotic pressure and cation balance. Chloride-sensitive crops grown in saline soils show leaf injury symptoms when the leaves accumulate between 0.3 to 0.5% Cl on a dry-weight basis. Of all the plant essential elements, it is unique in terrestrial systems in that it is rarely needed as a fertilizer. Its concentration in terrestrial soils ranges from 20 to 900 mg kg^{-1} with an average value of 100 mg kg^{-1} (Lindsay, 1979). Precipitation and KCl fertilizers are important sources of Cl to terrestrial soils. The chloride ion is weakly adsorbed to soils and it is readily leached.

The concentration of Cl in the lunar regolith ranges from 15 to 35 mg kg^{-1} (Jovanovic & Reed, 1974). Lunar igneous rocks have even lower concentrations, 1 to 15 mg kg^{-1} (Jovanovic & Reed, 1974), with a few samples having much higher values, such as anorthosite samples 66095 and 61016 (200–300 mg kg^{-1}) (Jovanovic & Reed, 1973). Up to two-thirds of the Cl in the lunar regolith materials is soluble in a 10-min exposure to hot water (Reed & Jovanovic, 1971; Jovanovic & Reed, 1974). Chlorine is strongly associated with a P-rich phase in lunar materials (Jovanovic & Reed, 1973). There should be no need for Cl to be added to the lunar soils as the plant requirement for this nutrient is small. Continuous recycling of waste and leachate water may require that the concentrations of Cl, Na, Ca, and sulfate be reduced in the irrigation water.

Manganese

The concentration of Mn in terrestrial soils is highly variable, with an average concentration of 600 mg kg^{-1} (Lindsay, 1979). The dominant mineral forms are pyrolsite (MnO_2) manganite (MnOOH), and hausmanite (Mn_3O_4) (McKenzie, 1977). The dominant form in solution is in the plus two oxidation state, where it can form complexes with organic and inorganic ligands, such as sulfate. The concentrations of soluble Mn in soil is highly variable and dependent on the soil pH and redox status. The solubility of Mn is highest in reduced soils. The Mn^{2+} ion is the form available to plants. Soil solution concentrations of 1.4 μmol L^{-1} are sufficient to supply all of the Mn to plants if the mass flow of soil water is the sole mechanism of uptake (Barber, 1984). Aerated soils in equilibrium with oxides of Mn at a pH lower than 6.5 have concentrations higher than this. Manganese is seldom deficient in most soil-plant systems unless the pH is alkaline.

The concentration of Mn in the lunar regolith ranges from 200 to 2800 mg kg^{-1} (Laul & Schmitt, 1973; Haskin et al., 1973; Helmke et al., 1973). The concentration of Mn in lunar materials is highly correlated with that of Fe, the ratio of Fe to Mn being 81, with the ratio nearly constant over a factor of 30 in Fe and Mn abundances (Laul & Schmitt, 1973). Most of the Mn in lunar materials is found in pyroxenes, olivines, ilmenite, and spinels instead of discrete Mn minerals. Manganese should not pose any special problems in a lunar soil-plant system. The susceptibility of pyroxenes to weathering plus the ability of plant roots to modify the soil to enhance Mn availability (Godo & Reisenauer, 1980) will probably supply sufficient Mn for plant growth. If not, the amounts of Mn required to correct deficiencies are small.

Iron

Iron is present in terrestrial soils at concentrations higher than that of any other nutrient. In the trivalent form it occurs as oxides and hydroxides, while in the divalent form it occurs as oxides and primary silicates, pyroxene, and olivine. Most of the dissolved Fe in aerated soils is Fe^{3+}, which also forms complexes with organic materials. The solubility of Fe is thought to be controlled by a form of ferric hydroxide that is slightly less soluble than freshly precipitated $Fe(OH)_3$ (Lindsay, 1979). The concentration of soluble Fe is much higher in reduced soils, where most of the dissolved Fe is divalent. In the lunar regolith, the average concentration of FeO ranges from 5.9% for the terra (lunar highlands) to 14.1% for the mare (Turkevich, 1973). Most of the Fe is divalent and in primary rock-forming minerals and glass. A small fraction is present as native Fe, usually along with Co, Cr, and Ni. The initial conditioning process of converting the lunar regolith to soils will consume O_2 as the native Fe and some of the divalent Fe is oxidized to Fe^{3+} and precipitated as $Fe(OH)_3$. Amorphous ferric hydroxide is slowly converted to less soluble crystalline forms such as maghemite, hematite, lepidocrocite, and goethite, but the reactions are so slow in soils that the concentration of Fe^{3+} is usually higher than those in equilibrium with any of these materials (Lindsay, 1979).

Small quantities of Fe are needed for plant growth, but Fe deficiencies occur in spite of its abundance in soils because so little of the element is in an available form. In aerated soils, the concentration of disolved Fe^{3+} is about 10^{-16} m at pH 6.5 (Lindsay, 1979). Research indicates that Fe^{3+} is reduced to Fe^{2+} before it is absorbed by the plant. Plants have been classified as either Fe efficient or Fe inefficient; plants classed as Fe efficient respond to Fe deficiency by releasing H^+ ions, reductants, and/or complexing agents into the rhizosphere, especially at the root tip (Bienfait, 1988; Chaney, 1988). Iron deficiencies are most likely to occur in alkaline soils (Murphy & Walsh, 1972). Deficiencies can be corrected by judicious selection of cultivars or foliar application of iron. Addition of ferrated organics to soil is also effective (Norvell, 1972).

Boron

The concentration of B in terrestrial soils ranges from 2 to 100 mg kg^{-1}, with an average concentration of 10 mg kg^{-1} (Lindsay, 1979). The highest concentrations are found in soils derived from sedimentary rocks of marine origin, while the lowest concentrations are found in soils from basic rocks. The most common B mineral is tourmaline. Boron in solution usually occurs as the undissociated acid H_3BO_3. It is weakly adsorbed by organic matter. Boron-deficient soils probably have <5 μmol L^{-1} of B in the soil solution (Barber, 1984). Unlike ionic species, B appears to be assimilated by plant roots by passive uptake. Boron in soil solution apparently diffuses into the root or moves with the water flow until the concentrations inside and outside the root are in equilibrium. Toxic levels of B are absorbed by plants when the soil solution concentrations are high.

Few measurements of B have been made in lunar material. The data available indicate that it is present at low concentrations, from <1 to 12 mg kg^{-1} (Wanke et al., 1970; Morrison et al., 1973). The highest values occur in KREEP-rich basalts (Morrison et al., 1973). For a lunar soil containing 1 mg kg^{-1} of B at a 20% moisture content, about 1% of the total B in the soil would have to be in the soil solution to prevent B-deficiency symptoms. Research is needed to determine the solubility of B in lunar materials, and its leaching and adsorption characteristics.

Molybdenum

The concentration of Mo in soils is lower than that of most other micronutrients. It ranges from 0.2 to 5 mg kg^{-1} with an average value of 2 mg kg^{-1} (Lindsay, 1979). The concentration of Mo in soil organic matter is usually several times higher than that of the mineral fraction. No Mo minerals have been identified that control the solution levels of Mo. The dominant form of Mo in solution is MoO_4^{2-}. Its concentration increases as the pH of the soil increases, unlike most other nutrients. The uptake of Mo by plants decreases with increasing pH, but the increase in solution Mo with increasing pH more than offsets this effect, so that the concentration of Mo in plants increases with increasing pH (Barber, 1984). Legumes require more Mo than most other plants because the N_2-fixing bacteria associated with legumes require Mo (Marschner, 1986). The few data available for Mo in lunar materials indicate that it is present in low concentrations, ranging from <0.03 to 0.7 mg kg^{-1} (Morrison et al., 1970; Turekain Kharkar, 1970).

Zinc

The mean concentration of Zn in soils is 50 mg kg^{-1} (Lindsay, 1979). Its concentration ranges from 300 mg kg^{-1} in soils with high clay and organic matter contents to low values for sandy soils. The mechanisms that control the concentration of soluble Zn in soils are not known but probably involve some combination of dissolution and sorption phenomena (Ellis & Knezek, 1972; Lindsay, 1979). Zinc is adsorbed by hydrous oxides, especially those of Fe, and it forms complexes with soil organic matter, some of which are soluble. The Zn^{2+} ion is the form available to plants. Typical values for soluble Zn in soil solution range from 0.4 to 0.8 μmol L^{-1} (Barber, 1984).

The concentrations of Zn in lunar materials range from <1 to over 90 mg kg^{-1} (Helmke et al., 1972; Haskin et al., 1973; Helmke et al., 1973; Baedecker et al., 1973). The concentration of Zn is generally higher in the lunar regolith than in the crystalline rocks. Sphalerite (ZnS) is found in association with troilite (FeS) in some samples (Taylor et al., 1973). Sulfide minerals should oxidize readily in contact with O_2 and this may be an initial source of Zn. There is sufficient Zn in the lunar regolith to supply Zn, but research is needed to determine if it is released in quantities and rates sufficient for plant growth.

Copper

The concentration of Cu in terrestrial soils ranges from 2 to 100 mg kg^{-1} (Lindsay, 1979). Like the case for Zn, the mechanisms that control the concentrations of soluble Cu in soils are not known. Copper is readily complexed by soil organic matter, and the soluble portion of these complexes comprises most of the Cu in the soil solution. The concentration of total soluble Cu and the percent complexed Cu decrease as the pH of the soil increases. The Cu^{2+} ion is the form available to plants. Plant uptake of Cu is lower than that of most other micronutrients (Barber, 1984), which is probably why Cu deficiencies are uncommon except in organic soils.

The concentration of Cu in lunar materials ranges from <0.1 to 18 mg kg^{-1} (Haskin et al., 1970; Cuttitta et al., 1973; Brunfelt et al., 1974). The higher concentrations tend to be in mare material. The amounts of Cu in the lunar material are sufficient for plant growth, but nothing is known about its release and availability in lunar soils.

Carbon

Carbon as CO_2 is not purposely managed in terrestrial soils, unless one accepts the premise that the increase in atmospheric CO_2 from combustion of fossil fuels is global management. The concentration in soils ranges up to 50 times that of the open atmosphere (Bolt & Bruggenwert, 1976). Carbon in the form of organic matter is an important component of typical soils, where it serves as a nutrient source for microorganisms and plays an important part in the recycling of nutrients and buffering their concentrations.

Carbon in lunar material is present as carbides, carbon monoxide and dioxide, and methane. Much of the C is the result of entrapment of the solar wind. Its concentration in the lunar regolith ranges from 14 to over 200 mg kg^{-1} (Moore et al., 1971; DesMarais et al., 1975). It is obvious that C must be significantly enriched to support lunar agriculture. One source of C is from the by-product lunar volatiles collected from the proposed extraction of He-3 from the lunar regolith. He-3 is such an attractive fuel for fusion energy that there are serious proposals to mine the trace amounts of He-3 trapped in the lunar regolith. The amounts of regolith that must be processed are large but not prohibitive, about 10^7 to 10^8 t of regolith per tonne of He-3 (Kulcinski, 1989). Each tonne of He-3 extracted will give sufficient CO_2 to support 22 000 people per year, assuming a 10% loss in the recycling through a life support system and a 1% leakage per day of the atmosphere to space (Kulcinski, 1989). Enough C will be available under this scenario that selected decisions can be made about what kinds and forms of waste are to be recycled in the soil-plant system. The methods of recycling wastes and the role of organic matter in lunar soils will require extensive research.

Nitrogen

Most of the N in terrestrial soils is present in the organic fraction. Atmospheric N can only be used by leguminous plants, which have symbiotic N_2-fixing microorganisms on their roots. Inorganic N occurs mostly as NH_4^+ and NO_3^-, with nitrate generally being the dominant form (Barber, 1984). Both forms are taken up by plants. Nitrate is readily leached from soils and the ammonium cation can be lost by volatilization of ammonia, especially at higher values of pH. Nitrogen fertilizers are needed in all intensively cropped systems.

The concentration of N in the lunar regolith ranges from 20 to 120 mg kg^{-1} (Moore et al., 1971; Muller, 1974). Additional amounts of N will be required for plant growth in lunar material. One source of N is as a by-product of He-3 extraction as in the case for C. Each tonne of He-3 recovered is estimated to provide sufficient N for 1400 people per year, again assuming a 10% loss in recycling through a life-support system and a 1% leakage per day of a N_2–O_2 atmosphere to space (Kulcinski, 1989). Organic matter from recycled wastes will help supply and buffer N in the lunar soil.

Toxic Elements

The lunar regolith does not contain any materials that pose potential toxicity to plants except the native Co, Cr, and Ni associated with the free Fe. This phase is expected to be oxidized quickly and the initial levels and availability of Co and Ni in a new lunar soil may result in phytotoxicity. The total concentrations in the lunar regolith (24–58 mg kg^{-1} Co and 185 to 370 mg kg^{-1} Ni) (Cuttitta et al., 1973; see Chapter 4 in this book) are within the range found for terrestrial soils, but the normal phases that fix these elements in terrestrial soils, hydrous oxides and organic matter, may be insufficient in the intial stages of developing a lunar soil. This Fe-rich phase can be removed magnetically from the lunar regolith if research shows that these elements are a problem. Cadmium is present at concentrations ranging from 0.03 to 0.3 mg kg^{-1} (Baedecker et al., 1973). This level will not pose any phytotoxicity problems, but research should be done to demonstrate that the human dose is within allowable limits. The levels of other potentially toxic elements are low enough they are unlikely to pose any problems at normal values of soil pH.

FUTURE RESEARCH

The amounts of lunar materials available for research are too small and precious to conduct the traditional chemical extraction, glasshouse, and field experiments commonly used to assess the fertility status of soils. We must improve our present base of knowledge of how the soil-water-air-plant system operates so that we can devise more sensitive and specific assessment procedures. The kinds of basic research needed have not been emphasized by the federal agencies that fund agricultural research, as evidenced by the absence of any federal competitive grants program for soils research. Such research, while imperative for the development of lunar soils, would also have tremendous benefits for terrestrial agriculture, improving the efficiency of scarce fertilizers and minimizing the contamination of groundwater and other parts of the environment.

Research Using Lunar Simulants

It will be necessary to synthetically produce materials similar to the lunar regolith from terrestrial materials because only small amounts of lunar materials are available for research. Research using a lunar simulant would serve as a stepping stone to confirmatory research using samples of the lunar regolith. A lunar simulant will allow techniques and conceptual understandings to be tested and improved with minimal consumption of lunar materials.

The fidelity of the simulant will depend upon the experimental design. Many factors will have to be considered, such as: the elemental composition of glass, lithic fragments, and agglutinates; mineralogy of the lithic fragments; particle-size distribution and porosity; oxidation state of Fe, Mn, etc.; implanted gases; and radiation and shock damage. Reducing conditions may be necessary during production of the simulant.

Optimal Physical Properties of Lunar Material for Plant Growth

The movement of water and gases through lunar soils in one sixth gravity can be estimated by scaling terrestrial data. Models of water movement through soils made from lunar simulant can be tested in research aircraft, where one sixth gravity can be maintained for a short time. Major areas of research include: (i) procedures to produce soils having optimum particle size; (ii) the effects of particle size and porosity on drainage; (iii) the potential for synthesis of soil materials or additives that have a contact wetting angle larger than zero, and if so what are their effects on chemical and biological reactions; (iv) the effectiveness of encouraging soil drainage by maintaining a partial vacuum at the bottom of lunar soils; (v) the development of models to predict rooting behavior of plants; and (vi) the functioning of microorganisms in lunar soils as related to waste recycling, plant nutrient budgets, and symbiotic relationships with plants.

Understanding the Geochemistry of the Lunar Regolith with Water

The reaction of the lunar soil with water will have to be understood to produce a highly productive soil for plant growth (see Chapter 16 in this book). The main areas of research will be: the reactions of the lunar regolith with water to produce a lunar agricultural soil; the reactions of the derived soil with nutrients and potential toxicants; and the interactions of the derived soil with microorganisms and plants.

The production of a lunar agricultural soil must include research on: the selection of which components of the lunar regolith are desirable to use for a lunar soil; the dissolution of lunar regolith and the formation of secondary minerals, including kinetic studies and the effects of elevated temperature and pressure; the stability of primary and secondard mineral phases or synthetic soil phases during projected agricultural practices; and the production of solid phases with suitable cation and anion exchange properties.

Maintenance of a soil solution that is optimal for plant growth will require research on: the nutrient release and buffering characteristics of the lunar soil; the reactions that control ionic strength, pH, and redox status of the soil solution; the development of analytical procedures to monitor the fertility status of the lunar soil; and an assessment of the potential toxicity of selected elements in the lunar regolith.

We advocate that future research on the above topics should use species-specific techniques and experimental designs that give information on reaction pathways and rates and identify the exact nature of the solid and liquid phases involved.

REFERENCES

Allen, H.E., R.H. Hall, and T.D. Brisbin. 1980. Metal speciation: effects on aquatic toxicity. Environ. Sci. Technol. 14:441–442.

Baedecker, P.A., C.-L. Chou, E.B. Grudewicz, and J.T. Wasson. 1973. Volatile and siderophilic trace elements in Apollo 15 samples: Geochemical implications and characterization of the long-lived and short-lived extralunar materials. Proc. Lunar Sci. Conf., 4th, 1973 2:1177–1195.

Barber, S.A. 1984. Soil nutrient bioavailability. Wiley-Interscience Publ., New York.

Baver, L.D., W.H. Gardner, and W.R. Gardner. 1972. Soil physics. 4th ed. John Wiley and Sons, New York.

Berner, R.A., and G.R. Holden, Jr. 1979. Mechanism of feldspar weathering. II. Observations of feldspars from soils. Geochim. Cosmochim. Acta 43:1173–1186.

Bienfait, H.F. 1988. Mechanisms in Fe-efficiency reactions of higher plants. J. Plant Nutr. 11:605–629.

Bohn, H.L., B.L. McNeal, and G.A. O'Connor. 1985. Soil chemistry. 2nd ed. John Wiley and Sons, New York.

Bolt, G.H., and M.G.M. Bruggenwert (ed.). 1976. Soil chemistry: A. Basic elements. Elsevier, Amsterdam.

Brunfelt, A.O., K.S. Heier, B. Nilssen, E. Steinnes, and B. Sundvoll. 1974. Elemental composition of Apollo 17 fines and rocks. Proc. 5th Lunar Sci. Conf., 5th, 1974 2:981–990.

Busenberg, E. 1978. The products of the interaction of feldspars with aqueous solutions at 25 °C. Geochim. Cosmochim. Acta 42:1679–1686.

Busenberg, E., and C.V. Clemency. 1976. The dissolution kinetics of feldspars at 25 °C and 1 atm CO_2 partial pressure. Geochim. Cosmochim. Acta 40:41–49.

Carrier, W.D. III, J.K. Mitchell, and A. Mahmoud. 1973. The nature of lunar soil. J. Soil Mech. Found. 99:813–832.

Carroll, D. 1970. Rock weathering. Plenum Press, New York.

Chaney, R.L. 1988. Recent progress and needed research in plant Fe nutrition. J. Plant Nutr. 11:1589–1603.

Chaney, R.L., J.C. Brown, and L.O. Tiffin. 1972. Obligatory reduction of ferric chelates in iron uptake of soybeans. Plant Physiol. 50:208–213.

Checkai, R.T., R.B. Corey, and P.A. Helmke. 1987. Effects of ionic and complexed metal concentrations on plant uptake of cadmium and micronutrient metals from solution. Plant Soil 99:335–345.

Cuttitta, F., H.J. Rose, Jr., C.S. Annell, M.K. Carron, R.P. Christian, D.T. Ligon, Jr., E.J. Dwornik, T.L. Wright, and L.P. Greenland. 1973. Chemistry of twenty-one igneous rocks and soils returned by the Apollo 15 mission. Proc. Lunar Sci. Conf., 4th, 1973 2:1081–1096.

DesMarais, D.J., A. Basu, J.M. Hayes, and W.G. Meinschein. 1975. Evolution of carbon isotopes, agglutinates, and the lunar regolith. Proc. 6th Lunar Sci. Conf., 6th, 1975 3:2353–2373.

Dreschel, T.W., R.P. Prince, C.R. Hinkle, and W.M. Knott. 1987. Porous membrane utilization in plant nutrient delivery. Paper 87-4025. Am. Soc. Agric. Engr. Summer Mtg., Baltimore. ASAE, St. Joseph, MI.

Ellis, B.G., and B.D. Knezek. 1972. Adsorption reactions of micronutrients in soils. p. 59–77. *In* J.J. Mortvedt et al., (ed.) Micronutrients in agriculture. SSSA, Madison, WI.

Gibson, E.K., Jr., and G.W. Moore. 1973. Carbon and sulfur distributions and abundances in lunar fines. Proc. Lunar Sci. Conf., 4th, 1973 2:1577–1586.

Godo, G.H., and H.M. Reisenauer. 1980. Plant effects on soil manganese availability. Soil Sci. Soc. Am. J. 44:993–995.

Goldhaber, M.B. 1983. Experimental study of metastable sulfur oxyanion during pyrite oxidation at pH 6–9 and 30°C. C. Am. J. Sci. 283:193–217.

Haskin, L.A., R.O. Allen, P.A. Helmke, T.P. Paster, and M.R. Anderson, R.L. Korotev, and K.A. Zweifel. 1970. Rare earths and other trace elements in Apollo 11 lunar samples. Proc. Apollo 11 Lunar Sci. Conf. 2:1213–1231.

Haskin, L.A., P.A. Helmke, D.P. Blanchard, J.W. Jacobs, and K. Telander. 1973. Major and trace element abundances in samples from the lunar highlands. Proc. Lunar Sci. Conf., 4th, 1973 2:1275–1296.

Helmke, P.A., L.A. Haskin. R.L. Korotev, and K.E. Ziege. 1972. Rare earths and other trace elements in Apollo 14 samples. Proc. Lunar Sci. Conf., 3rd, 1972 2:1275–1292.

Helmke, P.A., D.P. Blanchard, J.W. Jacobs, and L.A. Haskin. 1973. Rare earths, other trace elements and iron in Luna 20 samples. Geochim. Cosmochim. Acta 37:869–874.

Hewitt, E.J. 1966. Sand and water culture methods used in the study of plant nutrition. Commonw. Bur. Hort. Plant Crops Tech. Commun. 22 (Revised). 2nd ed. Eastern Press Ltd., London.

Holdren, G.R., Jr., and P.A. Speyer. 1985. Reaction rate-surface area relationships during the early stages of weathering—I. Initial observations. Geochim. Cosmochim. Acta 49:675–681.

Hubbard, N.J., and P.W. Gast. 1971. Chemical composition and origin of nonmare lunar basalts. Proc. Lunar Sci. Conf., 2nd, 1971 2:999–1020.

Jovanovic, S., and G.W. Reed, Jr. 1973. Volatile trace elements and the characterization of the Cayley Formation and the primitive lunar crust. Proc. Lunar Sci. Conf., 4th, 1973 2:1313–1324.

Jovanovic, S., and G.W. Reed, Jr. 1974. Labile and nonlabile element relationships among Apollo 17 samples. Proc. Lunar Sci. Conf., 5th, 1974 2:1685–1701.

Kaplan, I.R., and C. Petrowski. 1971. Carbon and sulfur isotope studies on Apollo 12 lunar samples. Proc. Lunar Sci. Conf., 2nd, 1971 2:1397–1406.

Kulcinski, G.L., E.N. Cameron, J.F. Santarius, I.N. Sviatoslavsky, L.J. Wittenberg, and H. Schmidt. 1989. Fusion energy from the Moon for the 21st century. *In* W.W. Mendell (ed.) Proc. 2nd Conf. on Lunar Bases and Space Activities of the 21st Century. Lunar and Planetary Inst., Houston. (In press.)

Laul, J.C., and R.A. Schmitt. 1973. Chemical composition of Apollo 15, 16, and 17 samples. Proc. Lunar Sci. Conf., 4th, 1973 2:1349–1367.

Lindsay, W.L. 1979. Chemical equilibria in soils. John Wiley and Sons, New York.

Marschner, H. 1986. Mineral nutrition of higher plants. Academic Press, London.

McKenzie, R.M. 1977. Manganese oxide and hydroxides. p. 181–194. *In* J.B. Dixon and S.B. Weed (ed.) Minerals in soil environments. SSSA, Madison, WI.

Ming, D.W., and G.E. Lofgren. 1989. Crystal morphologies of minerals formed by hydrothermal alterations of synthetic lunar basaltic glass. *In* L.A. Douglas (ed.) Proc. of the Soil Micromorphology Workshop. Elsevier, Amsterdam. (In press.)

Moore, C.B., C.F. Lewis, J.W. Larimer, F.M. Delles, R.C. Gooley, W. Nichiporuk, and E.K. Gibson, Jr. 1971. Total carbon and nitrogen abundances in Apollo 12 lunar samples. Proc. Lunar Sci. Conf., 2nd, 1971 2:1343–1350.

Morrison, G.H., J.T. Gererd, A.T. Kashuba, E.V. Gangadharam, A.M. Rothenberg, N.M. Potter, and G.B. Miller. 1970. Multielement analysis of lunar soil and rocks. Science 30:505–507.

Morrison, G.H., R.A. Nadkarni, J. Jaworski, R.I. Botto, and J.R. Roth. 1973. Elemental abundances of Apollo 16 samples. Proc. Lunar Sci. Conf., 4th, 1973 2:1399–1405.

Morrow, R.C., R.J. Bula, R.B. Corey, T.W. Tibbitts, and E.E. Richards. 1988. A porous tube nutrient delivery system for plant growth in space p. 32. *In* Am. Soc. for Gravitational and Spcae Biology, 4th Annu. Mtg., Programs & Abstracts. 20–23 Oct. Am. Soc. for Gravitational and Space Biology, Washington, DC.

Mortvedt, J.J., P.M. Giordano, and W.L. Lindsay. 1972. Micronutrients in agriculture. SSSA, Madison, WI.

Muller, O. 1974. Solar wind nitrogen and indigenous nitrogen in Apollo 17 lunar samples. Proc. Lunar Sci. Conf., 5th, 1974 2:1907–1918.

Murphy, L.S., and L.M. Walsh. 1972. Correction of micronutrient deficiencies with fertilizers. p. 347–388. *In* J.J. Mortvedt et al. (ed.) Micronutrients in agriculture. SSSA, Madison, WI.

Norvell, W.A. 1972. Equilibria of metal chelates in soil solution. p. 115–138. *In* J.J. Mortvedt et al. (ed.) Micronutrients in agriculture. SSSA, Madison, WI.

Page, A.L., R.H. Miller, and D.R. Keeney (ed.). 1982. Methods of soil analysis. Part 2. 2nd ed. Agronomy 9.

Papike, J.J., F.N. Hodges, A.E. Bence, M. Cameron, and J.M. Rhodes. 1976. Mare basalts: crystal chemistry, mineralogy and petrology. Rev. Geophys. Space Physics 14:475–540.

Petrowski, C., J.F. Kerridge, and I.R. Kaplan. 1974. Light element geochemistry of the Apollo 7 site. Proc. Lunar Sci. Conf., 5th, 1974 2:1939–1948.

Reed, G.W., and S. Jovanovic. 1971. The halogens and other trace elements in Apollo 12 samples and the implications of halides, platinum metals, and mercury on surfaces. Proc. Lunar Sci. Conf., 2nd, 1971 2:1261–1276.

Rose, H.J., Jr., P.A. Baedecker, S. Berman, R.P. Christian, E.J. Dwornik, R.P. Finkelman, and M.M. Schnepfe. 1975. Chemical composition of rocks and soils returned by the Apollo 15, 16, and 17 missions. Proc. Lunar Sci. Conf., 6th, 1975 3:1363–1373.

Schott, J., and R.A. Berner. 1983. X-ray photoelectrol studies of the mechanism of iron silicate dissolution during weathering. Geochim. Cosmochim. Acta 47:2233–2240.

Schott, J., R.A. Berner, and E.L. Sjoberg. 1981. Mechanism of pyroxene and amphibole weathering—I. Experimental studies of iron-free minerals. Geochim. Cosmochim. Acta 45:2123–2135.

Singer, P.C., and W. Stumm. 1970. Acidic mine drainage: The rate-determining step. Science 167:1121–1123.

Stites, W.D., and P.A. Helmke. 1989. Aqueous solubility of synthetic franklinite-magnetite solid solutions. (Submitted for publication).

Taylor, L.A., H.K. Mao, and P.M. Bell. 1973. "Rust" in the Apollo 16 rocks. Proc. Lunar Sci. Conf., 4th, 1973 1:829–839.

Turekian, K.K., and D.P. Kharkar. 1970. Neutron activation analysis of milligram quantities of Apollo 11 lunar rocks and soil. Proc. Apollo 11 Lunar Sci. Conf. 2:1659–1664.

Turkevich, A.L. 1973. The average chemical composition of the lunar surface. Proc. Lunar Sci. Conf., 4th, 1973 2:1159–1168.

Wallace, A., E. Frolich, and O.R. Lunt. 1966. Calcium requirements of higher plants. Nature (London) 209:634.

Wanke, H., R. Rieder, H. Baddenhausen, B. Spettel, F. Teschke, M. Quijano-Rico, and A. Balacescu. 1970. Major trace elements in lunar material. Proc. Apollo 11 Lunar Sci. Conf. 2:1719–1727.

Williams, R.J., and J.J. Jadwick (ed.). 1980. Handbook of lunar materials. NASA Ref. Publ. 1057. NASA Rep. RP-1057. Natl. Sci. Tech. Info. Office, Washington, DC.

16 Geochemistry of Soils for Lunar Base Agriculture: Future Research Needs

Gene Whitney

U.S. Geological Survey
Denver, Colorado

The successful use of lunar regolith as a plant growth medium will require a thorough understanding of many geochemical processes. Some of these processes have been studied in Earth soil environments, and several have been investigated experimentally. But lunar regolith differs substantially from Earth soils, and much of our knowledge is not directly transferable. The task at hand is to apply our current understanding of rock-water reactions to the interaction of lunar materials with water in an Earth-like (but artificial) atmosphere, and to expand our understanding of those processes through observation and experimentation. The purpose of this chapter is to outline some of the geochemical research topics that require investigation to develop soils for controlled ecological life support systems (CELSS) on the Moon.

Geochemical processes during rock-water interaction may be defined in a variety of ways, but a list of important geochemical processes to be considered must include:

1. Dissolution of primary minerals and glass.
2. Precipitation of colloids and secondary minerals.
3. Cation, anion, and ligand adsorption and exchange.
4. Redox change and control.
5. Metal translocation and partitioning.

In an operational CELSS environment, these processes will depend upon how the system works and, conversely, the operation of the system will depend on what we learn about these processes under different conditions. Therefore, each process should be investigated in several different ways. With the actual CELSS environment in mind, we must consider some of the following factors for each process: How open or closed will the plant growth environment be? The lunar base itself will be a totally contained system of course, but what materials will be added to or subtracted from the soil during its functional cycle? Will recycled fluids be treated between applications to remove dissolved components? What additives will be introduced? What is the effect of fluid flow rate? What will be the composition of beginning fluids? Will pure water be used, or is there some benefit to using organic or inorganic acids or other reagents? Is it advantageous to pretreat the regolith, either chemically or hydrothermally, prior to planting in order to optimize its usefulness? What kinds of physical or chemical pretreatment would help the plant growth processes? What solid materials should be used for meaningful experiments: pure minerals, phase mixtures, terrestrial lithologic analogs, or lunar soil simulants?

As the research progresses, some of these questions will be answered, but inevitably others will arise. The purpose of this chapter is simply to propose some starting points for research. After briefly reviewing some of the consequential properties of lunar materials, directions for research for each of the geochemical processes listed above will be outlined with the rationale for concentrating on certain aspects of each.

SUMMARY OF CRITICAL CHARACTERISTICS OF LUNAR MATERIALS

Mineralogic and Lithologic Characteristics

Lunar crustal rocks fall into two broad groups, the mare basalts and the highland lithologies. The mare basalts consist mainly of clinopyroxene, calcic plagioclase feldspar, olivine, ilmenite, spinel, and armalcolite (Papike et al., 1976), with lesser amounts of other phases, including some native metals (see Chapter 4 in this book). The lunar basalts are grossly similar to terrestrial basalts but differ in some important ways. The most abundant mineral in the basalts is pyroxene, generally within the augite-ferroaugite compositional range. Some mare basalts contain pigeonite, a relatively unusual pyroxene in terrestrial basalts, and pyroxferroite, a pyroxenoid not seen in terrestrial basalts. The pyroxenes are commonly compositionally zoned. Some of the lunar basalts are extremely enriched in ilmenite relative to basalts or gabbros from terrestrial oceanic basins (the rocks that they resemble most closely). Feldspars are dominantly anorthite (Ab_{4-25}, Or_{0-3}) (Papike et al., 1976). All of the lunar basalts are depleted in alkali feldspars relative to Earth basalts. Lunar mare basalts are enriched in S relative to terrestrial basalts, and the S generally occurs in troilite (FeS).

The highland rocks are predominantly polymict breccias resulting from meteorite impacts. Compositionally, they resemble anorthositic gabbro and are composed primarily of Ca-plagioclase, orthopyroxene, olivine, Mg-Al spinel, and minor clinopyroxene. A second lithology identified in the lunar highlands has been named KREEP because it is enriched in K, rare-earth elements, and P relative to other Moon rocks. The KREEP materials generally make up only a small portion of the brecciated highland rocks.

A critical lithologic characteristic of the lunar materials is the high proportion of glass. In addition to glassy mesostasis in the basalt parent rock, a significant amount of glass has been produced during impact events. Typical lunar regolith is composed of 25 to 60% agglutinates (described below) and glass (Heiken, 1975; Warner, 1975). In addition, breccia fragments may also include as much as 50% glass (Williams & Jadwick, 1980). There are no hydrous minerals in lunar materials, and the lunar rocks contain some minerals that are unknown on Earth.

Chemical Characteristics

In keeping with their distinct mineral content, the lunar mare basalts are FeO-rich (17–23 wt. %), depleted in alkali metals (<1 wt. % total Na_2O + K_2O), and mafic (15–35 wt. % plagioclase) relative to terrestrial basalts (Warner, 1975). The mare basalts are divided into a low-Ti (1–5 wt. % TiO_2) group, which is similar to terrestrial basalts, and a high-Ti (9–14 wt. % TiO_2) group, which is greatly enriched in Ti relative to Earth basalts. The highlands rocks are richer in Al and Ca and lower in Fe and Ti than the mare basalts, reflecting the greater abundance of Ca-plagioclase in the highlands. All lunar materials are uniformly depleted in volatile elements, but exhibit a large range of values (Burnett, 1975). There is no evidence for the presence of water in any lunar rocks. Lunar rocks are also depleted in siderophile elements (those that tend to concentrate in a metallic phase) by a factor of 10 to 100 compared to average solar-system material (Burnett, 1975). All lunar materials are extremely reduced (e.g., approximately 10^{-7} Pa. O_2 at 1200 °C for alkali-depleted mare basalts; Warner, 1975). The distinctive highlands material called KREEP, though it is enriched in K and P relative to the other lunar materials, contains less than a weight percent each of K_2O and P_2O_5 (Williams & Jadwick, 1980). The surface of the regolith has been exposed to the solar wind, and components of the solar wind (H, C, N, He, Ne, Ar, Kr, and Xe) are concentrated in the outer 1 μm relative to the bulk composition of the lunar material (Burnett, 1975).

Physical and Textural Characteristics

Unpulverized basaltic rocks from the mare exhibit a wide variety of primary igneous textures (Warner, 1975). Much of the regolith material, however, is dominated by impact-induced textures rather than by primary lithologic textures. Most surface materials consist of impact breccia or unconsolidated impact debris having an average grain size of 45 to 100 μm (Heiken, 1975). These materials are poorly sorted and contain particles ranging in size from <2 μm to boulders. Porosity for "soils" from the Apollo 11, 12, 14, and 15 sites averaged 43.3% (Heiken, 1975). The regolith may be several meters deep and is commonly layered, each layer consisting of ejecta from a large meteorite impact. The upper few millimeters of regolith has been subjected to intense micrometeorite bombardment that has produced intense crushing and melting on a small scale. The result of this process is the formation of small irregular clusters of soil particles bonded by vesicular glass. These glassy clusters are called *agglutinates*. Agglutinates are generally <1 mm across and may comprise a significant proportion of the regolith, depending on the maturity (time of exposure at the lunar surface) of the regolith (see Chapter 4 in this book). Glass spheres with diameters of 1 to 3 mm make up several percentage of some lunar regoliths and may have pitted or irregular surfaces formed by the impact of micrometeorites (see Chapter 4 in this book).

STRATEGY FOR RESEARCH ON GEOCHEMICAL PROCESSES

Because so much work has been done in the study of the geochemical processes under consideration, it is not possible to provide a complete review of all past work, nor an exhaustive list of all possible future studies. Rather, I will briefly describe some of the critical considerations for each of the major categories of geochemical processes and then summarize the research strategies that serve to address those issues. Many processes described here are related. For example, the dissolution of primary Fe phases, the oxidation of Fe and precipitation of secondary Fe-bearing phases cannot be viewed as separate or unrelated processes, but each aspect of the process will be examined separately.

Dissolution of Primary Minerals and Glass

Background and Important Considerations

Soil scientists and geologists have studied rock and mineral dissolution to understand the surface and subsurface processes that determine the compositions of soils, sediments, and their aqueous fluids (Lindsay, 1979; Stumm & Morgan, 1981). The literature on chemical weathering of minerals in natural environments is huge, and hundreds of weathering reactions have been written and studied. It has been known for 50 yr, for example, that olivine, pyroxene, and anorthite (the most common lunar minerals) are among the most easily dissolved silicate minerals under weathering conditions (Goldich, 1938). Likewise, high-quality experimental studies of mineral dissolution date back to Correns and von Engelhardt (1938) and continue widely today. Despite this long history, however, many important questions concerning terrestrial weathering and dissolution reactions remain unanswered. Furthermore, much of the information is not applicable to the lunar CELSS environment. For example, two important assumptions of many of the weathering or dissolution studies are that solutions are in equilibrium with atmospheric CO_2 and that the fluids contain dissolved O_2. In the lunar base, the available CO_2 and O_2 will come from the tightly controlled CELSS atmosphere. If soil alteration reactions consume large quantities of water, CO_2, and/or O_2, the designers of the CELSS must be able to quantify those processes and design for them.

The dissolution of some of the common lunar minerals has been studied experimentally. Pyroxene and olivine dissolution has been examined by Schott et al. (1981) and Schott and Berner (1983, 1985), who showed that solution is congruent for both minerals in an O_2-free environment (Fig. 16–1a) but that the presence of O_2 produces an Fe-oxide surface layer that retards dissolution of the Fe-bearing minerals (Fig. 16–1b). The dissolution of feldspars has been widely studied in inorganic systems (Busenberg & Clemency, 1976; Busenberg, 1978; Berner & Holdren, 1977, 1979; Holdren & Berner, 1979; Holdren & Speyer, 1985, 1986, 1987; Lagache, 1976; Petrovic et al., 1976; Wollast & Chou, 1985). Although most of these studies did not include anorthite, the most common lunar feldspar, Busenberg and Clemency (1976) and Holdren and Speyer (1987) found in their comparative studies that anorthite dissolves most rapidly of all the feldspars under most conditions. In general, feldspar dissolution proceeds in two steps: (i) exchange of structural cations for H^+ at the mineral surface, and (ii) attack of the silicate-

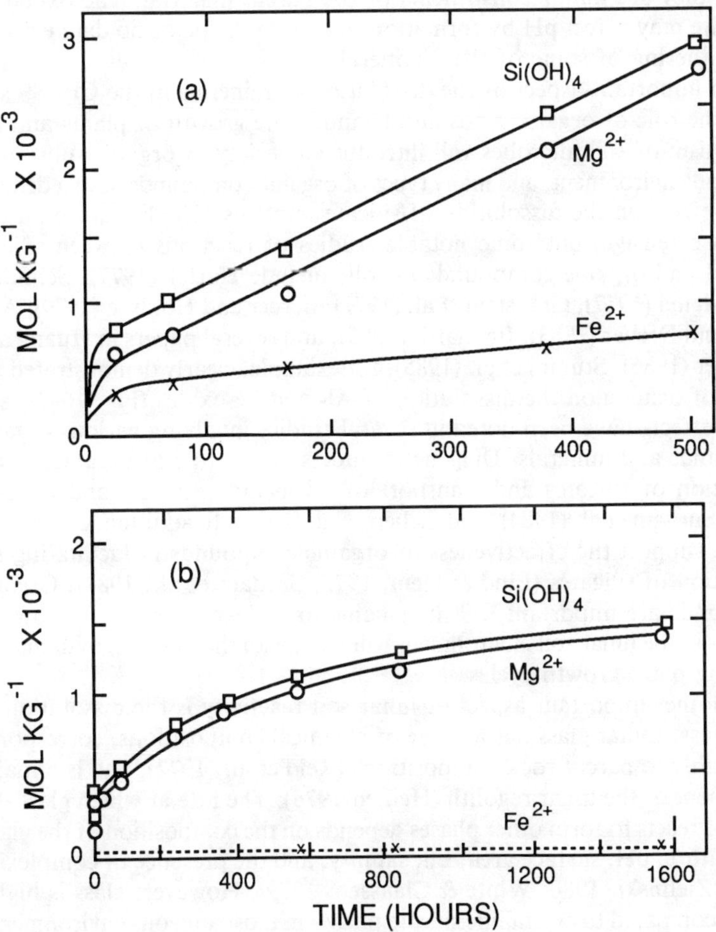

Fig. 16–1. Plot of released Si(OH)$_4$, Mg^{2+}, and Fe^{2+} as a function of time for the dissolution of bronzite (Ca-pyroxene) at pH 6 for (a) anoxic conditions and (b) oxic conditions. The formation of an Fe-oxide armor under oxic conditions greatly retards the dissolution of many Fe-bearing minerals. After Schott and Berner (1983).

aluminate framework of the mineral (Holdren & Speyer, 1986). These processes may be sensitive to extremes in pH (Helgeson et al., 1984), fluid composition, and temperature. Significant progress has been made through these experimental studies in understanding the mechanisms of dissolution, and reasonable kinetic models have been used to quantify the processes. Nevertheless, more work is needed to understand the dynamics of feldspar, pyroxene, and olivine weathering in aqueous solutions.

In addition to the abundant lunar minerals, the dissolution behavior of several minor minerals must also be understood. Some of these phases may be important in determining the geochemical characteristics of the plant growth medium that far exceeds their relative abundance. For example, ilmenite, pyroxferroite, and native Fe may control the Eh of the soil even though they are minor constituents of the parent material. The oxidation of troilite may affect pH by formation of sulfate. Little or no data exist on the weathering of some of these minerals.

An important aspect of the dissolution of minerals in the CELSS system is the role of organic acids and ligands. The growth of plants and the metabolism of soil microbes will introduce a variety of organic substances to the soil environment, and many types of organic compounds have a demonstrable effect on the dissolution of silicate minerals. The literature on this topic is extensive, but some notable studies of reactions between silicate material and organic compounds in soils include Harter (1977), Schnitzer and Kodama (1977), Graustein et al. (1977), Berner and Holdren (1979), Antweiler and Drever (1983), Eckhardt (1985), and several papers in Huang and Schnitzer (1986). Stumm et al. (1985), for example, clearly demonstrated the effects of oxalate on the dissolution of Al- and Fe-oxides (Fig. 16–2), and similar effects have been noted in several studies involving various organic compounds and minerals. Diagenetic studies on the role of organics in the dissolution of silicates and transport of Al include Surdam and Crossey (1985), Surdam et al. (1984), and Siebert et al. (1984). In addition, experimental data support the effectiveness of organic compounds in facilitating the dissolution of silicates (Lind & Hem, 1975; Surdam et al., 1984). Organic compounds are important, but it remains to be learned precisely how the organics in the lunar soil environment will react with the lunar regolith material during plant growth cycles.

Another important aspect of lunar soil reactivity is the dissolution of lunar glass. Lunar glass has a range of chemical compositions, corresponding roughly to parent rock compositions (Reid et al., 1972), and is a major component of the lunar regolith (Heiken,1975). The rate at which glass dissolves or reacts to form other phases depends on the composition of the glass, temperature, pH, surface area, Eh, salinity, and the presence of complexing agents (Zielinski, 1980; White & Claassen, 1979). However, glass is highly soluble compared to crystalline silicate phases in most aqueous environments. This fact, combined with the fine grain size of most lunar regolith, will render the regolith extremely reactive in water.

The physical characteristics of the minerals and glass in lunar regolith will play a major role in mineral and glass solubility. Reactions appear to proceed preferentially at sites of increased surface energy, such as crystalline defects (Berner & Holdren, 1979). Shock metamorphic effects are not as abundant as might be expected in pulverized lunar regolith (Heiken, 1975), but the mineral fragments in the regolith are locally fractured and brecciated and tend to be angular. This brecciation might be expected to produce weathering rates that are much faster than terrestrial soil-weathering rates (Velbel, 1985, 1986). The agglutinates generally exhibit elongate or branched morphology (see Chapter 4 in this book), thus making the surface area of

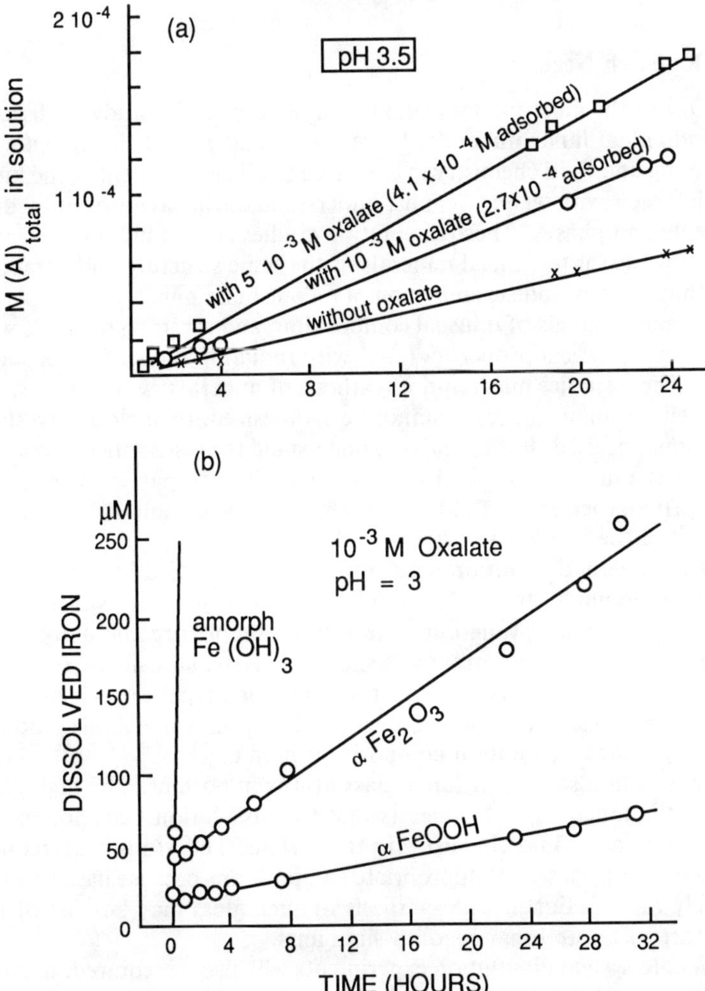

Fig. 16–2. (a) The effect of oxalate concentration on the dissolution of Al_2O_3 at pH 3.5. (b) The effect of oxalate on the dissolution of $Fe(OH)_3$, Fe_2O_3, and FeOOH, showing the difference in susceptibility of different Fe-oxide phases to the accelerating effects of the oxalate on dissolution. The solubility of these phases in the absence of the oxalate is negligible on the scale shown. After Stumm et al. (1985).

these glassy particles high. Furthermore, the outermost micron or so of the regolith particles are rendered amorphous by bombardment of solar wind over a long time. This process would also accelerate initial dissolution rates (Burnett, 1975). Considering these characteristics, combined with the fine grain size of most lunar regolith, it is difficult to imagine a natural silicate material more reactive in water than the lunar regolith. These important physical attributes make it imperative that researchers obtain experimental materials that are physically, as well as chemically and mineralogically, similar to the lunar regolith for dissolution experiments. The fabrication of lunar simulants for experimental work must assume primary importance.

Critical Research Needs

A. One of the most fundamental research needs is the study of dissolution of individual lunar minerals. If we are to understand the systematics of weathering and fluid chemistry in the CELSS soil environment as the lunar regolith interacts with water, we must understand the behavior of individual components and phases. These dissolution studies should build on existing data and models for terrestrial minerals of the same structure and composition, but must also include examination of unique lunar minerals (e.g., pyroxferroite), lunar minerals of unusual composition, and common minerals with uniquely lunar physical properties (e.g., with radiation-induced amorphous coatings). These studies may require synthesis of minerals, which is no small task in itself. If lunar minerals cannot be synthesized, then close terrestrial analogs should be used. It is essential to understand the systematics of mineral dissolution as a function of pH, Eh, presence of ligands, particle size or surface area, fluid/rock ratio, fluid flow rates, and wetting and drying cycles, to understand the whole soil system.

B. The mineral dissolution reactions should be studied in the presence of organic compounds appropriate to a CELSS environment in which plants and microbes grow and function. The role of simple organic anions and chelating agents will be important and should be systematically assessed. As work progresses and the specific plant and microbe types are selected, the mineral dissolution studies should incorporate the specific metabolic products of those organisms as solution composition parameters.

C. Dissolution studies of lunar glass also is important. The high glass content of the lunar regolith suggests that the dissolution behavior of the glass may dominate fluid chemistry in the system. This study may require synthesis of several glasses of appropriate compositions because basaltic glass is relatively rare on Earth. The synthesis of such glass may be part of the larger effort to create lunar regolith simulants.

D. Whole-system dissolution experiments will also be required, in addition to the study of individual minerals. If simulants or suitable regolith analogs are obtained, the combination of glass and minerals in a texture and particle-size distribution similar to that of the regolith will be important in determining the bulk chemical properties of the system. The drawback of such a study is that so many variables operate at once that it may be difficult to understand the specific processes producing the observed fluid and solid reaction products. Nevertheless, only in a whole-system approach can we see the overall behavior of the material in water. These studies should be conducted in both inorganic and in organic-bearing solutions.

E. The consumption of water, O_2, and CO_2 during the weathering and dissolution reactions must be monitored because these components will be consumed from the CELSS atmosphere. Eventually, a balance between consumption of O_2 by weathering processes and production of O_2 by plant metabolism may be achieved, but early stages of the alteration processes may take up significant amounts of O_2. Some of the CO_2 and water consumed during weathering may be recovered, but much of it will remain in the soil and soil solutions and should be measured.

Precipitation of Colloids and Secondary Minerals

Background and Important Variables

The study of precipitation of colloids and solids in the lunar soil environment is an extension of the study of the dissolution of primary phases. Even as primary minerals begin to react, secondary phases may form. If the lunar materials are as reactive as anticipated, then soil solutions will rapidly become supersaturated with respect to various secondary phases. In addition, the rates of dissolution reactions and attendant degrees of supersaturation may preclude the direct application of well-tested principles of mineral equilibrium, at least during the early stages of alteration, because of precipitation of metastable phases. The goal of this part of the research is to understand what initial phases are likely to form from lunar regolith under different conditions, and how they subsequently react. The experiments designed to explore these reactions must be tailored to the chemical and mineralogical conditions of the lunar soil, drawing also on the store of data in the literature (e.g., mine waste reclamation studies).

During the alteration experiments, it will be necessary to monitor solution composition as the primary phases dissolve, monitoring speciation, complexation, polymerization, nucleation, and crystal growth in secondary phases as fluid concentrations increase. Volumes have been written on the thermodynamics of these phenomena in terrestrial soils (e.g., Sposito, 1981). But the mixture of lunar regolith with water presents a unique chemical environment that must be studied and modeled. Because the lunar regolith is Fe-rich, highly reduced, and contains a large proportion of glassy material, the identity and behavior of dissolved species may be quite different from terrestrial soil environments. In addition, the solid/fluid ratios and fluid flow rates will alter the nature of the dissolution/precipitation processes (Dibble & Tiller, 1981). In general, the solution compositions will be a key to understanding the dissolution and precipitation processes.

A second approach is to monitor surface reactions on minerals and glass as they react with the primary fluid. In some rock/water environments, secondary phases form almost solely by replacement of primary minerals.

Sposito (1985) has demonstrated that the thermodynamic understanding of a soil system must be augmented by kinetic constraints. A thorough theoretical knowledge of stable assemblages may be accompanied by total ignorance of reaction paths, but reaction path is critical in the formation of soils in the artificial CELSS environment because of the relatively short time of reaction. For the lunar materials, the early stages of the reaction will probably precipitate unstable or metastable solids. The formation and reactivity of these materials may dominate solution chemistry. Furthermore, most crystalline solids in natural solutions form by heterogeneous nucleation; that is, they nucleate on the solid surface of another phase (Stumm et al., 1985). In a highly reactive environment such as the glassy lunar regolith, optimum nucleation surfaces may be scarce, thus affecting the nucleation of certain crystalline solids. It is important that study of the kinetics of glass and mineral dissolution be combined with an examination of the alteration processes on solid surfaces and the kinetics of nucleation and crystal growth to understand the overall alteration process.

The high Fe content of lunar materials will play an important role in the precipitation processes. As Fe^{2+} is released from the extremely Fe-rich and highly reduced lunar regolith, a suite of Fe-oxide minerals will form. The species, crystallinity, and chemistry of these secondary minerals will depend upon a number of environmental factors including temperature, moisture, organic matter, and pH (Schwertmann, 1985; Schwertmann & Taylor, 1977; several papers in Stucki et al., 1988). The Fe-oxide minerals also show strong kinetic effects, with persistent metastable phases common in many soils. The formation of Fe-hydroxides is accompanied by a drop in pH, and Fe^{2+} is soluble and may be easily transported under reducing and acidic conditions. If dissolved silica is present due to glass and minerals dissolution, Fe-silicate phases (such as nontronite) may form. It is important

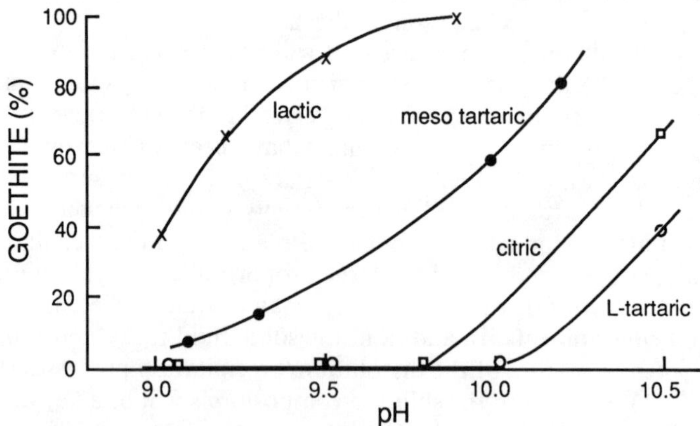

Fig. 16–3. Relationship between relative amounts of goethite and hematite formed in the presence of various organic acids at various pH (hematite = 100% − goethite). Hematite is favored at lower pH in all organic acids, and the type of organic acid present determines the relative abundance of the two minerals at a particular pH. After Cornell and Schwertmann (1979).

to understand how these Fe minerals form and react in the CELSS environment because hydrolysis of Fe may consume significant amounts of water and the oxidation of the Fe may consume significant quantities of O_2 from the CELSS atmosphere. In addition, the Fe-oxide minerals may produce an undesirable laterite-like soil under certain conditions. The geochemistry of the mobility, hydrolysis, oxidation, precipitation, and transformation of Fe and Fe-oxides is a critical factor in soil quality.

Organic matter also plays an important role in the precipitation of secondary soil minerals. The ability of organic material to poison the crystal growth of carbonate minerals is well documented (Suess, 1970) and may play a similar role in the precipitation of silicates. In addition, the role of organic matter in dissolving and transporting Al (Huang & Violante, 1986; Huang & Keller, 1971) and other metals affects secondary precipitation by altering the concentration and form of those metals in solution. Furthermore, the species of organic compounds present may determine which crystalline species forms, as Cornell and Schwertmann (1979) have shown for the Fe-oxides (Fig. 16–3). These processes are too complex to predict with confidence for the lunar materials; only careful experimentation will provide the data necessary to understand the entire system.

Research Needs

A. The primary need in studying the precipitation reactions in the CELSS soil environment is a thorough understanding of fluid chemistry during the interaction of starting materials with water. Part of this work can be combined with the dissolution studies outlined above. Concentrations of the various chemical components will be measured during dissolution, but the detailed analysis of speciation, polymerization, nucleation, and crystal growth will require analyses not generally performed during dissolution studies. As in the dissolution studies, the role of pH, Eh, ligands, fluid/rock ratio, and wetting and drying cycles will be important.

B. All secondary minerals and colloids must be identified and characterized. Colloids should be separated and analyzed by the appropriate analytical means, such as detailed optical, electron-optical, and surface analysis, as well as bulk x-ray diffraction analysis.

C. The role of appropriate organic compounds in facilitating or retarding the precipitation of secondary minerals or colloids must be thoroughly investigated.

D. Because the ultimate goal of these studies is to produce a soil for the CELSS enviroment that is optimum for plant growth, the precipitation of some colloids or minerals may be undesirable. Therefore, another avenue of study is the deliberate modification of soil processes by adjustment of solution chemistry. If the unmodified soil tends to become clogged with secondary Fe-oxides, for example, then we must understand how to inhibit the precipitation of the troublesome materials.

Cation, Anion, and Ligand Adsorption and Exchange

Background and Important Variables

The study of ion exchange in soils has an extensive empirical and theoretical basis (Sposito, 1981, 1984; Mott, 1981). Specific issues related to cation, anion, and ligand adsorption and exchange during the formation and use of lunar soils includes the ionic composition of fluids interacting with CELSS soils, the mineralogy and surface chemistry of secondary phases in the soils, the presence of organic species, and rates of fluid movement through the soil.

Because of the bulk composition of the lunar materials, Ca and Mg will dominate over the other dissolved alkali and alkaline-Earth cations in lunar soils. However, the selectivity of cations during adsorption, based on ionic radius, suggests that less abundant cations may still be present in significant quantities on mineral exchange sites. The importance of some of the less-abundant cations (e.g., K) for nutritional purposes suggests that careful exchange studies should be done specifically on lunar soil analogs to determine the availability of such cations.

Important anions in terrestrial soils include chloride, fluoride, borate, bicarbonate/carbonate, phosphate, selenite, silicate, molybdate, nitrate, sulfate, and organic anions (Mott, 1981). Most of these anions are absent or present in extremely small amounts in primary lunar materials, owing to the igneous origin of the parent material and extreme reducing conditions. Simple alteration of lunar regolith by oxygenated water may produce measurable amounts of sulfates and phosphates, but the other anions will become quantitatively important only if added to the soil. Nevertheless, because some of the other anions may be added, it is important to monitor their behavior in lunar soil analog.

The adsorption and exchange processes will depend not only on the ionic species present in solution, but also on the composition of the solid exchange surfaces. As mentioned above, the glassy lunar regolith may react rapidly and form amorphous or colloidal solids as immediate alteration products. If the soil becomes dominated by such materials, the adsorption and exchange dynamics may be different from most terrestrial soils and obviously should be investigated in detail.

Once plants are grown in the CELSS soil, a variety of soluble and insoluble organic acids and ligands may be present. These organics play a dual role in ion exchange (Sposito, 1984). First, they act as exchangers themselves, exhibiting exchange phenomena similar to the solid substrates (Frizado, 1977). The affinity of various organics for metals varies greatly, and their role in adsorbing or chelating metals may be negligible or may dominate the system, depending on the specific identity of the organic materials. In addition, the organics themselves may be adsorbed onto exchange sites of minerals, with or without their complexed metals. The study of organic-clay interactions has received much attention in both natural and experimental systems (Theng, 1974). In clay-organic interaction, the clays may play the role of passive adsorbants, or may actively catalyze reactions altering the organics.

This wide variability in ion-exchange behavior demonstrates how important it is to know what kinds of organic compounds will be present in the lunar soils and how they interact with the actual dissolved cations, minerals, amorphous solids, and colloids.

Research Needs

A. The first stage of this research must be the identification of adsorbed and exchangeable species: dissolved cations, anions, and ligands resulting from the interaction of the lunar regolith with water. Because the lunar materials are chemically different from terrestrial rocks, the dissolved ion composition will be different from terrestrial soil solutions. In addition, the change in composition with time, as the soil evolves, will be important. The composition must ultimately include organics derived from plant growth and microbial activity and any additives to the soil.

B. To understand and quantify the adsorption and exchange processes, the adsorption and exchange characteristics of the solid surfaces in the weathered lunar soil (including exchange capacities, ion selectivity, anion competition, and catalysis) should be investigated in detail, using the specific species likely to be present in the lunar soil. The effects of texture, surface area, ionic strength, and pH will be important in the lunar soil environment.

Redox Change and Control

Background and Important Variables

Lunar materials are extremely reduced, and some lithologies are Fe-rich. Lunar basalts commonly contain >20 wt. % FeO, locally contain metallic Fe and contain 0.2 to 0.3 wt. % MnO and 0.1 to 0.8 wt. % S (Williams & Jadwick, 1980). Highland anorthositic lithologies are much lower in Fe, Mn, and S, but even highland regolith contains significant brecciated mafic material. Thus, the role and importance of redox processes in the CELSS soil will be somewhat dependent upon starting lithology. For a particular starting composition, three of the most important variables that will determine redox conditions in the lunar soil during plant growth are: (i) the relative stabilities of the various oxides and oxy-hydroxides of Fe, Mn, and other metals (e.g., Ni, Cr, and Cu), (ii) the kinetics of the oxidation reactions for various species and phases, and (iii) the effects of organic matter on these processes. Other factors, such as S speciation or the chemistry of soil additives, may also play important roles depending upon specific concentrations.

The relative stabilities of Fe-oxides and oxy-hydroxides in terrestrial environments have been successfully described in Eh-pH space (Garrels & Christ, 1965; Stumm & Morgan, 1981; Krauskopf, 1979). Because of the unique character of the lunar soil environment, the CELSS soil system may experience excursions in Eh and pH beyond what is normally observed in natural terrestrial environments (Fig. 16–4). The starting Eh for the water-regolith system will probably be well below values normally encountered. During the subsequent hydrolysis of Fe producing hydroxy-Fe complexes and Fe-oxides, the pH of the system may fall. However, this reaction may be coupled with or offset by the silicate hydrolysis reactions that tend to raise the system pH (Krauskopf, 1979). Clearly, describing or predicting stable mineral assemblages in such a system will require some creative geochemical modeling. An experimental approach using lunar simulants may be the only way to know what the products will be.

This problem is complicated by the kinetic constraints on oxidation reactions, especially where compounded with the kinetic problems associated with the silicate glass and mineral dissolution processes. Many redox reactions are slow and are pH dependent. For example, we know that oxidation proceeds faster at higher pH (Fig. 16-5a) and that the rate of oxidation may vary considerably within a fairly narrow pH range (Fig. 16-5b). Many re-

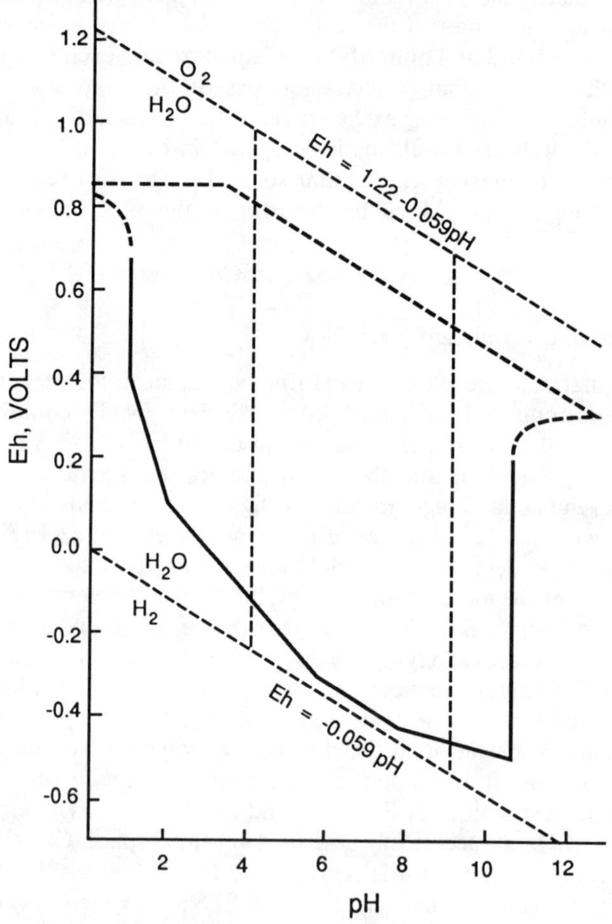

Fig. 16-4. Range of Eh and pH values, including extreme values, measured in natural environments. From Baas Becking et al. (1960). Central parallelogram defining the limits within which most natural Eh and pH values fall (Krauskopf, 1979). The lunar regolith is so reducing that CELSS soils may fall at the lower limit of these fields.

dox processes do not couple readily with one another, and several different oxidation states may persist within a limited volume, locally producing partial redox equilibrium (Stumm & Morgan, 1981). These behavior problems not only make it difficult to predict Eh-pH conditions in the CELSS soil system but will make determinative experiments difficult. In fact, simply finding a means of measuring and reporting meaningful Eh values in such a complex system presents significant problems. Furthermore, rates of aeration of the soils will play a critical role in rates of oxidation.

A third key factor in the dynamics and control of Eh is the role of organic matter and organisms. Even less is known about the redox processes in and among organic materials than those involving the inorganic redox couples. Organic matter in soil environments is heterogeneous and difficult to characterize. Furthermore, the meaning of oxidizing or reducing strength in covalently bonded materials is unclear. Although some of the most reduced natural materials are inorganic matter, quantifying the reducing power of organics is extremely difficult. For these reasons, the organic matter that will enter the lunar soils as plant material or by-products must be assessed specifically for its redox influence on the system.

In most natural terrestrial environments, redox reactions are mediated by microorganisms. Although they cannot force a redox reaction that would not ultimately proceed anyway (Krauskopf, 1979), these microbes play an important role in determining the rates of redox reactions. Some of these organisms thrive under specific environmental conditions. For example, $T.$ $ferrooxidans$ oxidizes Fe^{2+} to Fe^{3+} but functions best at low pH (Rowell,

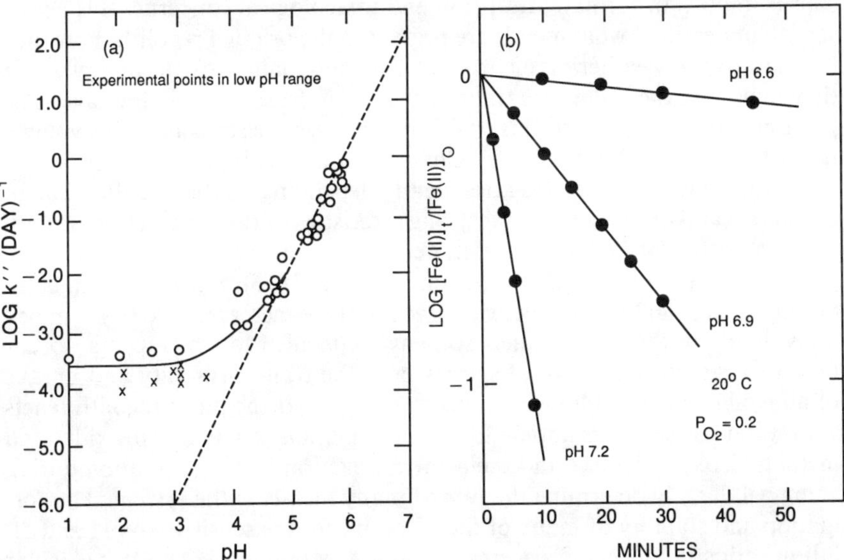

Fig. 16–5. (a) Oxidation rate of Fe^{2+} as a function of pH. The rate of oxidation is highly dependent upon pH at all but the low pH values. From Singer and Stumm (1970). (b) The ratio dissolved Fe^{2+} to Fe^{3+} as a function of time in O_2-bearing fluids at different pH values. The rate of oxidation of Fe^{2+} may be highly sensitive to pH even within a narrow pH range. After Stumm and Morgan (1981).

1981). The selection of soil microbes and their functioning in the specific chemical conditions of the lunar soil should be investigated because their influence on redox conditions will influence many of the inorganic reactions that occur in the system.

Research Needs

A. Basic research is needed on the redox processes involving highly reduced minerals and glass that interact with fluids in equilibrium with a CELSS-like atmosphere. No terrestrial rocks as reduced as the lunar regolith exist, so artificial materials may need to be used in these experiments. The synthesis of such materials is no small task. The kinetics of oxidation and hydrolysis; the formation of oxide or hydroxide coatings; the long-term availability of reduced Fe and other metals; and the consumption of water, CO_2, and O_2 should be assessed for a lunar soil analog.

B. The role of soil organic matter in the CELSS soil in controlling the redox conditions must be investigated. As the soil evolves and organic plant by-products and recycled waste enter the soil, the Eh conditions may change dramatically. The long-term effects of these processes may be great and should be assessed.

Metal Translocation and Partitioning

The mobilization, transportation, and final residence of metals in lunar soils is important for plant nutrition and toxicological concerns. It is necessary to understand what metals are present in the original regolith, how they respond to the weathering environment during plant growth, and whether they remain in the soil or are transported from the soil in soil drainage. The plant nutritional aspects of this problem are covered elsewhere in this volume (see Chapter 6 in this book; see Chapter 10 in this book).

Mobilization of metals varies greatly depending on the specific element of interest and on several chemical characteristics of the fluid. The ease with which a particular metal is released depends on the overall solubility of the parent solid, the Eh and pH of the solution (Fig. 16–6), the presence of complexing agents, and biologic activity. Each of these factors should be examined individually in the unique lunar soil environment. Most metals may occur in aqueous systems in several species, and the transportability and uptake of a metal depends on how it occurs in solution. If the lunar regolith reacts with water as readily as anticipated, the formation of metal hydroxides and metal hydroxo polymers and their polymerization and precipitation during aging will largely determine the fate of many metals in the system. The formation and stability of many of these complexes are sensitive to pH and dilution ratios in inorganic systems (Stumm & Morgan, 1981). Of particular importance in lunar soils are the metals that occur in especially high concentrations, such as Ti. Although the mobility of Ti has been studied (Hutton, 1977), its behavior in the geochemically unusual lunar environment must be examined.

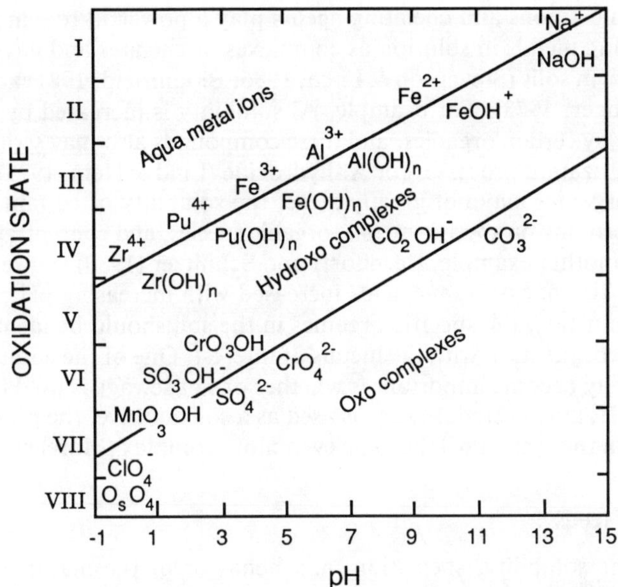

Fig. 16–6. The predominant pH range for the occurrence of aquo, hydroxo, hydroxo-oxo, and oxo complexes for several metals for various oxidation states. This simplified diagram shows only general relationships and does not include all possible complexes, but demonstrates the pH dependence of complexation of metals. After Stumm and Morgan (1981).

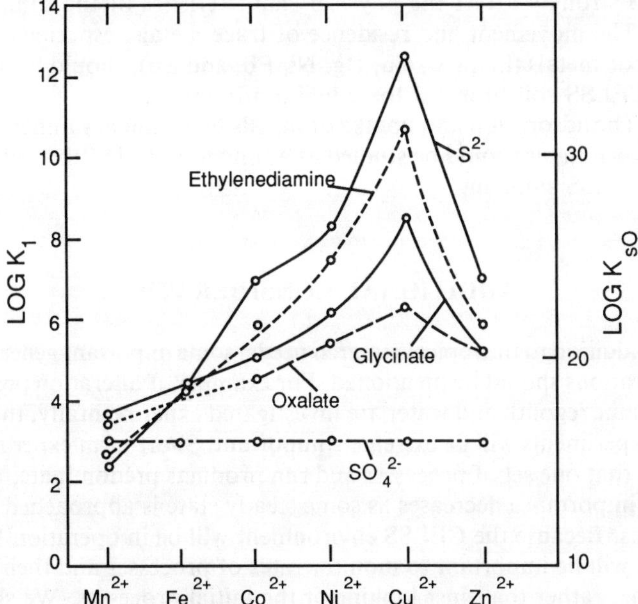

Fig. 16–7. The stability constants of several transition metals, demonstrating that the stability of organo-metal complexes depends on the specific metal and the specific organic compound involved in the complex. After Stumm and Morgan (1981).

Organic anions and chelating agents play a powerful role in liberating or stabilizing metals in solution as complexes or chelates and have received much study in soils (Stevenson & Fitch, 1986; Bloomfield, 1981) and natural waters (Singer, 1973). For example, Al solubility is increased by orders of magnitude by certain organics, and these compounds also may seriously alter the polymerization processes for Al-hydroxide (Lind & Hem, 1975). In other systems, organics may not greatly affect the solubility of certain trace elements, depending on specific metals, organic species, and concentrations (Fig. 16-7). In another example, Kerndorff and Schnitzer (1980) showed that the sorption efficiency of humic acids increased with increasing pH. For lunar soil environments, the specific organics in the soil should be identified and studied in conjunction with the metals of interest. One of the additional factors that may become important is whether or not sewage is used in soil fertilization. If, as expected, sewage is used as a soil additive, the geochemistry of metals in the system will become even more complex (Morel et al., 1975).

Research Needs

A. The solubility, speciation, and behavior of the major metal constituents of the lunar regolith must be studied during alteration and plant growth. Iron and Ti are enriched in many lunar materials far above normal terrestrial ranges, and the dissolution, transport, and precipitation behavior must be understood. The movement and precipitation of these metals, along with Al will not only determine much of the geochemistry of the system but will also strongly affect the physical characteristics of the soil.

B. The movement and residence of trace metals, especially the potentially toxic metals (i.e., Cd, Cu, Hg, Ni, Pb, and Zn), should be understood in the CELSS soil to avoid toxic buildup in crops.

C. The adsorption and uptake of metals by secondary minerals and colloidal precipitates should be studied as a function of pH, Eh, and fluid flow in the CELSS simulant.

ADDITIONAL CONSIDERATIONS

In addition to the topical research needs, some important general research considerations should be mentioned. For example, if alteration processes between lunar regolith and water are investigated experimentally, the duration of the experiments will be extremely important. Short-term experiments may indicate that one set of processes and run products predominate, while their relative importance decreases as some steady state is approached after long run times. Because the CELSS environment will be in operation for several years, it will be important to monitor rates of processes and their evolution with time, rather than just looking at the initial processes. We should also know, as early as possible, what types of plants and microbes are likely to be used in the CELSS soil so that the appropriate organic compounds can be incorporated into the ongoing experiments.

The emphasis in this chapter has been on the geochemical aspects of the production and use of CELSS soil, but the chemical aspects are inseparable from the physical aspects. The water/rock ratio, drainage and fluid flow rates, and frequency of tilling will play important roles in the geochemical processes. A frequently tilled soil will experience much more interaction with the atmosphere than will an untilled soil, and many of the geochemical processes under consideration will be dramatically affected by physical processes. In addition, physical analysis of experimental run products should be an integral part of the analyses. Textural information gained by scanning electron microscopy (SEM), transmission electron microscopy, or petrographic microscopy will be invaluable in understanding the formation of coatings, the clogging of pores with colloidal material, or transport of organic and inorganic materials through the soil.

An aspect of this research related to the physical processes is the production and use of a lunar regolith simulant. Terrestrial analogs of the lunar regolith that exhibit all of the necessary chemical and physical characteristics are simply not available. In an attempt to manufacture a lunar regolith simulant material, the chemical composition is obviously important. Equally important is the physical nature of the analog material. Because of the fine grain size and the unusual textures of the lunar breccias and agglutinates, reaction rates and the nature of certain reactions will not be faithfully duplicated in materials having different physical characteristics. The duplication of the physical characteristics of the lunar regolith must, therefore, be a top priority for those attempting to produce a simulant material. Studies of mine waste reclamation may be extremely useful in understanding the problems associated with growing plants in crushed rock, perhaps the closest terrestrial analog to lunar regolith.

A further consideration for studies of lunar soil formation and development is the potential use of artificially modified materials for soil. This topic has been covered in detail in another chapter of this volume (see Chapter 7 of this book). If plans are made to chemically or hydrothermally pretreat regolith materials prior to use as CELSS soil, the studies suggested in this chapter should be applied to the altered material in the same manner as suggested for the unaltered material. Many of the potential uncertainties and problems in using raw lunar regolith as starting material for the CELSS soil may be overcome by pretreatment, and the behavior of various materials must be examined if such a plan is possible.

Finally, analytical techniques will play an important role in the experimentation leading to a thorough understanding of the geochemical processes in lunar soils. To gain the greatest profit from the intensive experimental work involved in this project, investigators should analyze experimental products by all available techniques. As mentioned above, the bulk concentration of elements in solution during dissolution studies tells only part of the geochemical story. Detailed analysis of speciation and complexation of elements and organic compounds from a small number of experiments may reveal more about the soil processes than large numbers of concentration values. The analysis of the solids by a wide variety of available sophisticated techniques (analytical electron microscopy, Mossbauer spectroscopy, x-ray photoelectron spectroscopy, electron spin resonance, nuclear magnetic resonance) as well as the more conventional analytical methods (x-ray diffraction, thermal analysis, UV-vis-IR spectrophotometry, SEM, optical microscopy, atomic absorption spectrophotometry, and x-ray fluorescence spectrometry) will yield a wealth of information about the soil-forming processes that may otherwise go undetected. It is unlikely that this variety of analytical techniques is practiced in any single laboratory, thus making collaboration between experimentalists and analysts highly desirable. Such collaboration within a group of soil scientists and geochemists may produce enormous advances in the understanding of lunar soil alteration, but will also benefit the science of rock alteration and soil formation in general.

CONCLUSION

Although geochemical processes have received an enormous amount of study in terrestrial soils, the soils formed from lunar regolith in the CELSS environment will constitute a large experimental system in which many variables and processes cannot be understood in advance. It is possible, however, with well conceived and well-designed experiments and with close observation of natural analogs, to gain a level of understanding that will allow us to design and develop a soil enviroment for a lunar base that will fulfill the needs of the CELSS inhabitants. The studies required for this understanding can be performed with available technology. As the results from early work become available and as decisions are made about plant species to be planted and harvested, the studies will evolve into closer approximations of the CELSS environment. Ultimately, the data derived from these studies will not only serve to solve many of the problems for the lunar base, but will also contribute to the understanding of soil-forming processes in terrestrial environments.

REFERENCES

Antweiler, R.C., and J.I. Drever. 1983. The weathering of a Late Tertiary volcanic ash: Importance of organic solutes. Geochim. Cosmichim. Acta 47:623–629.

Baas Becking, L.G.M., I.R. Kaplan, and D. Moore. 1960. Limits of the natural environment in terms of pH and oxidation-reduction potentials. J. Geol. 68:243–284.

Berner, R.A., and G.R. Holdren, Jr. 1977. Mechanism of feldspar weathering: Some observational evidence. Geology 5:369–372.

Berner, R.A., and G.R. Holdren, Jr. 1979. Mechanism of feldspar weathering. II. Observations of feldspars from soils. Geochim. Cosmochim. Acta 43:1173–1186.

Bloomfield, C. 1981. The translocation of metals in soils. p. 463–504. In D.J. Greenland and M.B.H. Hayes (ed.) The chemistry of soil processes. John Wiley and Sons, Chichester, England.

Burnett, D.S. 1975. Lunar science: The Apollo legacy. Rev. Geophys. Space Phys. 13:13–34.

Busenberg, E. 1978. The products of the interaction of feldspars with aqueous solutions at 25 °C. Geochim. Cosmochim. Acta 42:1679–1686.

Busenberg, E., and C.V. Clemency. 1976. The dissolution kinetics of feldspars at 25 °C and 1 atm. CO_2 partial pressure. Geochim. Cosmochim. Acta 40:41–49.

Cornell, R.M., and U. Schwertmann. 1979. Influence of organic anions on the crystallization of ferrihydrite. Clays Clay Miner. 27:402–410.

Correns, C.W., and W. von Engelhardt. 1938. Neue Untersuchungen uber die Verwitterung des Kalifeldspates. Chem. Erde 12:1–22.

Dibble, W.E., and W.A. Tiller. 1981. Non-equilibrium water/rock interactions—I. Model for interface-controlled reactions. Geochim. Cosmochim. Acta 45:79–92.

Eckhardt, F.E.W. 1985. Solubilization, transport, and deposition of mineral cations by microorganisms—Efficient rock weathering agents. p. 161–173. *In* J.I. Drever (ed.) The chemistry of weathering. Reidel Publ. Co., Dordrecht, Netherlands.

Frizado, J.P. 1977. Ion exchange on humic materials—A regular solution approach. p. 133–146. *In* E.A. Jenne (ed.) Chemical modeling in aqueous systems. Am. Chem. Soc., Washington, DC.

Garrels, R.M., and C.L. Christ. 1965. Solutions, minerals, and equilibria. Freeman, Cooper and Co., San Francisco.

Goldich, S.S., 1938. A study in rock-weathering. J. Geol. 46:17–58.

Graustein, W.C., J. Cromack, Jr., and P. Sollins. 1977. Calcium oxalate: occurrence in soils and effect on nutrient and geochemical cycles. Science 198:1252–1254.

Harter, R.D. 1977. Reactions of minerals with organic compounds in the soil. p. 709–740. *In* J.B. Dixon and S.B. Weed (ed.) Minerals in soil environments. SSSA, Madison, WI.

Heiken, G. 1975. Petrology of lunar soils. Rev. Geophys. Space Phys. 13:567–587.

Helgeson, H.C., W.M. Murphy, and P. Aagaard. 1984. Thermodynamic and kinetic constraints on reaction rates among minerals and aqueous solutions. II. Rate constants, effective surface area, and the hydrolysis of feldspar. Geochim. Cosmochim. Acta 48:2405–2432.

Holdren, G.R., Jr., and R.A. Berner. 1979. Mechanism of feldspar weathering. I. Experimental studies. Geochim. Cosmochim. Acta 43:1161–1171.

Holdren, G.R., Jr., and P.M. Speyer. 1985. Reaction-rate surface area relationships for an alkali feldspar during the early stages of weathering. Geochim. Cosmochim. Acta 49:675–681.

Holdren, G.R., Jr., and P.M. Speyer. 1986. Stoichiometry of alkali feldspar dissolution at room temperature and various pH values. p. 61–81. *In* S.M. Colman and D.P. Dethier (ed.) Rates of chemical weathering of rocks and minerals. Academic Press, Orlando, FL.

Holdren, G.R., Jr., and P.M. Speyer. 1987. Reaction rate-surface area relationships during the early stages of weathering. II. Data on eight additional feldspars. Geochim. Cosmochim. Acta 51:2311–2318.

Huang, P.M., and M. Schnitzer (ed.). 1986. Interactions of soil minerals with natural organics and microbes. SSSA Spec. Publ. 17. SSSA, Madison, WI.

Huang, P.M., and A. Violante. 1986. Influence of organic acids on crystallization and surface properties of precipitation products of aluminum. p. 159–222. *In* P.M. Huang and M. Schnitzer (ed.) Interactions of soil minerals with natural organics and microbes. SSSA Spec. Publ. 17. SSSA, Madison, WI.

Huang, W.H., and W.D. Keller. 1971. Dissolution of clay minerals in dilute organic acids at room temperature. Am. Mineral. 56:1082–1095.

Hutton, J.T. 1977. Titanium and zirconium minerals. p. 673–688. *In* J.B. Dixon and S.B. Weed (ed.) Minerals in soil environments. SSSA, Madison, WI.

Kerndorff, H., and M. Schnitzer. 1980. Sorption of metals on humic acid. Geochim. Cosmochim. Acta 44:1701–1708.

Krauskopf, K.B. 1979. Introduction to geochemistry. McGraw-Hill Book Co., New York.

Lagache, M. 1976. New data on the kinetics of the dissolution of alkali feldspars at 200 °C in CO_2 charged water. Geochim. Cosmochim. Acta 40:157–161.

Lind, C., and J. Hem. 1975. Effects of organic solutes on chemical reactions of aluminum. U.S. Geol. Surv. Water Supply Paper 1827-G. U.S. Gov. Print. Office, Washington, DC.

Lindsay, W.L. 1979. Chemical equilibria in soils. Wiley-Interscience, New York.

Morel, F.M., J.C. Westall, C.R. O'Melia, and J.J. Morgan. 1975. Fate of trace metals in Los Angeles County wastewater discharge. Environ. Sci. Technol. 9:756–761.

Mott, C.J.B. 1981. Anion and ligand exchange. p. 179–220. *In* D.J. Greenland and M.H.B. Hayes (ed.) The chemistry of soil processes. John Wiley and Sons, Chichester, England.

Papike, J.J., F.N. Hodges, A.E. Bence, M. Cameron, and J.M. Rhodes. 1976. Mare basalts: Crystal chemistry, mineralogy and petrology. Rev. Geophys. Space Phys. 14:475–540.

Petrovic, R., R.A. Berner, and M.B. Goldhaber. 1976. Rate control in the dissolution of alkali feldspars. I. Study of residual feldspar grains by x-ray photoelectron spectroscopy. Geochim. Cosmochim. Acta 40:537–548.

Reid, A.M., J. Warner, W.I. Ridley, and R.W. Brown. 1972. Major element composition of glasses in three Apollo 15 soils. Meteoritics 7:395–415.

Rowell, D.L. 1981. Oxidation and reduction. p. 401–461. *In* D.J. Greenland and M.H.B. Hayes (ed.) The chemistry of soil processes. John Wiley and Sons, Chichester, England.

Schnitzer, M., and H. Kodama. 1977. Reactions of minerals with soil humic substances. p. 741–796. *In* J.B. Dixon and S.B. Weed (ed.) Minerals in soil environments. SSSA, Madison, WI.

Schott, J., and R.A. Berner. 1983. X-ray photoelectron studies of the mechanism of iron silicate dissolution during weathering. Geochim. Cosmochim. Acta 47:2233–2240.

Schott, J., and R.A. Berner. 1985. Dissolution mechanisms of pyroxenes and olivines during weathering. p. 35–53. *In* J.I. Drever (ed.) The chemistry of weathering. Reidel Publ. Co., Dordrecht, Netherlands.

Schott, J., R.A. Berner, and E.L. Sjoberg. 1981. Mechanisms of pyroxene and amphibole weathering—I. Experimental studies of iron-free minerals. Geochim. Cosmochim. Acta 45:2133–2135.

Schwertmann, U. 1985. Formation of secondary iron oxides in various environments. p. 19–20. *In* J.I. Drever (ed.) The chemistry of weathering. Reidel Publ. Co., Dordrecht, Netherlands.

Schwertmann, U., and R.M. Taylor. 1977. Iron oxides. p. 145–180. *In* J.B. Dixon and S.B. Weed (ed.) Minerals in soil environments. SSSA, Madison, WI.

Siebert, R.M., G.K. Moncure, and R.W. Lahann. 1984. A theory of framework grain dissolution. p. 163–176. *In* D.A. McDonald and R.C. Surdam (ed.) Clastic diagenesis. Am. Assoc. of Petroleum Geol., Tulsa, OK.

Singer, P.C. (ed.). 1973. Trace metals and metal-organic interactions in natural waters. Ann Arbor Sci. Publ., Ann Arbor, MI.

Singer, P.C., and W. Stumm. 1970. Acidic mine drainage: The rate-determining step. Science 167:1121–1123.

Sposito, G. 1981. The thermodynamics of soil solutions. Oxford Univ. Press, New York.

Sposito, G. 1984. The surface chemistry of soils. Oxford Univ. Press, New York.

Sposito, G. 1985. Chemical models of weathering in soils. p. 1–18. *In* J.I. Drever (ed.) The chemistry of weathering. Reidel Publ. Co., Dordrecht, Netherlands.

Stevenson, F.J., and A. Fitch. 1986. Chemistry of complexation of metal ions with soil solution organics. p. 29–58. *In* P.M. Huang and M. Schnitzer (ed.) Interactions of soil minerals with natural organics and microbes. SSSA Spec. Publ. 17. SSSA, Madison, WI.

Stucki, J.W., B.A. Goodman, and U. Schwertmann (ed.). 1988. Iron in soils and clay minerals. NATO ASI Ser. Reidel Publ. Co., Dordrecht, Netherlands.

Stumm, W., and J.J. Morgan. 1981. Aquatic chemistry. John Wiley and Sons, New York.

Stumm, W., G. Furrer, E. Wieland, and B. Zinder. 1985. The effects of complex-forming ligands on the dissolution of oxides and aluminosilicates. p. 55–74. *In* J.I. Drever (ed.) The chemistry of weathering. Reidel Publ. Co., Dordrecht, Netherlands.

Suess, E. 1970. Interaction of organic compounds with calcium carbonate. I. Association phenomena and geochemical implications. Geochim. Cosmochim. Acta 34:157–168.

Surdam, R.C., S.W. Boese, and L.J. Crossey. 1984. The chemistry of secondary porosity. p. 127–150. *In* D.A. McDonald and R.C. Surdam (ed.) Clastic diagenesis. Am. Assoc. of Petroleum Geol., Tulsa, OK.

Surdam, R.C., and L.J. Crossey. 1985. Mechanisms of organic/inorganic interactions in sandstone/shale sequences. p. 177–233. *In* Society of Economic Paleontologists and Mineralogists Short Course 17 Lecture Notes: Relationship of organic matter and mineral diagenesis. Soc. of Economic Paleontologists and Mineralogists, Tulsa, OK.

Theng, B.K.G. 1974. The chemistry of clay-organic reactions. John Wiley and Sons, New York.

Velbel, M.A. 1985. Geochemical mass balances and weathering rates in forested watersheds of the Southern Blue Ridge. Am. J. Sci. 285:904–930.

Velbel, M.A. 1986. Influence of surface area, surface characteristics, and solution composition on feldspar weathering rates. p. 615–634. *In* J.A. Davis and K.F. Hayes (ed.) Geochemical processes at mineral surfaces. ACS Symp. Ser. 323. Am. Chem. Soc., Washington, DC.

Warner, J. 1975. Mineralogy, petrology, and geochemistry of the lunar samples. Rev. Geophys. Space Phys. 13:107–113.

White, A.F., and H.C. Claassen. 1979. Dissolution kinetics of silicate rocks—Application to solute modeling. p. 447–473. *In* E.A. Jenne (ed.) Chemical modeling in aqueous systems. Am. Chem. Soc., Washington, DC.

Williams, R.J., and J.J. Jadwick. 1980. Handbook of lunar materials. NASA Ref. Publ. 1057. NTIS, Springfield, VA.

Wollast, R., and L. Chou. 1985. Kinetic study of the dissolution of albite with a continuous flow-through fluidized bed reactor. p.75–96. *In* J.I. Drever (ed.) The chemistry of weathering. Reidel Publ. Co., Dordrecht, Netherlands.

Zielinski, R.A.1980. Stability of glass in geologic environment: Some evidence from studies of natural silicate glasses. Nucl. Technol. 51:197–200.

17 Plant Considerations for Lunar Base Agriculture

T. W. Tibbitts

University of Wisconsin
Madison, Wisconsin

A major emphasis is being placed on growth of higher plants for life support in controlled ecological life support systems (CELSS) for lunar and other space bases. The success of CELSS will depend upon using plants with a maximum of efficiency and dependability. Effective plant use in CELSS will require a significant research effort to understand how to maintain production in closed recycling systems under essentially completely automated procedures. The research needs are large and the challenges involve the expertise of a wide variety of plant scientists. For this chapter on plant considerations for lunar bases, research needs have been divided into four general categories to emphasize the unique and different area of research facing scientists working on CELSS. These four categories are (i) plant productivity, (ii) closed environments, (iii) automation and robotics, and (iv) space environment.

PLANT PRODUCTIVITY

Plant productivity is being investigated by several different scientists both in the USA and USSR. The amount of research, however, is quite limited and only a few species are being studied. The research for establishing productivity needs in CELSS, based on $g\ m^{-2}\ d^{-1}$, are significantly different than for studying productivity in traditional agricultural systems. In CELSS, there is the opportunity to manipulate different parameters of the environment, including temperature, photoperiod, irradiance, humidity, and CO_2, through a wide range of levels, to determine the range and maximum response for each cultivar of each species and to find the interaction of conditions that provide optimum levels of crop production and quality. This potential is in contrast to traditional agriculture where the environmental conditions of a particular producing area have distinct limits and research is directed toward developing cultivars (or breeding cultivars) that provide maximum crop production and quality under the environment conditions prevailing in a specified producing area. Thus, research efforts in CELSS are quite different and will require imagination by scientists to manipulate the environment to extract the productive potential of different genetic strains.

The research needs in plant productivity can be broken down into the following areas.

Range of Response to Environment

There is a need to determine the range of plant response to varying levels of all controllable environmental parameters. Determination of the range of response is a more critical need than establishing maximum response levels. The range-of-response data are needed to calculate the most cost-effective environment in different types of bases requiring life support. The cost of providing environmental control will vary in different bases in space primarily because of different costs for providing irradiation to the plants and removal of heat from the growing areas. Thus, plant response information needs to be obtained for radiation level, radiation duration, temperature, humidity, and CO_2 level, and then response models developed to predict the productivity at all interacting levels of these factors.

The complexity of this effort is multiplied by the fact that such data should be obtained for varied levels of the environmental conditions during both the light and dark period and for different stages of plant development.

The response data obtained from the plants should include:

1. Yield of edible product in mass per unit area per day.
2. Proportion of non-edible to edible biomass.
3. Proportion and nutritional quality of the edible biomass that is digestible.
4. Size of plants.
5. Amount of water transpired.

Germplasm Screening and Plant Breeding

There is a need for extensive screening of lines and cultivars of species from all over the world. There is the possibility that species adapted to environments in other locations of the world may have much greater potential in a CELSS than commonly available cultivars adapted to locations in the USA. It has been found that a potato (*Solanum tuberosum* L.) cv. Denali, obtained from Alaska has the highest productivity of cultivars screened. This highest productivity has been obtained in environmental rooms using continuous irradiation. Thus, the selection pressure of long days in Alaska likely contributed to this high-yielding capability. There are large collections of cultivars and species in Plant Introduction Stations in many different locations of the world that can provide a wealth of germplasm for CELSS research. The need for research to develop plant germplasm that is adapted to the CELSS environment will be required as the environmental constraints of space bases become more clearly defined.

Reproduction Methods

There is a need to develop specific procedures for reproduction of each different CELSS candidate species that will have minimum cost and yet maintain disease-free, true-to-type progeny. For some species, as with potato, the development of effective procedures using vegetative propagation likely will be preferred to seed propagation to avoid genetic variability. Various micro-propagation procedures should be studied as this will allow the rapid multiplication of vigorous plantlets (Evans et al., 1983; Zimmerman et al., 1986). Reproduction research is not a critical need at this time but will become more important as CELSS are being developed for specific space bases.

Plant Spacing

There is a need for determining the productivity of plants with different spacing to minimize the number of transplants or seeds that would have to be used on a given area (Bugbee & Salisbury, 1986). This determination is complicated by the fact that yield response to spacing will vary with the environmental conditions under which plants are grown.

CLOSED ENVIRONMENT

Water, nutrients, and atmospheric gases within a bioregenerative life support system must be recycled in an essentially closed loop to minimize resupply needs from Earth. With a "closed" CELSS there is a significant potential for any toxic compounds present, or produced within the unit, to accumulate and have significant effects on growing plants and to humans. Thus, there is a significant research effort needed to establish what toxic components will accumulate in closed systems and to develop effective procedures for their control. Little effort in this direction has been required in growing systems on Earth because of the availability of large quantities of fresh atmospheric gases and water for dilution of toxicants. Only in a few situations, as in submarines, and in recent years with closure of homes and other buildings to conserve energy, has this problem been of some concern.

Toxic Components

Toxicity must be determined for components that accumulate in both the atmosphere and in the recycled liquid nutrient solution provided to the roots. Many different sources must be evaluated including (i) materials used in the construction of the environment rooms and for the plant growing system, (ii) compounds released from the plants themselves as volatile gases into the atmosphere and as exudations from the roots into the nutrient solution; and (iii) compounds produced by microorganisms contaminating the nutrient solution and released into the nutrient solution or released as volatile gases into the atmosphere.

Different plant species may have quite different chemical emissions and will likely be hosts for varying types of microorganisms, thus this research effort must involve evaluation of each separate species that is a candidate for food production in CELSS.

The potential toxicity of atmospheres and nutrient solutions must be evaluated through use of plant bioassays as the chemical composition of the toxic compounds generaly will not be known. Thus, analytical procedures cannot be relied upon solely for detection. The selection of the plants to be used for bioassays poses a significant problem, for different species, and even cultivars within species, have different sensitivity to toxic substances (Tibbitts, 1978). Thus, a group of different plant lines should be used for bioassays including strains or cultivars that are recognized to be sensitive to air pollutants and nutrient inbalances. At least one cultivar of each candidate species for CELSS should be used in the bioassays.

Plant Removal of Contaminants

There is a significant potential for plants to be used for removing unwanted contaminants from the atmosphere and from liquid solutions (Bennett & Hiel, 1973; Wolverton et al., 1983). Research should be undertaken to quantify the amount of toxic gases or nutrient solution contaminants that can be removed by the different candidate species. Absorption of volatile components from the atmosphere by plants will likely be closely proportional to the extent of leaf stomatal opening. Thus, plants growing in an environment that provides rapid photosynthetic rates will have open stomates and will have maximum effectiveness for removal of volatile organics. Similarly, high photosynthetic rates produce rapid growth and will likely promote the uptake of contaminants that may have accumulated in the nutrient solution.

Waste Recycling

A major research need is to develop procedures for decomposing inedible plant parts and human wastes so that the resultant breakdown products can be reused by the plants and humans. An extensive effort needs to be initiated to recover the C, O_2, N, and other elements required for plant growth and human nutrition in forms that can be reassimilated by the plant. Research is needed to determine what partially decomposed compounds can be absorbed and assimilated by plants to reduce the cost of waste recycling. Each separate species may have different potential for utilization of partially decomposed compounds.

AUTOMATION AND ROBOTICS

The culture and harvesting procedures for plant production in space should be automated as much as possible to reduce the need for providing man support to maintain crops. This will require that plant development must be precisely defined and that plants are effectively monitored on a regular basis so that all growing and harvesting procedures can be effectively performed and automated. Automated procedures will need to be developed for the following culture operations:

1. Planting and transplanting operations.
2. Supply of nutrient and water to the plants.
3. Maintenance of proper nutrient concentrations and pH.
4. Monitoring of plant size, maturity stage, and for stress or disease on the plants.
5. Harvesting of the plant material.
6. Separation and transport of edible and nonedible plant parts to appropriate compartments.

SPACE ENVIRONMENT

Of primary concern in the space environment is the plant response to reduced gravity levels. This will be of primary concern in orbiting space bases where there will be no significant gravity force in a particular direction. Gravitational effects will likely be of less significance on Moon bases where a significant, but reduced, gravity level will be present. However, even on these bases there is a need for research to establish to what extent the reduced gravity levels affect plant growth and productivity.

Productivity

The level of plant productivity under reduced gravity is a major research need that can only be determined through experiments undertaken in space. On the basis of current information, it is not known whether reduced gravity may result in stress to the plant, with a consequent reduction in productivity, or whether reduced gravity may stimulate the plant and thereby increase productivity. The productivity of each separate candidate species should be evaluated under space conditions because photosynthate accumulation in seeds, (e.g., as in wheat, *Triticum aestivum* L.) may be affected differently than accumulation in edible tubers and root tissue [e.g., white potato or sweet potato, *Ipomoea batatas* (L.) Lam.].

Nutrient Movement

Systems need to be developed to move nutrient solutions under weightlessness for study of plant growth in orbiting space craft. These systems should be functional and useful under the one-sixth gravity of Moon bases. The systems for weightlessness need to keep the nutrient solution effectively contained and provided with adequate O_2 levels for the roots. Systems are being evaluated that involve keeping nutrient solutions under a suction within porous plates or in porous tubes as it circulates to the root system (Wright, 1984).

Plant Orientation

It is anticipated that blue-wavelenths of light can be used for orientation of plants in space (Schmidt, 1984). More research is needed to quantify the amount of blue photons required to provide effective orientation under different growing conditions. Separate species may have varying blue light requirements.

Plant Support

Specific types of flexible materials need to be identified to provide support and separation of seedlings and plants in the growing trays. These materials should be of light weight, have no toxic emmanation and maintain integrity following sterilization, cleaning, and reuse for many different growth cycles.

Atmospheric Composition and Pressure

Research is needed to evaluate the use of modified atmospheres for plant growing and food storage. This is needed to make the most effective utilization of gases required for pressurizing a CELSS. The effects of partial pressures of all gases on plant growth and productivity should be studied to establish the feasibility of maintaining growing units at a reduced atmospheric pressure. Human-occupied space craft have consistently been maintained with CO_2 concentrations that exceeded 5000 μmol mol^{-1}. These high partial pressures of CO_2 must be studied to determine if they accelerate production or cause significant injury to plants. Of interest also is the substitution of He for N_2 as the inert gas in the atmosphere of a CELSS because He is only 30% as heavy as N_2 and significant quantities may be available from the lunar regolith (R.J. Bula, 1988, personal communication).

SUMMARY

It is obvious that a tremendous research effort is required if plants are to be effectively used in life support systems in space. Some research is underway but there are many critical research needs that need to be addressed in the near future if CELSS is to become a reality within the time frame outlined for development of space bases. The capability and enthusiasm are present in the plant-science community. A significantly greater commitment from NASA and the general populus is needed to mobilize this interest for the development of effective and useful CELSS.

REFERENCES

Bennett, J.H., and A.C. Hill. 1973. Absorption of gaseous air pollutants by a standardized plant canopy. J. Air Pollut. Control Assoc. 23(3):203–206.

Bubgee, B.G., and F.B. Salisbury. 1986. Studies on the maximum yield of wheat for the controlled environments of space. p. 447–485. *In* R.D. MacElroy (ed.) Controlled ecological life support systems: CELSS '85. NASA TM-851388. Ames Res. Ctr., Moffett Field, CA.

Evans, D.A., W.R. Sharp, P.V. Ammirato, and Y. Yamada. 1983. Handbook of plant cell culture. Vol. 1. Techniques for propagation and breeding. Macmillan Publ. Co., New York.

Schmidt, W. 1984. Blue light physiology. BioScience 34(11) 698:704.

Tibbitts, T.W. 1978. Air contaminants. p. 101–116. *In* R.W. Langhans (ed.) A growth chamber manual. Cornell Univ. Press, Ithaca, NY.

Wolverton, B.C., R.C. McDonald, and E.A. Watkins, Jr. 1983. Foliage plants for removing indoor air pollutants from energy-efficient homes. Econ. Bot. 38(2):224–228.

Wrigh, B.D. 1984. A hydroponic method for plant growth in microgravity. IAF-ST-8405. Am. Inst. of Aeronautic and Astronautics, New York.

Zimmerman, R.H., R.J. Griesbach, F.A. Hammerschlag, and R.H. Lawson. 1986. Tissue culture as a plant production system for horticultural crops. Nijhoff, Netherlands.

18 Microbiological Considerations for Lunar-Derived Soils

D. B. Alexander and D. A. Zuberer

Texas A&M University
College Station, Texas

D. H. Hubbell

University of Florida
Gainesville, Florida

Terrestrial soils are composed of inorganic particles, water, air, organic matter, and living organisms (Alexander, 1977; Brady, 1974). Lunar regolith lacks all but one of these essential components: mineral matter. It contains no living organisms, no water, no organic matter, and only vanishingly small concentrations of the gases that comprise the major portion of the Earth's atmosphere—N_2 and O_2 (Williams & Jadwick, 1980). To produce from this material a "soil" that will support plant growth, these components must be introduced into the lunar environment. In this chapter we will discuss some of the contributions that will be required of microorganisms to make the development of such a soil possible.

The biotic portion of a terrestrial soil contains a vast array of microscopic and macroscopic organisms, including fungi, bacteria, actinomycetes, algae, protists, viruses, insects, and other small invertebrates. Together these organisms, particularly the microscopic forms, play a major role in the formation of soils, the maintenance of soil structure, and nutrient cycling in terrestrial environments (Coleman, 1985; Lynch, 1983; Alexander, 1977). Microbial respiration produces CO_2 that dissolves in soil water to form a dilute solution of carbonic acid. Carbonic acid and organic acids that are exuded as waste products of metabolism contribute to the disintegration of rocks and minerals into smaller particles, facilitating the formation of soils from parent materials (Ugolini & Edmonds, 1983). Microorganisms also play a major role in the dissolution of primary minerals, the mobilization of elements, and the precipitation of secondary minerals. Polysaccharides and other organic polymers produced by these organisms cement individual soil particles together into stable aggregates that improve soil structure. Fungal mycelia contribute to aggregate stability by entangling soil particles among their filamentous hyphae (Lynch, 1983).

Perhaps the most important role of microorganisms in soil is to function as decomposers. These organisms are responsible for the breakdown of the complex organic molecules in plant and animal residues and waste products to simple compounds that can be recycled and used by other organisms. Complete oxidation of these compounds leads to the formation of CO_2 that can be recaptured by plants for photosynthesis. Inorganic nutrients such as N, S, and P are also released through the decomposition of organic matter in forms that are available for utilization by plants and other primary producers.

Microorganisms are responsible for the conversion of nutrient elements, notably N and S, from one oxidation state to another, thereby regulating the availability of the nutrients to plants. These redox reactions often result in a change of the physical state of the element, such as the volatilization of soluble ions or the conversion of gases into cellular constituents. Microorganisms contribute to the dissolution of inorganic nutrients such as P and Fe from insoluble phases through the production of enzymes, organic acids, and chelators (e.g., siderophores) (Lynch, 1983).

All of these microbial activities must be incorporated into lunar regolith to produce a "soil" that will support plant growth and serve to recycle wastes. Ehrlich (see Chapter 10 in this book) has touched on the role that soil microorganisms play in converting P and S to forms that are available for assimilation by plants, including descriptions of the organisms involved and possible sources of these nutrients in lunar materials. He has also provided an extensive list of microbes that contribute to the cycling of micronutrients in soils. Some of these micronutrients become toxic with increasing concentration. Stotsky (see Chapter 9 in this book) has discussed how microorganisms may be used to extract such elements from lunar regolith, thereby reducing the risk of the regolith becoming toxic to plant life. The focus of this chapter will be on the role microbes play in cycling the major nutrients, C and N, required for plant production.

CARBON CYCLING IN LUNAR-DERIVED SOILS

The primary factor limiting the activity of heterotrophic organisms (those requiring performed organic nutrients) in terrestrial soils is the availability of organic substrates (Richards, 1987; Alexander, 1977). The same will undoubtedly be true in soils derived from lunar materials. Lunar regolith contains little C, none of it organic. The highest concentrations of C detected in samples of lunar regolith collected during the Apollo missions are on the order of 100 to 200 mg kg^{-1} (Williams & Jadwick, 1980). These are mostly gaseous forms (methane, CO, CO_2) that have been implanted in the crystal lattices of lunar minerals by solar wind and solar flare bombardment. No carbonate minerals or graphite have been found in these samples.

Initially organic matter will have to be added to lunar regolith to establish a heterogeneous community of heterotrophic microorganisms. Research will be needed to determine what types of organic amendments are best suited for this purpose. Suitable amendments should include a combination of simple substrates that are immediately available as sources of C and energy, organic residues containing a low C/N content for enhanced mineralization, complex substrates which will provide a long-term supply of organic C, and recalcitrant forms that will form a stable organic fraction (humus) in a lunar soil. Ground plant residues and organic wastes are likely sources of these types of materials. The long-term goal is to achieve a regenerative, self-sustaining ecosystem in which unused plant residues and human wastes produced at the lunar base are returned to the soil to recycle nutrients, water, and gases (Henninger et al., 1987; MacElroy et al., 1985).

A variety of microorganisms must be introduced into the lunar regolith to achieve adequate cycling of carbonaceous materials (Fig. 18–1). Photoautotrophs and chemoautotrophs will serve to add organic C to the soil using

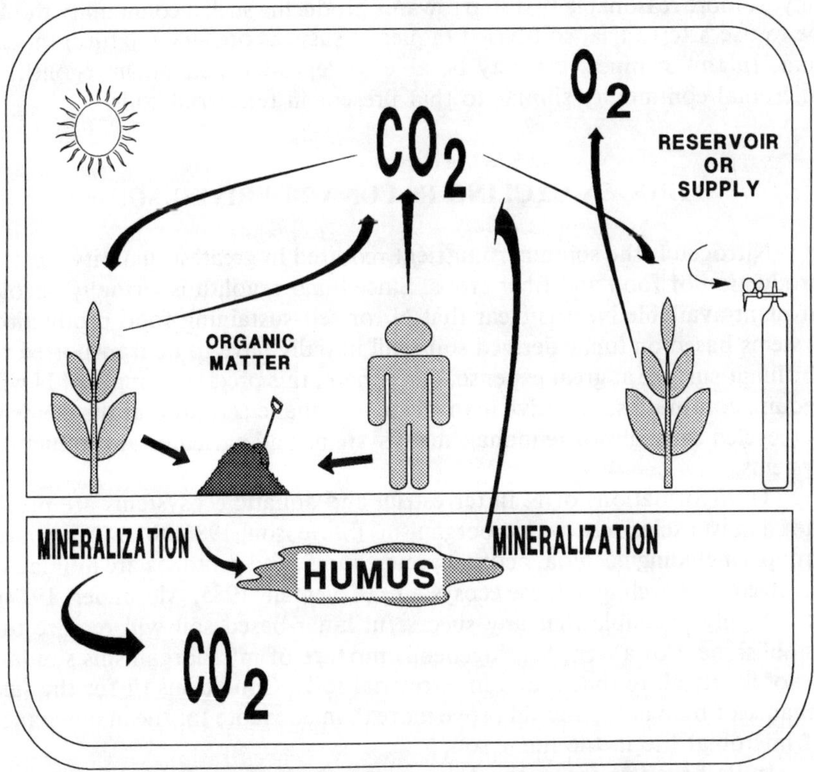

Fig. 18–1. Simplified C cycle operating in a plant production module of a closed environmental life support system containing facilities for "lunar agriculture." The cycle will be critical for balancing CO_2 and O_2 contents of the atmosphere of the closed system. Major imbalances, should they result, would be corrected by adjusting the levels of gases through addition or removal of the appropriate gases using reservoirs or supply sources. The ultimate goal is to recycle as much C as possible through biomass and soil organic matter to achieve a balanced atmosphere. Initially, some source of organic matter will need to be transported to the lunar surface. Later, plant residues and human wastes will be processed for recycle through the plant production module to conserve nutrients.

energy from light and energy obtained through the oxidation of inorganic minerals. They will also complement the system by recycling CO_2 which is evolved by respiration. A vast array of heterotrophic decomposers will be needed to break down the many different forms of organic C (carbohydrates, proteins, lipids, and aromatic compounds) contained in the residues and waste materials that will be produced at a lunar base. It would not be reasonable to attempt to construct a suitable combination by inoculating with a few, or even several, selected species. Because we are presently unable to culture all indigenous soil microorganisms in vitro or to measure all of their diverse metabolic activities in situ, our knowledge regarding the types of microorganisms that are present in soils and their ecological interactions with one another and with their environment is far from complete. If we were to attempt to construct a defined inoculum, we might fail to include the diverse population of organisms needed to achieve a stable, self-regulating community. A more reasonable first step towards producing such a community would be to use a terrestrial soil-based (aqueous suspension, dry mixture) inoculum. In this manner, we may be able to reproduce, in lunar regolith, a microbial community similar to that present in terrestrial soils.

NITROGEN CYCLING IN LUNAR-DERIVED SOILS

Nitrogen is the soil macronutrient required in greatest quantity for the production of food and fiber crops. Since lunar regolith is virtually devoid of plant-available N, it is clear that N for self-sustaining food production systems based on lunar-derived soils will initially have to be transported to the lunar surface at great expense. Once there, this precious supply of N will require continuous, effective management to make certain that the element is recycled through waste management systems and agricultural production systems.

Transformations of N in terrestrial and aquatic ecosystems are mediated almost exclusively by microorganisms (Stevenson, 1986). Many microbial groups, including bacteria, actinomycetes, fungi, and protozoa are intimately involved in N cycling in these ecosystems (Coleman, 1985; Alexander, 1977). It is highly probable that any successful lunar-based soil will require the establishment of a rich, heterogeneous mixture of microorganisms similar, if not identical, to that found in terrestrial soils. Conditions fit for the sustenance of human life should prove more than adequate for the maintenance of microbial life in the lunar soil.

In its broadest sense, the N cycle represents the balance between the production of gaseous forms of N and their recapture into fixed forms (Fig. 18–2). It is this balance between "capture" (N_2 fixation) and volatilization (denitrification, etc.) that will require careful maintenance in a closed life support system. As on Earth, the excessive loss of N from the lunar soil through volatilization could have substantial negative effects on plant production. On the other hand, though not likely, it might be possible to deplete the supply of N_2 from the atmosphere of the closed system by fixation.

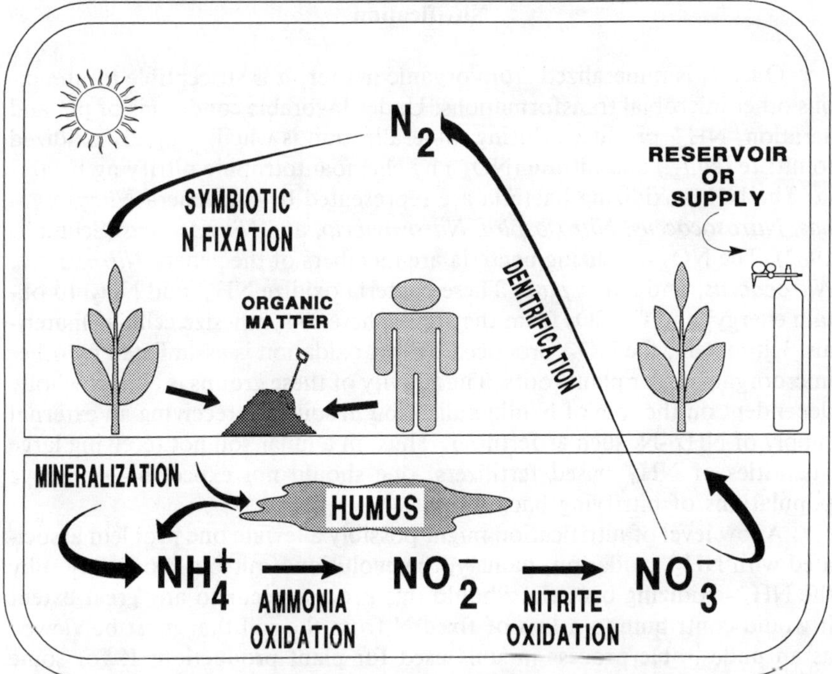

Fig. 18-2. An abbreviated N cycle that might be operative in a plant production module of a closed environmental life support system. Conservation of N will be extremely important in any system used to produce food and biomass for other purposes. While combined N must be brought to the lunar surface to establish an agricultural base, biological N$_2$ fixation will be useful in recycling any N$_2$ that might be lost from the lunar soil through denitrification. Appropriate legumes could serve this purpose in addition to producing high-quality foodstuffs and plant residues. The mineralization of crop residues and humus can provide the fertility buffer that is lacking in many hydroponic systems. Nitrification of NH$_4^+$ to NO$_3^-$ could be easily controlled with an appropriate inhibitor thereby reducing leaching of NO$_3^-$ as well as denitrification.

Nitrogen Mineralization

Between the volatilization and fixation of N lies a range of transformations of paramount importance for the continued cycling of N and the well-being of plant and microbial communities. The process of N mineralization, the liberation of ammonium (NH$_4^+$) from organic nitrogenous compounds, is essential for the return of N from plant and animal remains. Only through this process can the N in organic forms such as the protein, amino acids, and nucleic acids of animal, plant, and microbial remains be made available for subsequent populations of plants and microbes (Stevenson, 1982, 1986; Alexander, 1977). Fortunately, the process of N mineralization is universal among microorganisms and occurs wherever sufficient nitrogenous organic materials are available and environmental conditions are not adverse to microbial life. The only probable limitation to N mineralization in a lunar soil would be the availability of organic materials to support microbial populations.

Nitrification

Once N is mineralized from organic matter, it is susceptible to numerous other microbial transformations. Under favorable conditions of pH and aeration, NH_4^+ produced during mineralization is usually rapidly oxidized to nitrite (NO_2^-) and nitrate (NO_3^-) by chemoautotrophic nitrifying bacteria. The NH_4^+-oxidizing bacteria are represented by the genera *Nitrosomonas, Nitrosococcus, Nitrosospira, Nitrosovibrio,* and *Nitrosolobus* (Schmidt, 1982). The NO_2^--oxidizing bacteria are members of the genera *Nitrobacter, Nitrococcus,* and *Nitrospina.* These bacteria oxidize NH_4^+ and NO_2^- to obtain energy, and fix CO_2 from the atmosphere to synthesize cellular materials. Ultimately, the NO_3^- produced by this oxidation is assimilated by other microorganisms or plant roots. The activity of these groups is almost wholly dependent on the rate of N mineralization in soils not receiving an external supply of NH_4^+-N, such as fertilizer. Thus, in a lunar soil not receiving large quantities of NH_4^+-based fertilizers, one should not expect highly active populations of nitrifying bacteria.

A low level of nitrification might possibly alleviate one problem associated with NH_4^+ oxidation, namely, the evolutionof nitrous oxide (N_2O) by the NH_4^+-oxidizing bacteria. Should this process occur to any great extent it would contribute to a loss of fixed N from the soil that must be viewed as an undesirable process in soils used for plant production. If for some reason, nitrification was viewed as an undesirable process in the lunar-based system it might be feasible to control or eliminate the process through the use of a nitrification inhibitor such as nitrapyrin (Schmidt, 1982). It may be feasible to transport such chemicals to the lunar surface because the application rates would be small and large quantities would not have to be transported. Alternatively, it may be possible to establish a microflora free of nitrifying bacteria.

Nitrate produced during nitrification is taken up by plants and other organisms or lost from the soil through leaching or volatilization. Since the lunar system will be closed, leaching of NO_3^- from the soil would not be a major problem because any leachate would be reapplied to the soil or otherwise purified for reuse, during which process N, other nutrients, and water would be recovered. In fact, in a closed system such as the models proposed for a lunar base it would be desirable to use the soil and vegetation as a means for removal of N from water which could be used for other purposes.

Denitrification

If leaching is manageable, denitrification represents a possible problem that could be more difficult to manage. Numerous genera of bacteria are capable of denitrification. Among these are species of *Pseudomonas, Alcaligenes, Agrobacterium, Bacillus,* and others. All of these are common soil microorganisms that are typically aerobic heterotrophs that switch to anaerobic respiration when soil O_2 becomes limiting (Firestone, 1982; Tiedje, 1982). Denitrification occurs when there is insufficient O_2 for aerobic respiration, sufficient available organic energy sources, and sufficient NO_3^- to support an active population of denitrifying bacteria. Many of these bacteria would be transported to a lunar base with its human occupants and other biological materials, or be present in any natural inoculum used to establish the lunar soil microflora, therefore their eventual presence seems inevitable.

Since the loss of combined N from a lunar-based agricultural soil represents loss of perhaps the most essential element for plant production, denitrification must be viewed as a potential problem in these systems. Such a problem could be managed in these systems in several ways. If there is abundant O_2 available at the lunar base, such as might be recovered from the lunar regolith (Gibson & Knudsen, 1985), some form of artificial aeration might be imposed to minimize denitrification. From a more practical standpoint, it is likely that denitrification might be limited by lack of sufficient oxidizable organic matter in the lunar-derived soils. If nitrification inhibitors are used in these systems, the production of NO_3^-, the substrate for denitrification, would be greatly reduced or eliminated.

Dinitrogen Fixation

Nitrogen is returned to terrestrial soils through abiotic mechanisms such as rainfall deposition, and through biological N_2 fixation. In the latter process, prokaryotic microorganisms convert N_2 into NH_3, most of which is incorporated into microbial cells. Numerous bacterial genera contain species capable of N_2 fixation. Included among these are a wide range of free-living (nonsymbiotic) N_2 fixers as well as symbiotic N_2-fixing bacteria.

The root nodule symbionts *Rhizobium* and *Bradyrhizobium* produce N_2-fixing root nodules on a vast array of legume species. In the legume root nodule, the bacteria are amply supplied with photosynthate to drive the energy-intensive N_2-fixing enzyme complex. Dinitrogen fixation by free-living bacteria is largely restricted by the availability of C sources in soil. For this reason, these bacteria are frequently found in association with the roots of many different plant species. It is widely recognized that N_2-fixing bacteria such as *Azospirillum, Azotobacter, Bacillus, Klebsiella, Enterobacter* and others are closely associated with roots and rhizospheres of many non-leguminous plants including most of the cereals (Döbereiner & Pedrosa, 1987). Current evidence indicates that root-associated bacteria fix N_2 and that some of the fixed N is transferred to the host plant.

In a closed system such as a lunar base, it is quite probable that N_2-fixing plant/microbe associations could play a vital role in plant production as well as in biological systems designed to modulate the atmosphere of the facility. Thus, there is a clear role for N_2 fixation in the replenishment of combined N lost from the lunar-derived soil through biological volatilization reactions such as those described above. Since C sources would initially be scarce in the lunar-derived soil, legume use for food production and other biomass seems highly desirable. Through the use of appropriate legume species, the lunar soil could be enriched with N derived from the atmosphere as well as organic matter produced by the roots, thereby reducing the need for transported nitrogenous fertilizers.

Since these soils are likely to exhibit some unusual properties in comparison with terrestrial soils, it will be necessary to gain an understanding of the survival and functioning of these bacteria and others in such habitats. As Ehrlich (see Chapter 10 in this book) has pointed out, Mo may be in short supply in the lunar regolith. Since Mo is essential for the N_2-fixing enzyme complex, nitrogenase, and for plant and microbial NO_3^- reductases, it appears there may be a need to transport Mo to the lunar base. Experiments conducted with lunar regolith or simulants on Earth will provide valuable information regarding the behavior of plants and microorganisms in these materials. If, as Stotsky (see Chapter 9 in this book) has pointed out, it becomes feasible to transport clay minerals for amelioration of the lunar regolith, appropriate materials could be chosen or prepared to ensure that adequate Mo as well as other critical micronutrients, such as B, (see Chapter 10 in this book) were present in the lunar-derived soil. Alternatively, zeolites might be tailor-made for this purpose (see Chapter 7 in this book; Henninger et al., 1987).

It should be indicated that given a sufficient supply of Earth-derived fixed N for agricultural use, the process of biological N_2 fixation is not critical. If N is in perennial short supply, however, as it is in many of the soils on Earth, N_2 fixation will be a vital process in these lunar-based systems. An ample supply of N gas would most likely be brought in with the atmosphere used to establish a lunar base.

At the other extreme of the subject of supply of atmospheric N_2, there might be concern that an overactive population of N_2-fixing microbes might deplete the atmosphere of N. This seems a remote possibility, however, since the energy supply for the bacteria would be a limiting factor and should the soil become sufficiently enriched with fixed N the enzyme system would be turned off. There is little reason to assume that the same set of checks and balances that controls the reactions on Earth would not regulate them in the closed system of the lunar base.

It should be clear from the considerations above that establishing an active, balanced, N-cycling microflora in lunar-derived soils will be essential for the production of long-term sustainable plant production systems not dependent on costly transported materials. Such would be the case in any production scheme using solid-rooting media and probably in hydroponic systems as well. It is also likely that the N-cycling microflora will play a critical role in environmental aspects of the lunar bases.

Just as life on Earth would cease without the continual activities of the soil microflora, so would life in a closed system elsewhere be more perilous without the activities of a diverse, metabolically stable microflora such as that found in all arable soils on Earth.

THE RE-CYCLE PROCESS

Human life in a lunar environment is totally dependent on in situ sustainable agriculture, which in turn is dependent on near-perfect recycling of organic and inorganic nutrients. Loss of such materials from the recycling process would be a fatal luxury. Virtually every substance brought to the lunar surface, if not intended for "permanent" function in its arrival form, should have the potential to enter readily into the nutrient recycling process. They must be susceptible to total biodegradation to products that can be used in the recycle process. Each of these materials must be a substrate. But matter cannot be created; there must be a base level of substrate to start the process. All materials brought to the lunar surface, if not designed for permanence, should be so formulated chemically that they can easily enter the nutrient cycle at some point, as nutrients for plants or microbes. Materials of biological origin are an obvious final choice. An alternative might be to use custom-designed packaging materials composed of artificial polymers of C. These would be readily biodegradable synthetic materials that might incorporate elements, such as N and P required by biological systems.

SOURCES OF MICROORGANISMS FOR LUNAR-DERIVED SOILS

What should be the source of microorganisms used to stock the newly developed lunar-derived soil? It might be argued that the preparation of a lunar-derived soil provides the unique opportunity to establish some type of special soil microflora, perhaps one derived from a collection of genetically modified bacteria and fungi endowed with special properties for rapid organic waste degradation, high rates of N_2 fixation, or any of a multitude of other beneficial traits. Such an approach, while rather exciting and challenging for a microbial ecologist, might be somewhat short-sighted, for such a well-contrived system would probably lack one of the outstanding attributes of most terrestrial soil microbial populations, that of stability. It is a well-established principal that species diversity provides stability in ecosystems. The soil population is no different in this respect.

There are countless varieties of microbes, representing unknown numbers of genera, in the typical mineral soil on Earth. Many of these bacteria have never been cultured and so we are completely ignorant of their role(s) as members of the soil community. Yet they must have some function and must be persistent to some degree because they are regularly observed in soils (Alexander, 1977). Until we understand the ecological interactions of individual microbial species and microbial groups in the soil much more clearly it seems premature to plan a "designer microflora" for a lunar soil. Perhaps we will be able to eliminate undesirable microbes from a "lunar soil microbial inoculant," but it is clear that we should attempt to establish a diverse population of microbes in these soils. There is no reason why soil from a fertile field (one with low-disease incidence) would not provide the necessary "starter culture" to establish a vital soil microflora in a lunar-derived soil.

Comparatively recent and intensive investigation of the microbiology of groundwaters, prompted by recognition of frequent contamination of aquifers with a wide variety of xenobiotics, suggests another possible source of mixed inoculum for lunar soils. Evidence indicates that microbes found in deep groundwaters, though few in number due to substrate limitations, are for the same reason extremely versatile metabolically. Through selective pressure, only those organisms having this versatility have survived. An inoculum consisting of a mixed population of groundwater microbes might prove highly efficient in terms of its capacity for recycling a wide variety of substrates. It is conceivable, although by no means proven, that such a mixture might be largely, if not entirely, devoid of plant and animal pathogens.

Numerous other sources could be used as well, including composted sludge. After all, as years of experience have shown us, and as the Dutch microbiologist, Martinus Beijerinck, stated long ago, "everything is everywhere and the *milieu* selects." Any population established in a lunar soil would soon come to some equilibrium based on the selective pressures provided by the environment. The challenge is to ensure that the lunar soil environment remains favorable for plant and microbial growth and resistant to rapid changes in its stability. Such a challenge requires intensive investigations by soil chemists, soil microbiologists, and soil and plant scientists working together to understand the chemical, physical, and biological oddities that may result from the growth of plants in a soil derived from lunar regolith.

A final point: until the experiments with lunar materials are actually begun, this contribution, like others in this collection, is largely speculative. The experiments will assuredly be exciting!

REFERENCES

Alexander, M. 1977. Introduction to soil microbiology, 2nd ed. John Wiley and Sons, New York.

Brady, N.C. 1974. The nature and properties of soils, 8th ed. Macmillan Publ. Co., New York.

Coleman, D.C. 1985. Through a ped darkly: an ecological assessment of root-soil-microbial-faunal interactions. p. 1–21. *In* A.H. Fitter (ed.) Ecological interactions in soil: Plants, microbes and animals. Blackwell Sci. Publ., Oxford, England.

Döbereiner, J., and F.O. Pedrosa. 1987. Nitrogen-fixing bacteria in nonleguminous crop plants. Science Tech Publ., Madison, WI.

Firestone, M.K. 1982. Biological denitrification. p. 289–326. *In* F.J. Stevenson (ed.) Nitrogen in agricultural soils. ASA, Madison, WI.

Gibson, M.A., and C. Knudsen. 1985. Lunar oxygen production from ilmenite. p. 543–550. *In* W.W. Mendell (ed.) Lunar bases and space activities of the 21st Century. Lunar and Planetary Inst., Houston.

Henninger, D.L., C.W. Lagle, and D.W. Ming. 1987. A lunar derived "soil" for the growth of higher plants. p. 1–20. *In* G.M. Andrus (ed.) The first lunar development symposium. Lunar Develop. Counc., Pitman, NJ.

Lynch, J.M. 1983. Soil biotechnology: Microbiological factors in crop productivity. Blackwell Sci. Publ., Oxford, England.

MacElroy, R.D., H.P. Klein, and M.M. Averner. 1985. The evolution of CELSS for lunar bases. p. 623–633. *In* W.W. Mendell (ed.) Lunar bases and space activities of the 21st Century. Lunar and Planetary Inst., Houston.

Richards, B.N. 1987. The microbiology of terrestrial ecosystems. Longman Sci. and Tech., Essex, England.

Schmidt, E.L. 1982. Nitrification in soil. p. 253–288. *In* F.J. Stevenson (ed.) Nitrogen in agricultural soils. ASA, Madison, WI.

Stevenson, F.J. (ed.). 1982. Nitrogen in agricultural soils. ASA, Madison, WI.

Stevenson, F.J. 1986. Cycles of soil. John Wiley and Sons, New York.

Tiedje, J.M. 1982. Denitrification. p. 1011–1026. *In* A.L. Page et al. (ed.) Methods of soil analysis. Part 2. ASA, Madison, WI.

Ugolini, F.C., and R.L. Edmonds. 1983. Soil biology. p. 193–231. *In* L.P. Wilding (ed.) Pedogenesis and soil taxonomy I. Concepts and interactions. Elsevier Publ., New York.

Williams, R.J., and J.J. Jadwick (ed.). 1980. Handbook of lunar materials. NASA Rep. RP-1057. U.S. Gov. Print. Office, Washington, DC.